Lecture Notes in Computer Science 2574

Edited by G. Goos, J. Hartmanis, and J. van Leeuwen

Springer
Berlin
Heidelberg
New York
Barcelona
Hong Kong
London
Milan
Paris
Tokyo

Ming-Syan Chen
Panos K. Chrysanthis
Morris Sloman
Arkady Zaslavsky (Eds.)

Mobile Data Management

4th International Conference, MDM 2003
Melbourne, Australia, January 21-24, 2003
Proceedings

Springer

Volume Editors

Ming-Syan Chen
National Taiwan University, No. 1, Sec. 4, Roosevelt Road, Taipei, Taiwan
E-mail: mschen@cc.ee.ntu.edu.tw

Panos K. Chrysanthis
University of Pittsburgh, Dept. of Computer Science
Sennott Square Building, 210 S. Bouquet Street, Pittsburgh, PA 15260, USA
E-mail: panos@cs.pitt.edu

Morris Sloman
Imperial College of Science Technology and Medicine, Department of Computing
180 Queen's Gate, London SE7 2BZ, U.K.
E-mail: m.sloman@doc.ic.ac.uk

Arkady Zaslavsky
Monash University, School of Computer Science and Software Engineering
900 Dandenong Road, Caulfield East, Vic 3145, Melbourne, Australia
E-mail: a.zaslavsky@monash.edu.au

Cataloging-in-Publication Data applied for

A catalog record for this book is available from the Library of Congress.

Bibliographic information published by Die Deutsche Bibliothek
Die Deutsche Bibliothek lists this publication in the Deutsche Nationalbibliografie; detailed bibliographic data is available in the Internet at <http://dnb.ddb.de>.

CR Subject Classification (1998): C.2, C.5.3, C.3, D.2, D.4, H.5, H.4, H.3

ISSN 0302-9743
ISBN 3-540-00393-2 Springer-Verlag Berlin Heidelberg New York

Springer-Verlag Berlin Heidelberg New York
a member of BertelsmannSpringer Science+Business Media GmbH

http://www.springer.de

Printed in Germany

Typesetting: Camera-ready by author, data conversion by Boller Mediendesign
Printed on acid-free paper SPIN: 10872310 06/3142 5 4 3 2 1 0

Table of Contents

Storage Management

Storing and Accessing User Context 1
Stéphanie Riché, Gavin Brebner (HP Labs, France)

Cooperative Caching in Ad Hoc Networks 13
Françoise Sailhan, Valérie Issarny (INRIA, France)

Investigation of Cache Maintenance Strategies for Multi-cell Environments 29
Zhijun Wang, Mohan Kumar, Sajal K. Das, Huaping Shen (University of Texas at Arlington, USA)

Resilient Data-Centric Storage in Wireless Ad-Hoc Sensor Networks 45
Abhishek Ghose, Jens Grossklags, John Chuang (University of California at Berkeley, USA)

Location Tracking

Shape-Based Similarity Query for Trajectory of Mobile Objects 63
Yutaka Yanagisawa, Jun-ichi Akahani, Tetsuji Satoh (NTT Communication Science Laboratories, Japan)

An Efficient Spatiotemporal Indexing Method for Moving Objects in Mobile Communication Environments 78
Hyun Kyoo Park (KAIST, Korea), Jin Hyun Son (Hanyang University, Korea), Myoung Ho Kim (KAIST, Korea)

DynaMark: A Benchmark for Dynamic Spatial Indexing 92
Jussi Myllymaki, James Kaufman (IBM Almaden Research Center, USA)

Information Management

Using Separate Processing for Read-Only Transactions in Mobile Environment 106
Eddie Y.M. Chan, Victor C.S. Lee, Kwok-Wa Lam (City University of Hong Kong, China)

Publish/Subscribe Tree Construction in Wireless Ad-Hoc Networks 122
Yongqiang Huang, Hector Garcia-Molina (Stanford University, USA)

Personal Workflows: Modeling and Management 141
San-Yih Hwang, Ya-Fan Chen (National Sun Yat-sen University, Taiwan)

Location-Aware Services

Architectural Support for Global Smart Spaces 153
Alan Dearle, Graham Kirby, Ron Morrison, Andrew McCarthy, Kevin Mullen, Yanyan Yang (University of St Andrews, UK), Richard Connor, Paula Welen, Andy Wilson (University of Strathclyde, UK)

FATES: Finding A Time dEpendent Shortest path 165
Hae Don Chon (Samsung Electronics, USA), Divyakant Agrawal, Amr El Abbadi (University of California at Santa Barbara, USA)

Search K Nearest Neighbors on Air 181
Baihua Zheng (Hong Kong University of Science & Technology, China), Wang-Chien Lee (Penn State University, USA), Dik Lun Lee (Hong Kong University of Science & Technology, China)

Adaptive Location Management in Mobile Environments 196
Ratul kr. Majumdar, Krithi Ramamritham (Indian Institute of Technology, India), Ming Xiong (Lucent Bell Labs, USA)

Context-Aware Services

Policy-Driven Binding to Information Resources in Mobility-Enabled Scenarios 212
Paolo Bellavista, Antonio Corradi, Rebecca Montanari (University of Bologna, Italy), Cesare Stefanelli (University of Ferrara, Italy)

Constructing Environment-Aware Mobile Applications Adaptive to Small, Networked Appliances in Ubiquitous Computing Environment 230
Kazunori Takashio (Keio University, Japan), Masakazu Mori, Masataka Funayama (The University of Electro-Communications, Japan), Hideyuki Tokuda (Keio University, Japan)

Experiences in Using CC/PP in Context-Aware Systems 247
Jadwiga Indulska, Ricky Robinson, Andry Rakotonirainy, Karen Henricksen (University of Queensland, Australia)

Document Visualization on Small Displays 262
Ka Kit Hoi, Dik Lun Lee, Jianliang Xu (Hong Kong University of Science & Technology, China)

Resource Discovery

Towards Autonomous Services for Smart Mobile Devices 279
Thomas Strang (German Aerospace Center, Germany)

Nomad: Application Participation in a Global Location Service 294
Kris Bubendorfer, John H. Hine (Victoria University of Wellington, New Zealand)

Mobiscope: A Scalable Spatial Discovery Service for Mobile Network Resources 307
Matthew Denny, Michael J. Franklin (University of California at Berkeley, USA), Paul Castro, Apratim Purakayastha (IBM T.J. Watson Research Center, USA)

Location Management

Presence, Location, and Instant Messaging in a Context-Aware Application Framework 325
Arjan J.H. Peddemors, Marc M. Lankhorst, Johan de Heer (Telematica Instituut, The Netherlands)

CAMEL: A Moving Object Database Approach for Intelligent Location Aware Services 331
Ying Chen, Fangyan Rao, Xiulan Yu, Dong Liu (IBM China Research Lab, China)

Using Hashing and Caching for Location Management in Wireless Mobile Systems 335
Weiping He, Athman Bouguettaya (Virginia Tech., USA)

SEB-tree: An Approach to Index Continuously Moving Objects 340
Zhexuan Song, Nick Roussopoulos (University of Maryland, USA)

Best Movement of Mobile Agent in Mobile Computing Systems 345
Chao-Chun Chen, Chiang Lee (National Cheng-Kung University, Taiwan), Chih-Horng Ke (Chang Jung University, Taiwan)

Storage Management and Query Processing

Clique: A Transparent, Peer-to-Peer Replicated File System 351
Bruno Richard, Donal Mac Nioclais, Denis Chalon (HP Laboratories, France)

Transactional Peer-to-Peer Information Processing: The AMOR Approach 356
Klaus Haller (ETH, Switzerland), Heiko Schuldt, Hans-Jörg Schek (ETH, Switzerland and University for Health Informatics and Technology Tyrol, Austria)

Performance Evaluation of Transcoding-Enabled Streaming Media Caching System ... 363
Bo Shen, Sung-Ju Lee, Sujoy Basu (Hewlett-Packard Labs, USA)

Adaptive File Cache Management for Mobile Computing ... 369
Jiangmei Mei, Rick Bunt (University of Saskatchewan, Canada)

Adaptive Power-Aware Prefetching Schemes for Mobile Broadcast Environments ... 374
Haibo Hu, Jianliang Xu, Dik Lun Lee (Hong Kong University of Science & Technology, China)

A Multi-layered Database Model for Mobile Environment ... 381
Sanjay Kumar Madria, Yongjian Fu (University of Missouri-Rolla, USA), Sourav Bhowmick (Nanyang Technological University, Singapore)

Context-Aware Information Services

A Task Oriented Approach to Delivery in Mobile Environments ... 386
François Paradis, Francis Crimmins, Nadine Ozkan (CSIRO, Australia)

Picturing the Future Personal Navigation Products and Services by Means of Scenarios ... 391
Olli Sotamaa (University of Tampere, Finland), Veikko Ikonen (VTT Information Technology, Finland)

Personal Digest System for Professional Baseball Programs in Mobile Environment ... 396
Takako Hashimoto, Takashi Katooka, Atsushi Iizawa (Ricoh Company, Japan)

Handling Client Mobility and Intermittent Connectivity in Mobile Web Accesses ... 401
Purushottam Kulkarni, Prashant Shenoy (University of Massachusetts Amherst, USA), Krithi Ramamritham (University of Massachusetts Amherst, USA and Indian Institute of Technology, India)

Enabling Web-Based Location-Dependent Information Services in Mobile Environments ... 408
Yongbo Niu (Lucent Technologies, China), Shanping Li (Zhejiang Univerisity, China), Jianliang Xu (Hong Kong University of Science & Technology, China)

Author Index ... 413

{stephanie.riche, gavin.brebner}@hp.com

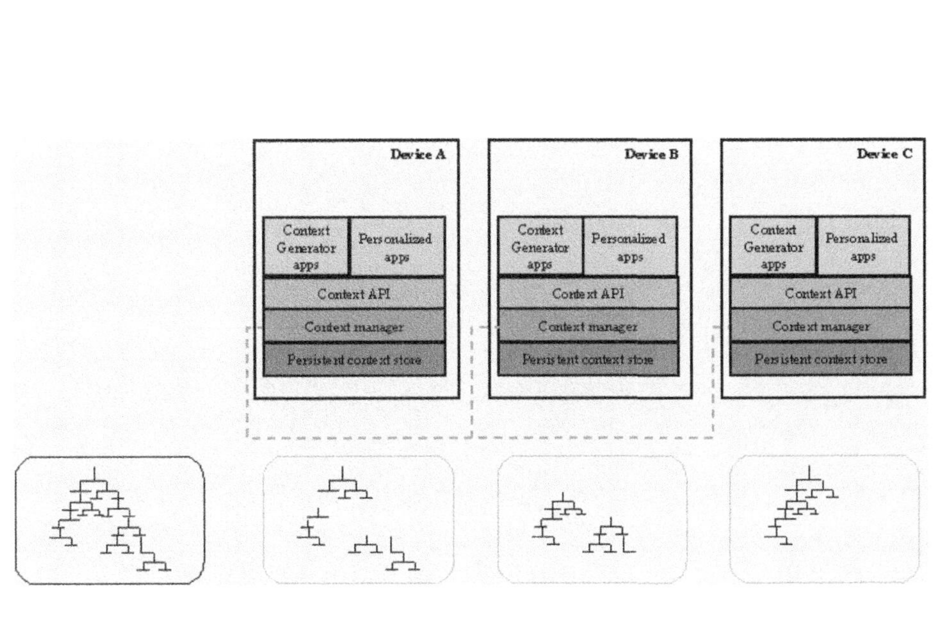

Device A
Context Generator apps
Personalized apps
Context API
Context manager
Persistent context store
Device B
Context Generator apps
Personalized apps
Context API
Context manager
Persistent context store
Device C
Context Generator apps
Personalized apps
Context API
Context manager
Persistent context store

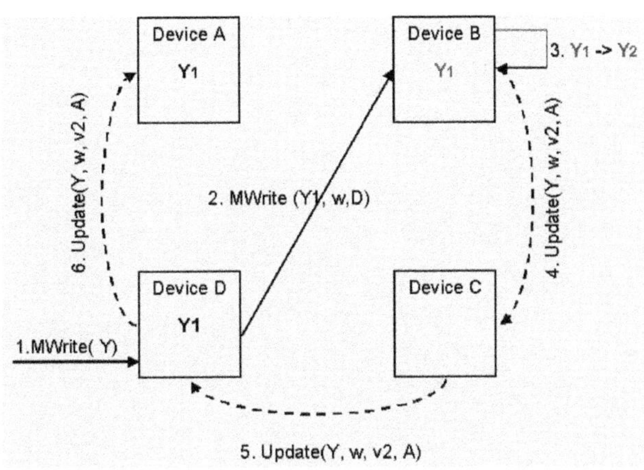

Device A
Y1
Device B
Y1
3. Y1 -> Y2
6. Update(Y, w, v2, A)
2. MWrite (Y1, w,D)
4. Update(Y, w, v2, A)
Device D
Y1
Device C
1.MWrite(Y)
5. Update(Y, w, v2, A)

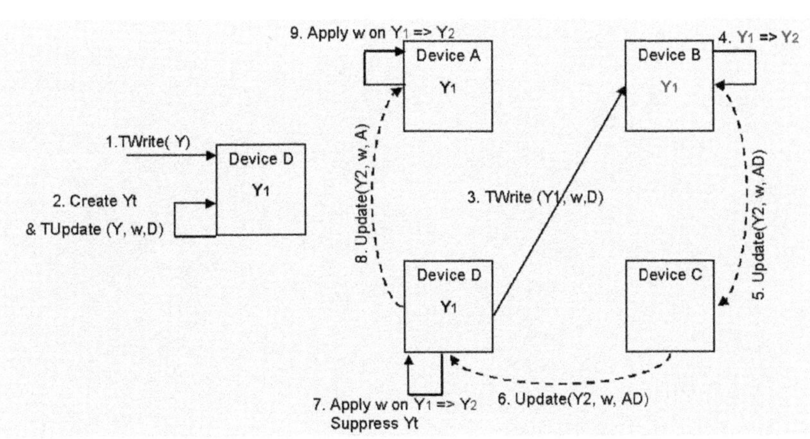

9. Apply w on Y1 => Y2
Device A
Y1
Device B
Y1
4. Y1 => Y2
1.TWrite(Y)
Device D
Y1
2. Create Yt
& TUpdate (Y, w,D)
8. Update(Y2, w, A)
3. TWrite (Y1, w,D)
5. Update(Y2, w, AD)
Device D
Y1
Device C
7. Apply w on Y1 => Y2
Suppress Yt
6. Update(Y2, w, AD)

Cooperative Caching in Ad Hoc Networks

Françoise Sailhan and Valérie Issarny

Inria-Rocquencourt, Domaine de Voluceau, Rocquencourt,
BP 105, 78153, Le Chesnay Cédex, France
{Francoise.Sailhan, Valerie.Issarny,}@inria.fr

Abstract. Terminal's latency, connectivity, energy and memory are the main characteristics of today's mobile environments whose performance may be improved by caching. In this paper, we present an adaptive scheme for mobile Web data caching, which accounts for congestion of the wireless network and energy limitation of mobile terminals. Our main design objective is to minimize the energy cost of peer-to-peer communication among mobile terminals so as to allow for unexpensive Web access when a fixed access point is not available in the communication range of the mobile terminal. We propose a collaborative cache management strategy among mobile terminals interacting via an ad hoc network. We further provide evaluation of the proposed solution in terms of energy consumption on mobile devices.

1 Introduction

The last decade has seen the rapid convergence of two pervasive technologies: wireless communication and the Internet. The resulting mobile Internet a priori enables users to easily access information anytime, anywhere. However, we have not yet reached the point where anywhere, anytime Internet access is actually offered. This paper addresses the above issue, concentrating more specifically on Web caching in a mobile environment to allow for Web access, without requiring availability of an infrastructure in the nearby environment.

There exist two different ways of configuring a mobile network: *infrastructure-based* and *ad-hoc-based*. The former type of network structure is the most prominent, as it is in particular used in both Wireless LANs (e.g., IEEE 802.11) and global wireless networks (e.g., GSM, GPRS, UMTS). An infrastructure-based wireless network uses fixed network access points (known as *base stations*) with which mobile terminals interact for communicating, i.e., a base station forwards messages that are sent/received by mobile terminals. One limitation of the infrastructure-based configuration is that base stations constitute bottlenecks. In addition, it requires that any mobile terminal be in the communication range of a base station. These shortcomings may be solved through the deployment of a sufficient number of base stations. However, this comes at a high cost for network providers. The ad-hoc-based network structure alleviates this problem by enabling mobile terminals to cooperatively form a dynamic network without any pre-existing infrastructure. In general, ad hoc and infrastructure-based networking should be seen as complementary rather than as competitive. Ad hoc

M.-S. Chen et al. (Eds.): MDM 2003, LNCS 2574, pp. 13–28, 2003.

networking is much convenient for accessing information available in the local area, and possibly reaching a WLAN base station, which comes at no cost for users. Ultimately, the user may decide to pay for communication using wireless global networking facility. This paper concentrates on improving the Web latency using a WLAN, exploiting both the ad hoc and infrastructure-based capabilities of the network.

The issue that we are addressing is on setting an ad hoc network of mobile terminals that cooperate to exchange Web pages. One application of our work is, e.g, multimedia guided tour from PDAs. Indeed, several cities and museums have proposed such services over the last few years but the limited number of base stations due to a prohibitive financial cost and the limited communication range of mobile terminals prevent tourists from accessing data anywhere. In addition, the limited memory capacity of PDAs requires constant interaction with the base station to gain access to the rich set of available information. Our proposal allows users to access information when a base station is not reachable in one hop and then offers better scalability at lower financial cost.

In the above context, it is crucial to account for the specifics of mobile terminals. Mobile terminals that will soon be available will embed powerful hardware (e.g., LCD screens, accelerated 3D graphics, high performance processor), and an increasing number of devices (e.g., DVD, CD) [13]. However, the capacity of batteries goes up slowly and all these powerful components reduce battery life. Thus, it is mandatory to devise adequate solutions to energy saving on the mobile terminals, for all the constituents of the mobile environment, i.e., application software, network operating system, and hardware. In particular, communication is one of the major sources of energy consumption, and hence a number of wireless communication protocols have been designed to reduce energy consumption [12, 5, 18]. However, most of the communication protocols that have been proposed for ad hoc networks are addressed in terms of bandwidth usage and not energy consumption [9]. In addition, it is mandatory for these protocols to be coupled with distributed application software that are designed so as to minimize energy consumption.

This paper introduces such a distributed application software, which implements ad hoc cooperative Web caching among mobile terminals[1]. The proposed solution aims at improving the Web latency on mobile terminals while optimizing associated energy consumption. The solution accounts for both the capacities of mobile terminals and the network features; it comprises: (i) a cooperative caching protocol among mobile terminals that builds upon an existing ad hoc routing protocol, and (ii) a local caching strategy for the mobile terminal. The next section discusses related work, addressing background in the area of cooperative Web caching and Web access from mobile terminals. Section 3 then introduces the proposed ad hoc cooperative Web caching protocol, and is followed in Section 4 by the presentation of the caching strategy implemented on terminals. Section 5 provides an evaluation of our proposal, giving the energy

[1] In the following we do not address the issue of Web Data conversion for these devices for which a number of base solutions exists.

consumption associated with cooperative caching. Finally, Section 6 concludes with a summary of our contribution, and our current and future work.

2 Related Work

The ever growing popularity of the Web and the resulting poor latency for users have given rise to huge effort on improving the Web latency, which mainly lies in the introduction of Web caching protocols. Due to the network topology, the idea of making network caches (also referred to as *proxy caches* or simply *proxies*) cooperate has emerged. The Harvest project [2] introduced the notion of hierarchical caching, which is pioneering and lies in introducing a cache on every network node, the system's hierarchical structure coming from the national networks' hierarchical organization. A cache then locates a missing requested object by issuing a request to the cache at the hierarchy's upper level. The process is iterated until either the object is found or the root cache is reached, which may ultimately involve contacting the object's server. The object is then copied in all the caches contacted as it is returned to the client.

A transversal system enriches the hierarchy by integrating a set of sibling caches that are close in terms of latency time, at each hierarchical level. On a cache miss, a cache not only contacts its ancestral cache but also its siblings. A number of transversal cooperative caching protocols have been proposed in order to minimize the number of messages that are exchanged among sibling caches to retrieve cached objects. Proposed solutions all amount to maintain a partial knowledge of the objects that are cached on siblings. With Summary Cache [8], each proxy keeps a compact summary of the cache content of other sibling proxies. The Scoop protocol [17] proposes to distribute this knowledge among the cache composing the transversal system and the Cache Array Routing Protocol (CARP) [22] partitions the URL space among sibling proxies.

In the context of mobility, proxy caches are used not only to cache and retrieve documents but also to manage user mobility. These proxies are access points to the Internet for mobile terminals. Proxies implement functionalities dedicated to the transfer of data to mobile terminals, such as compression, filtering, format conversion [16][1]. It is further mandatory to account for the disconnection of mobile terminals, as their connectivity is highly dynamic in nature. Typically, a mobile terminal can be in either of the three following connection modes, shifting from one to another depending on the user's location: strongly connected (i.e., use of a fixed LAN), less strongly connected (i.e., use of a wireless network), or disconnected (i.e., due to the absence of connectivity or to save energy). Solutions to the above issue include the one offered by the ARTour Web Express software [4], which queues user requests that are made during an unexpected disconnection. These requests are later processed as a background task when the connection is restored. In addition, the user may act upon local cache management by requesting to keep in the cache the pages that he/she marks using the cache browsing tool. However, not all the disconnections are unexpected, and some may be anticipated by the user. In this case, the user may indicate the kind of Web pages she will be willing to access (e.g., subject of interest, URLs). The information is then exploited, possibly in combination with the user's profile, to prefetch a number of pages. Similar to our solution, the one of [3] in the area of service discovery and composition or [19] in the area of databases use a decentralized model in an ad hoc environment. Complementing proxy solutions with ad-hoc-based ones will allow for both enhanced connectivity and Web access at low cost. In that context, the caches of the mobile terminals cooperate in a way similar to proxy caches in transversal cooperative cache systems. However, the cooperative scheme must be devised so as to account for: the continuously changing set of cooperating terminals, the heterogeneity of terminals in terms of resource availability, and energy saving.

3 Cooperative Caching in Ad Hoc Networks

A number of ad hoc routing protocols are implemented over a base WLAN. Basically, these protocols maintain a *routing table* and differ in the way they manage it (§ 3.1). Using the ad hoc network protocol that offers the best trade-offs in terms of energy consumption and response time, we propose a specialization of the protocol that is specifically aimed at handling remote access to Web pages (§ 3.2). We then introduce our ad hoc cooperative caching protocol, which has primarily been designed to minimize energy consumption (§ 3.3).

3.1 Ad Hoc Networking

The main issue to be addressed in the design of an ad-hoc (network) routing protocol is to compute an optimal *communication path* between any two mobile terminals. This computation must minimize the number of control messages exchanged in order to minimize network congestion and energy consumption [7].

There exist two base types of ad hoc protocols: *proactive* and *reactive*. Proactive protocols (e.g., OLSR [6]) update their routing table periodically. Compared to proactive protocols, reactive protocols (e.g., AODV [20], DSR [15]) check the validity of, and possibly compute, the communication path between any two mobile terminals only when a communication is requested between the two. ZRP [10] is a hybrid protocol that combines the reactive and proactive modes. The design rationale of ZRP is that it is considered advantageous to accurately know the *neighbours* of any mobile terminal (i.e., mobile terminals that are accessible in a fixed number of hops), since they are close. Hence, communicating with neighbours is less expensive and neighbours are most likely to take part in the routing of the messages sent from the terminal. As a result, ZRP implements: (i) a proactive protocol for communication with mobile terminals in the neighbourhood, and (ii) a reactive protocol for communication with the other terminals. With respect to a given mobile terminal, its neighbourhood is referred to as its *zone*. Notice that it would be interesting to dynamically set the number of hops used to define a zone, according to the number of mobile terminals that are present in the perimeter. ZRP is currently among the most efficient ad hoc routing protocols in terms of network load due to control messages and hence of associated energy consumption on mobile terminals. We thus use ZRP over IEEE 802.11, as the base ad hoc protocol for realizing ad hoc cooperative caching among mobile terminals. Mobile terminals belonging to the zone of a given terminal then form a cooperative cache system for this terminal since the cost for communicating with them is low both in terms of energy consumption and message exchanges [11]. However, cooperative caching must not be restricted to the mobile terminals belonging to the zone: low-cost reachability of a base station must be accounted for as well as knowledge of a terminal that does not belong to the zone but that is likely to store a requested Web document given commonalities in performed Web accesses.

3.2 Ad Hoc Communication for Web Caching

A mobile terminal may get Web data that are not cached locally through two communication paths: (i) using the infrastructure-based mode, the terminal may interact with the nearby base station, which forwards the request to the Web, (ii) using the ad-hoc-based mode, the terminal may request for the data to the mobile terminals in its communication range (i.e., accessible in one hop in a base WLAN or in a number of hops using some ad hoc network routing protocol). The former is the most efficient in terms of energy consumption; the mobile terminal sends a single message to a base station that has an infinite energy budget. However, the ad-hoc-based mode must be enabled for the case where a base station is not reachable in one hop. In this case, a base station can still be reachable in a number of hops, thanks to mobile terminals forwarding the requests. For instance, UMTS supports such a mechanism to extend the range of the infrastructure[2]. The number of hops that are necessary to access a base station depends on the terminal's location. Let N be this number, then any mobile terminal that is at a distance greater or equal to N is not contacted to get a document. Figure 1 depicts the case where the mobile terminal A reaches the base station D in 3 hops, using the mobile terminals B and C for routing the request. Then, if either a mobile terminal in the zone of A (e.g., B belonging to the path leading to the base station or any other terminal in the zone) or a known mobile terminal located outside the zone but at a lower distance than the base station D (e.g., C that is in the path leading to D or E that does not belong to the path) holds the requested document in its cache, it returns it to A. Otherwise, the request reaches D, and D forwards it to the Web. We get the following ad hoc communication protocol over ZRP, to retrieve a remote Web object W, with respect to a given mobile terminal A:

- *In-zone communication:*
  If a base station is in the zone of A, then A requests for W through the base station only. Otherwise, A broadcasts the request message for W to the mobile terminals in the zone of A, incurring a low energy cost since the routing table contains the necessary information.

[2] http://www.3gpp.org

- *Peer-to-peer communication:*
  If W is not cached by any of the mobile terminals in the zone of A and if there is no base station in the zone of A, then a peer-to-peer communication (or point to point messaging) scheme is realized with mobile terminals that are known to share interests with A (see § 3.3) and that are at a distance that is less than the one between A and the nearest base station. Mobile terminals outside the zone of A are basically known through two ways: (i) they belong to the path used to reach the nearest base station, (ii) they were previously either in the zone or in the path used to reach the base station. The request for W is ultimately forwarded to the nearest base station.

Based on the above, the communication cost and hence the energy cost associated with getting a Web object is kept to a minimum: (i) broadcast is within a zone and is thus unexpensive by construction of ZRP, (ii) peer-to-peer communication occurs only with mobile terminals that are both the most likely to store a requested object and closer than a base station.

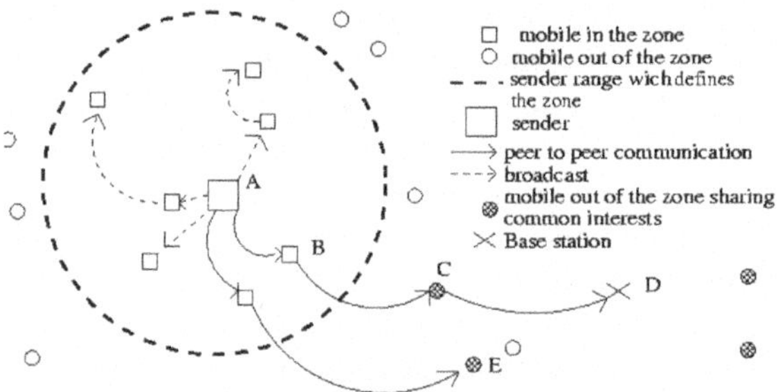

Fig. 1. Getting Web data

3.3 Ad Hoc Cooperative Caching

Without a proxy-type architecture that centralizes requests, local statistics are relied on for a mobile terminal A to identify mobile terminals that are likely to store a Web object requested on A. Such statistics are maintained on A through a *terminal profile* for every mobile terminal with which A interacts. The terminal profile is characterized by a value that counts the number of times the corresponding mobile terminal either is known to cache an object requested by A or requested for an object to A that A had in its cache. The value of the terminal profile is used to identify the mobile terminals with which peer-to-peer communication is undertaken. The list of known mobile terminals outside the zone and that are at a distance less than a base station are then weighed according to the value of $\mathcal{F} = terminal\ profile \div hops$ where the number of hops, *hops*, is obtained from the routing table. Mobile terminals for which the value of $\mathcal{F}$ is the greatest are first contacted and the process is iterated until a *hit* message is received or there are no more mobile terminals eligible for the request. In addition to the management of *terminal profiles* to identify mobile terminals that share common interests, we must account for the heterogeneity of the terminals' capacity (i.e., battery, processing, storage, communication). In particular, for two mobile terminals that are equally likely to store a requested object, it is better to contact the one that has the greatest capacity.

Ignoring the case where a base station is accessible in the zone and given the ad hoc communication protocol aimed at Web caching that was discussed in the previous subsection, the request for a Web object W that is not cached locally, from a mobile terminal A is handled as follows: (i) A first broadcasts the requests for W within the zone; (ii) if W is not retrieved then the retrieval protocol iterates on sending the request for W to known mobile terminals outside the zone and at a distance less than the base station according to the maximization of $\mathcal{F}$; (iii) the base station is ultimately contacted. The processing of requests is further handled as follows. A mobile terminal that receives the request for W and caches it increments its local value of A's terminal profile. If the terminal is further willing to cooperate (e.g., absence of energy safeguarding or of security policy enforcement), it returns a *hit* message, which embeds:

- (i) TTL that gives the Time To Live field of the document.
- (ii) *Capacity* that characterizes the capacity of the terminal to handle requests, whose value is in the range [0..1], 1 denoting the highest capacity. Currently, we use a simple scheme to set the value of *Capacity*; it is equal to the percentage of the energy budget that is left. It is part of our future work to investigate a more accurate way of computing *Capacity*.

For every *hit* message that it receives, A increments the *terminal profile* of the sender. Among the mobile terminals that replied by a *hit* message, A selects the terminal from which W should be obtained, that is the one that maximizes the following function[3]: $\mathcal{R} = Capacity \times (\lambda \times TTL + \mu \times hops)$ where λ and μ are set so as to favour communication with the closest nodes. More precisely, the value of R is computed as follows. The metrics used for consistency (duration given by TTL) and for communication cost (number of hops) are distinct and should be made comparable. As a dispersion measure of TTL and hops, we use the standard deviation. Thus, considering that A received N hits messages for W, A computes the mean m_{TTL} (resp. m_{hop}) of the TTL (resp. hops) and the standard deviation σ_{TTL} (resp. σ_{hop}):

$$m_{TTL} = \frac{1}{N}\sum_{i=1}^{N} TTL_{mobile_i}, \quad \sigma_{TTL} = \sqrt{\frac{1}{N}\sum_{i=1}^{n}[TTL_{mobile_i} - m_{TTL}]^2}$$

$$m_{hop} = \frac{1}{N}\sum_{i=1}^{N} hop_{mobile_i}, \quad \sigma_{hop} = \sqrt{\frac{1}{N}\sum_{i=1}^{n}[hop_{mobile_i} - m_{hop}]^2}$$

[3] For the case where the selected terminal is no longer accessible, e.g., due to energy safeguarding, the request will be sent to the next eligible terminal and will be so until the page is received.

For illustration, Figures 2 gives a set of TTL and hop values considering reception of hit messages from terminals [B,..,G]. We distinguish four intervals for the value of TTL: level $I1_{TTL}$ =]-∞, m_{TTL}-σ], level $I2_{TTL}$ = [m_{TTL}-σ, m_{TTL}], level $I3_{TTL}$ =[m_{TTL}, m_{TTL}+σ], level $I4_{TTL}$ = [m_{TTL}+σ, +∞].

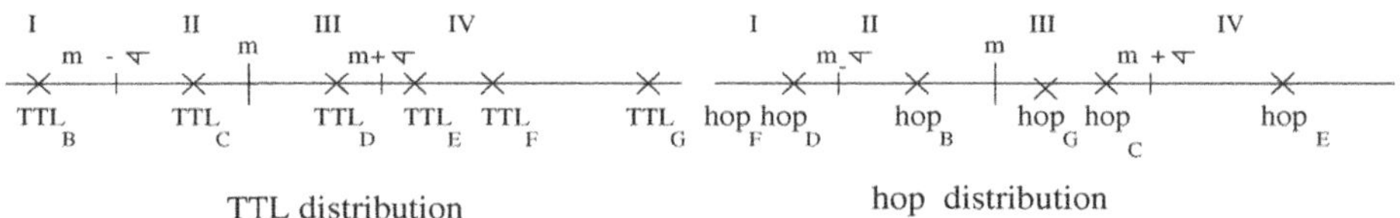

Fig. 2. Energy consumption.

The value of TTL is then mapped into one of the ranges [0..1], [1..2], [2..3] and [3..4], if it belongs to $I1_{TTL}$, $I2_{TTL}$, $I3_{TTL}$ and $I4_{TTL}$, respectively. The value taken in the target range is then set proportionnally to the value of the TTL in the initial range, Ij_{TTL}, j ∈ [1..4], and is the one used in the computation of R. For illustration, Table 1 gives the initial TTL and hops values, and values under the unified metric, considering the distribution given in Figure 2. Then the values of λ and μ are set so that $\lambda+\mu = 1$ and will be chosen according to the respective weighing factors for the consistency and communication cost. Figure 3 gives the value of $\frac{\mathcal{R}}{capacity}$ for the mobile terminals B,..,G considered in Figure 2, according to the value of λ, ranging from 1 (i.e, μ=0 and the communication cost is ignored) to 0 (i.e, μ=1 and the communication cost is the only selective factor).

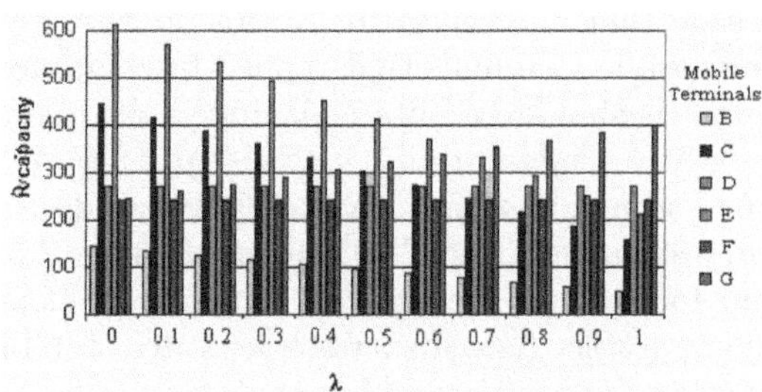

Fig. 3. R/capacity function.

We do not use *miss* messages for mobile terminals to notify that they do not cache a requested object. This is to minimize both network load and energy consumption. Hence, we use timeouts to detect the absence of a requested object. The value of the timeout is set according to the greatest number of hops that are involved to interact with the mobile terminals to which the object is requested,

Table 1. TTL and hop metric

mobile	TTL	TTL	TTL metric	hop	hop metric
b	$\frac{1}{2}$	$(m-\sigma)-\frac{1}{2}\sigma$	$1-\frac{1}{2}=\frac{1}{2}$	4	$1+\frac{1}{\sqrt{5}}$
c	$\frac{4}{7}$	$m-\frac{3}{7}\sigma$	$2-\frac{3}{7}=\frac{11}{7}$	8	$\frac{25}{7}+\frac{2}{\sqrt{5}}$
d	$\frac{27}{10}$	$(m+\sigma)-\frac{3}{10}\sigma$	$3-\frac{3}{10}=\frac{27}{10}$	2	$\frac{27}{10}$
e	$\frac{13}{4}$	$(m+\sigma)+\frac{1}{4}\sigma$	$\frac{17}{8}$	10	$\frac{49}{8}$
f	$\frac{23}{6}$	$(m+\sigma)+\frac{5}{6}\sigma$	$\frac{29}{12}$	2	$\frac{29}{12}$
g	5	$(m+\sigma)+2\sigma$	4	6	$2+\frac{1}{2\sqrt{5}}$

together with the current network load. Upon expiration of the timeout, if *hit* messages have been received, the Web object will be requested to the mobile terminal that maximizes $\mathcal{R}$. Otherwise, the next iteration of the cooperative caching protocol is processed (i.e., from the broadcast step to the peer-to-peer iterative step). For the case where a *hit* message is received after timeout expiration, while the object is still not retrieved, the message is accounted for in the current step of the protocol.

4 Local Caching

The ad hoc cooperative caching protocol introduced in the previous section is complemented by a local caching strategy, which is adaptive according to the current capacity of the terminal. Local cache management mainly amounts to the implementation of a replacement algorithm that is run when the cache gets full. We weigh every cached document according to both its probability of being accessed in the future and the energy cost associated with getting remotely the document. Documents with the lowest weighs are those that are removed from the cache. The document weigh is computed according to the following criteria:

- **Popularity**. The *Popularity* value serves approximating the probability of future access, both on the terminal and from remote terminals, as enabled by the cooperative caching protocol. The probability is approximated according to the number of times the document has been requested since it has been cached.
- **AccessCost**. The *AccessCost* value gives an estimate of the energy cost associated with getting the document remotely if it is to be removed from the cache. This cost varies depending on whether: a base station is accessible in the zone of the terminal, the document is cached on a mobile in the zone of the terminal, communication out of the zone is required to retrieve the document. In the first two cases, the cost is quite low, while it may be quite high in the third case. The value of the access cost is computed according to the energy consumption associated with intra-zone and inter-zone communication (see § 5). Intra-zone communication holds if a base station is known to be in the zone. It is further assumed if the document

was obtained from a terminal that is still in the zone, as identified using the routing table.

- **Coherency**. A document is valid for a limited lifetime, which is known using the TTL field. However, when the energy remaining on the terminal is low, it is better to favour energy saving over the accuracy of the document. Hence, the value of $Coherency$ is equal to $\nu \times TTL$ where ν increases as the available energy decreases.

We get the following function to compute the document weigh:
$\mathcal{W} = \alpha \times Popularity + \beta \times AccessCost + \gamma \times Coherency + \delta \times size$
The values of α, β, γ and δ are set so as make the values of $AccessCost$, $Popularity$, $Size$, and $Coherency$, decreasingly prominent factors for deciding whether a document should be kept in the cache. The metric used to compare the differents parameters is computed in a way similar to the one described in Section 3 for the function $\mathcal{R}$. Notice that our $\mathcal{W}$ function offers similarities with the one used by hybrid replacement algorithms that were proposed in the literature for Web caching on stationary hosts (e.g., [14]). In the same way, we use a function that accounts for a combination of criteria in the replacement decision. However, our function differs in that energy saving is a prominent criterion.

5 Evaluation

Control messages generated by ZRP and messages associated with ad-hoc cooperative caching affect the network traffic and energy consumption on mobile terminals. In this section, we examinate the energy consumption induced by:

- Maintening the routing table on each mobile terminal with ZRP.
- Using the ad-hoc cooperative strategy to identify terminals that are likely to store the requested Web data.
- Retrieving the Web data.

The above includes the energy cost due to both communication and computation. However, we do not consider the latter cost since it is induced by any local cache management, and it is negligible compared to the energy cost of communication: the energy cost of transmitting 1Kb is approximately the same as the energy consumed to execute 3 million intructions [21]. We further use the following wireless interface for our evaluation: 2.4Ghz DSSS lucent IEEE 802.11 WaveLan PC "Bronze" 2Mbps on the terminal.

5.1 Energy Consumption of ZRP Control Messages

Performance of ZRP has been evaluated in [11, 10] using event-driven simulation; this evaluation is gauged by considering the control traffic generated by ZRP. Figure 4 gives the energy consumption generated by the reception of ZRP traffic. Messages are used to maintain the zone routing tables with a beacon broadcast period of 0.2/s. To obtain the path to a mobile terminal out of the zone, additional control messages are exchanged only on-demand.

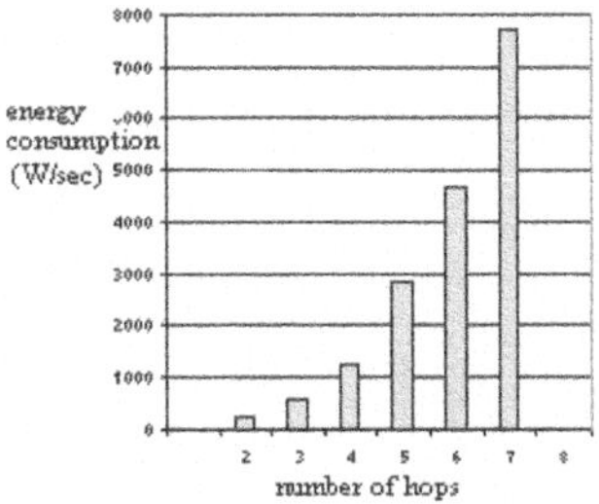

Fig. 4. Energy consumption generated by the reception of ZRP traffic in a zone per /s

Given the energy consumption generated by ZRP, we evaluate energy consumption associated with ad-hoc cooperative caching, as the sum of the energy consumption induced for the various mobile terminals that are involved (both in and outside the zone) in the cooperation, which adds to the energy cost induced by ZRP.

5.2 Energy Consumption of Ad-Hoc Networking

Focusing on the energy cost associated with communication, the cost associated with the emission of one message is the sum of:

- The cost of emission for the sender node.
- The cost of reception for the destination node.
- The cost of reception and emission for mobile nodes forwarding the message.
- The cost of reception for non-destination nodes (i.e., terminals that receive control messages due to their location with respect to the aforementioned nodes, although they are not involved in the routing of the message).

To minimize collisions in point-to-point traffic, the IEEE 802.11 protocol provides the following collision avoidance mechanism. Prior to any point-to-point transmission, the sender broadcasts a RTS (Request To Send) control message, which specifies the destination node and the data size (for duration). The sender then waits for a CTS (Clear To Send) message from the destination node. Once it receives the CTS, the sender sends the data message. Finally, the destination node sends an ACK message upon the reception of the data message. Therefore, the energy consumed by any mobile terminal for sending, receiving or discarding a message is given by the linear equation [9]: $\varepsilon = m * size + p$; where *size* is the message size, and m (resp. p) denotes the incremental (resp. fixed) energy cost associated with the message. The value of ε is detailed hereafter, for the various mobile terminals affected by the message transmission. Table 2 gives the energy cost, relative to the size of the message, for the sender and destination nodes (i.e., the actual sender and destination nodes and the forwarding nodes that act as both sender and destination nodes). For the sender X, the high value of the

Table 2. Energy consumption on nodes for point to point communication

Mobile	Energy consumption ($\mu W.sec$)	m ($\mu W.sec$)	p ($\mu W.sec$)
Sender X	$\varepsilon_{send} = m_{send} * size + p_{send}$	$m_{send} = 1.9$	$p_{send} = 454$
Destination A	$\varepsilon_{dest} = m_{dest} * size + p_{dest}$	$m_{dest} = 0.5$	$p_{dest} = 356$
Non-destination nodes			
in range of sender X	$\varepsilon_{AX}=m_{AX} * size + p_{AX}$	$m_{AX} = -0.22$	$p_{AX} = 210$
in range of sender X	$\varepsilon_{X}=m_{X} * size + p_{X}$	$m_{X} = -0.04$	$p_{X} = 90$
in range of destination A	$\varepsilon_{A}=m_{A} * size + p_{A}$	$m_{A} = 0$	$p_{A} = 119$

incremental cost m_{send} is due to the emission of the data message. The fixed cost p_{send} results from the reception of two control messages (CTS and ACK) and from the emission of the RTS message. For the destination node A, the fixed cost p_{dest} is due to the emission of two control messages (CTS and ACK) and to the reception of the RTS message. The value of m_{dest} for A follows from the reception of the data message. Table 2 also gives the energy cost, relative to the size of the message, for non-destination nodes. Non-destination nodes in the range of the sender receive RTS messages and thus enter a reduced energy consumption mode during data emission; this leads to have a negative value for m_{AX} and m_X since the energy consumption is less than the one in the idle mode. The fixed cost p_{AX} for non-destination nodes in the range of both the sender and the receiver is greater than the fixed cost p_X for non-destination nodes in the range of the sender only, because the latter do not receive the CTS and ACK messages. Finally, non-destination nodes in the range of the destination node but not the sender receive the CTS and ACK messages. On the other hand, they do not receive the RTS message, and thus cannot enter in the reduced energy consumption mode; this leads to have the incremental cost m_A equal to 0.

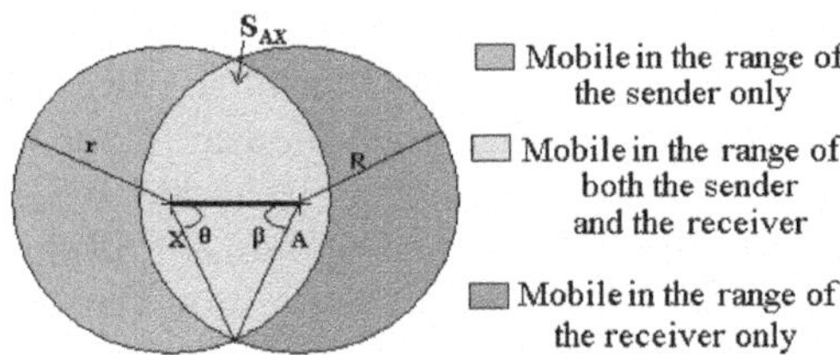

Fig. 5. Mobile terminals in the range of the sender and of the destination node.

Consider now a network of 500 mobile terminals whose communication range is of about 250m, that is such that the mobile terminals are uniformly distributed in the network surface S with $S = 4000[m] \times 4000[m]$ [4]. In a zone (see Figure 5), all the mobile terminals consume the same energy. Thus, the overall energy

[4] This network is taken as an example, as it is used in [11] for the evaluation of ZRP.

Table 3. Energy consumption in a zone.

Mobile terminal in range of	Surface	Number of mobile terminals	Total energy for a zone
the sender X & the receiver A	$S_{AX} = \theta(r^2 - a_1^2) + \beta(R^2 - a_2^2)$	$n_{AX} = \frac{NS_{AX}}{S}$	$n_{AX}\varepsilon_{AX}$
the sender X	$S_X = \pi r^2 - S_{AX}$	$n_X = \frac{NS_X}{S} = \pi r^2 - \theta r^2 \sin^2\theta + \beta R^2 \sin^2\beta$	$n_X\varepsilon_X$
the receiver A	$S_A = \pi R^2 - S_{AX}$	$n_A = \frac{NS_A}{S} = \pi R^2 - \theta r^2 \sin^2\theta + \beta R^2 \sin^2\beta$	$n_A\varepsilon_A$

consumption within a zone is the sum of the energy consumed by every mobile terminal in this zone (see Table 3). Then, the energy consumption of the overall network is the sum of the energy consumed per zone that is traversed. Figure 6 gives the energy consumption associated with the delivery of a message of 1 Kb to the destination node according to the number of hops, which is the sum of the energy consumed on all the nodes involved in the communication. The figure clearly demonstrates that the energy consumption increases with the number of hops. This directly follows from the resulting increase of both mobile terminals forwarding the message and non-destination nodes receiving control messages. Table 4 further evaluates the impact of the network density on energy consumption. For a constant number of hops (=3), we see that increased density of mobile terminals results in additional energy consumption for non-destination nodes. But, the ratio $\frac{\varepsilon_{send}+\varepsilon_{dest}+\varepsilon_{forwarding\ mobile\ terminals}}{\varepsilon_{overall\ non-destination\ mobile\ terminals}}$ highlights the weak impact of message reception on non-destination nodes, on the overall energy consumption. Indeed, for 600 mobile terminals and a destination node at 4 hops of the sender, the energy consumed by non-destination nodes is about 6 times less than the energy consumed by the sender, the receiver and the 3 forwarding mobile terminals.

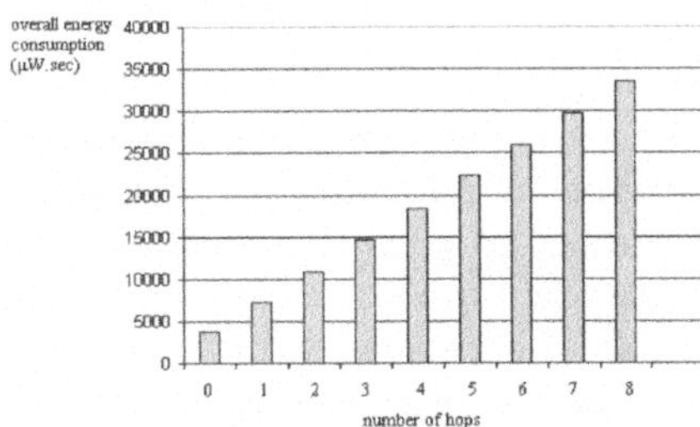

Fig. 6. Energy consumption in the network for the retrieval of Web data.

5.3 Energy Consumption of Ad-Hoc Copperative Caching

Having examined the energy consumption associated with data delivery, we now focus on the energy consumption induced by our ad-hoc cooperative caching

Table 4. Energy consumption according to the network density.

Number of Mobile terminals in the network surface	500	600	700
Energy consumption of non-destination mobile terminals (μ W.sec)	1389	18959	2136
ratio	9	7	6.11

protocol, which retrieves a requested Web page among mobile terminals. A request message for a Web page includes broadcasting (with mobile terminals in the requester's zone) and peer-to-peer communication (with mobile terminals sharing the same interests). Before broadcasting a message, the sender listens to the channel; if no traffic is detected, the message is broadcasted. Table 5 gives the energy consumption induced by a broadcast while the one associated with peer-to-peer communication is given in Table 2. The difference between the energy consumption associated with broadcasting and the energy consumption associated with peer-to-peer communication is due to the emission of control messages (RTS, CTS and ACK messages) in the case of peer-to-peer communication. Figure 7 gives the energy consumed by cooperative caching according

Table 5. Energy consumption on the sender and destination nodes for broadcast.

Mobile	Energy consumption ($\mu W.sec$)	m ($\mu W.sec/byte$)	p ($\mu W.sec/byte$)
Sender X	$\varepsilon_{brsend} = m_{brsend}$ *size+p_{brsend}	$m_{brsend} =$ 1.9	$p_{brsend} =$ 266
Destination A	$\varepsilon_{brdest} = m_{brdest}$ *size+p_{brdest}	$m_{brdest} =$ 0.5	$p_{brdest} =$ 56

to the number of hops that defines the zone. Precisely, Figure 7 gives the cost associated with broadcasting and peer to peer communication at a distance that adds one hop to the number of hops that defines the zone. As expected, the overall energy consumption of both broadcast and peer-to-peer communication increases with the number of hops that defines the zone (because more mobiles are in the zone). Comparing energy consumption of broadcasting with the one of peer-to-peer communication, we find that the cost added by broadcasting is weak compared to the number of mobile terminals contacted, and hence supports our approach.

6 Conclusion

Mobile technology has reached a stage that enables foreseeing easy access to information technology anywhere, anytime. However, it is necessary to devise

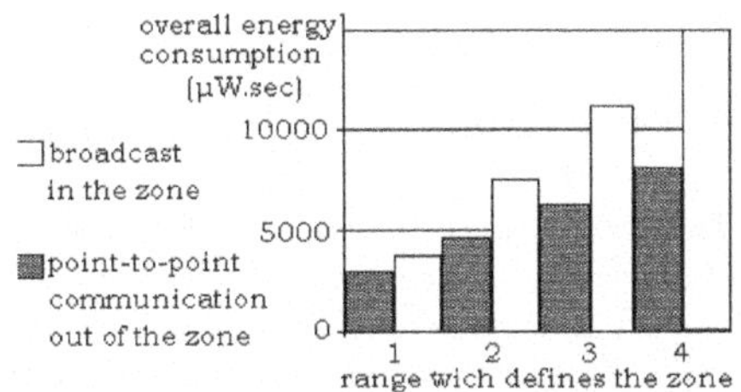

Fig. 7. Energy consumption in the network for the broadcast and the peer-to-peer communication.

adequate software and hardware solutions to mask unstable connectivity and limited energy. This paper has addressed one such solution, which focuses on enabling Web accesses from mobile terminals. Currently, Web access is easy to realize if the mobile terminal is in the communication range of a base station of either a WLAN or a global wireless network. However, this cannot always be assumed due to the financial cost associated with the deployment of the underlying infrastructure. In addition, the systematic use of a global wireless network is quite costly for users. Instead of relying on a base station for accessing the Web, an alternative solution is to exploit ad hoc networking, which allows for remote communication at no financial cost and also reaching a base station of a WLAN in a number of hops. In that context, we have proposed a cooperative Web caching system for ad hoc networks, which enables mobile terminals to share Web pages, while minimizing the resulting energy cost. Our system lies in implementing on each mobile terminal: (i) an ad hoc cooperative caching protocol that selects the mobile terminals from which a requested page should be retrieved, in a way that minimizes both energy consumption and network load, (ii) a caching strategy that maintains the local cache in a way that minimizes resource consumption. Such a system is well suited for location-dependent data such as multimedia guided tours. It is further a cheap solution compared to existing ones since it is not based on the infrastructure mode. Preliminary assessment of our proposal has been addressed by providing the energy cost that our system incurs for mobile terminals. We are currently working on further assessment of our proposal, which lies in the implementation of a simulator to thoroughly examine the behaviour of our system. Experiment using the simulator will in particular serves tuning the various weighing functions that we use for cooperative caching. We are also implementing our proposition for multimedia guided tours on the Web.

Acknowledgments:
This work has been partially funded by the OZONE European IST project[5].

[5] http://www.extra.research.philips.com/euprojects/ozone/

References

[1] C. Bennett. *Practical Wap, Developing applications for the Wireless Web.* Cambridge University Press, 2001.

[2] C.M. Bowman, P.B. Danzig, and D. Hardy. The Harvest information discovery and access system. *Computer Networks and ISDN Systems*, 28(1–2), 1995.

[3] D. Chakraborty, F. Perich, A. Joshi, T. Finin, and Y. Yesha. A reactive service composition architecture for pervasive computing environments. In *proc. of PWC*, 2002.

[4] H. Chang, C. Tait, N. Cohen, and M. Shapiro. Web browsing in a wireless environment: disconnected and asynchronous operation in ARTour Web express. In *proc. of MOBICOM*, 1997.

[5] J.H. Chang and L. Tassiulas. Energy conserving routing in wireless ad-hoc networks. In *proc. of INFOCOM (1)*, 2000.

[6] T. Clausen, P. Jacquet, A. Laouiti, P. Muhlethaler, A. Qayyum, and L. Viennot. Optimized link state routing protocol. In *proc. of INMIC*, 2001.

[7] S.R. Das, C.E. Perkins, and E.E. Royer. Performance comparison of two on-demand routing protocols for ad hoc networks. In *proc. of INFOCOM (1)*, 2000.

[8] L. Fan, P. Cao, J. Almeida, and A.Z. Broder. Summary cache: a scalable wide-area Web cache sharing protocol. *IEEE/ACM Transactions on Networking*, 8(3), 2000.

[9] L. Feeney and M. Nilsson. Investigating the energy consumption of a wireless network interface in ad hoc networking environment. *In proc. of INFOCOM (5)*, 2001.

[10] Z. Haas. A new routing protocol for the reconfigurable wireless networks. In proc. of ICUP, 1997.

[11] Z. J. Haas and M.R. Pearlman. The performance of query control schemes for the zone routing protocol. ACM/IEEE Transactions on Networking 9(4), 2001.

[12] W. Heinzelman, A. Chandrakasan, and H. Balakrisnan. Energy-efficient communication protocol for wireless microsensor networks. In proc. of HICSS, 2000.

[13] Intel. Mobile power guide v1.00. http://developer.intel.com/design/mobile/intelpower/, 2000.

[14] Z. Jinag and L. Kleinrock. Web prefetching in a mobile environment. *IEEE Personal Communications*, 5(8), 1998.

[15] D. Johnson and D Maltz. Dynamic source routing in ad hoc wireless networks. In Imielinski and Korth, editors, *Mobile Computing*. 1996.

[16] E. Markatos and C. Chronaki. A top-10 approach to prefetching on the Web. In *proc. of INET*, 1998.

[17] J.M. Menaud, V. Issarny, and M. Banatre. A scalable and efficient cooperative system for Web caches. In IEEE Concurrency. 8(3), 2000.

[18] K. Nakano and S. Olariu. Energy-efficient randomized routing in radio networks. In Proc. of DIALM, 2000.

[19] F. Perich, S. Avancha, D. Chakraborty, A. Joshi, and Y. Yesha. Profile driven data management for pervasive environments. In *proc. of DEXA*, 2002.

[20] C. Perkins. Ad-hoc on-demand distance vector routing. In Proc. of WMCSA, 1999.

[21] G.J. Pottie and W.J. Kaiser. Wireless integrated network sensors. Communications of the ACM, 2000.

[22] V. Valloppillil and K.W. Ross. Cache array routing protocol, Internet draft. http://ircache.nlanr.net/Cache/ICP/carp.txt, 1998.

Investigation of Cache Maintenance Strategies for Multi-cell Environments*

Zhijun Wang, Mohan Kumar, Sajal K. Das, and Huaping Shen

Center for Research in Wireless Mobility and Networking (CReWMaN)
Department of Computer Science and Engineering
The University of Texas at Arlington, Arlington, TX 76019, USA
zwang, kumar, das, hpshen@cse.uta.edu

Abstract. In this paper, three types of cache consistency maintenance strategies involving invalidation reports (IRs) are proposed for application in mobile environments. The three strategies: homogeneous IR, inhomogeneous IR without roaming check, and inhomogeneous IR with roaming check are applied to Timestamp (TS) and Scalable Asynchronous Cache Consistency Scheme (SACCS). Simulation based performance evaluations show that SACCS performs better than TS. Further investigation of SACCS reveals the following two cache system design guidelines: (1) Homogeneous IR strategy should be used for fast intra-roaming mobile users (MUs) and small systems. (2) Inhomogeneous SACCS should be used for inter-roaming MUs while SACCS with roaming check is better if power is not a consideration.

1 Introduction

A wireless data communication system contains a wired network, multiple mobile switching centers (MSCs), base stations (BSs) and mobile users (MUs). Figure 1 shows the architecture of such a system. In the system, An MSC is connected to a wired network and controls multiple BSs through wired links. A BS serves multiple MUs through wireless channels. Each MSC has a cache to store some retrieved data objects. When an MU requests a data object through a BS, the BS will retrieve the data object from its MSC, then broadcasts the data object to the cell. An MSC is also in charge of handoff and cache consistency maintenance between the origin servers and MU caches.

Based on the architecture of Figure 1, there are two types of MU roaming (we use *roaming* to indicate the movement of an MU from one cell to another cell). The first type is called *intra-roaming*, in which an MU roams among different cells which are all controlled by the same MSC. The movement of MU (A) from cell 2 to cell 3 is an example of intra-roaming. The other type is called *inter-roaming*, in which an MU roams between two cells that are belong to different MSCs. Movement of MU (B) from cell 4 to cell 3 is an example of this kind of roaming.

* Acknowledgement: This work is supported by a grant from Texas Advanced Research Program, TXARP Grant Number: 14-771032.

M.-S. Chen et al. (Eds.): MDM 2003, LNCS 2574, pp. 29–44, 2003.

To improve the system performance, some data objects are cached in MU's local buffers. Due to frequent disconnectedness and mobility of MUs, cache consistency is a critical issue for wireless data communication systems. Several cache consistency maintenance strategies have been proposed [1]-[9] to improve the system performance. In the literature, there are two types of cache consistency maintenance algorithms for wireless mobile computing environments: stateless and stateful. In stateful approaches [8], an MSC maintains a state for each MU's cache. When an MU roams to a new MSC, the corresponding cache state is also transferred. Hence the roaming impacts are minimized. But the MSC incurs nontrivial overhead to maintain MU's cache states.

In stateless approaches [1]-[7], an MSC has no knowledge of MU caches. Even though stateless approaches employ simple database management, their ability to support MU's mobility is poor. Typically, in stateless approaches, BSs broadcast the same invalidation reports (IR) to MUs. These IRs invalidate stale copies of cached objects in MUs, regardless of MU's location. However, when the number of BSs is large, such systems are not scalable. To the best of our knowledge, this is first study on the impact of MU's mobility on the cache consistency strategies in multi-cell environments, especially for stateless approaches.

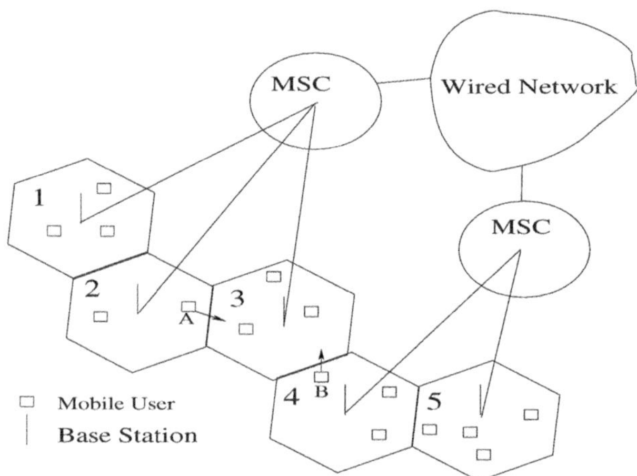

Fig. 1. Wireless Data Communication System Architecture

In this paper, we study the impact of MU's mobility for different cache consistency maintenance strategies for a read-only system. Three types of cache consistency maintenance strategies are proposed for roaming MUs: homogeneous IR, inhomogeneous IR without roaming check, and inhomogeneous IR with roaming check. These three types of strategies are applied to a stateless cache consistency scheme, Time Stamp (TS) [1] and a hybrid scheme, Scalable Asynchronous Cache Consistency Scheme (SACCS) [9] to evaluate their performance. Our study focuses on the cache consistency maintenance for intra- and inter-roaming MUs rather than on MU's handoff and in-session data transaction. The study of the effects of MU's handoff and in-session data transaction are beyond of this paper.

The first set of strategies is named as *homogeneous IR*. In the strategies, all BSs broadcast the same IRs. Therefore, there is no difference between a roaming MU and a static MU for cache consistency maintenance. The cache consistency maintenance for a roaming MU is the same way as that for an MU in a single cell. The strategy is useful only for intra-roaming MUs, because there is a very remote likelihood for all cells belonging to different MSCs to broadcast the same IR. The corresponding strategies for TS and SACCS are called HMTS and HMSACCS.

The second set of strategies, is called as *inhomogeneous IR without roaming check*. In this set of strategies, broadcast IR varies among different cells. In IMTS, the variation of TS, an MH simply drops all cached data objects when it roams to a new cell. In SACCS based strategy, called as IMSACCS, an MU forces a sleep-wakeup event. In other words, it sets all valid cache entries into an uncertain state whenever it roams. The cache consistency is maintained because an uncertain entry must be refreshed or checked before its usage.

The third set of strategies is called as *inhomogeneous IR with roaming check*. In this set of strategies, after an MU roams to a new cell, it needs immediately to send a valid data identifier (ID) list to MSC through its new BS to maintain cache consistency. In this case, the broadcast IR is treated the same way as in the second scheme. The corresponding strategies for TS and SACCS are said to be CKTS and CKSACCS, respectively.

Among these three types of strategies, inhomogeneous IR strategies can be used for both intra- and inter-roaming MUs, while homogeneous IR strategies can only be used by intra-roaming MUs.

The rest of the paper is organized as follows. Section II gives a brief description of TS and SACCS. Three types of extended cache consistency maintenance strategies based on TS and SACCS are proposed for intra- and inter-roaming MUs in Section III. Extensive simulation results are shown in Section IV to evaluate the different schemes. Section V draws some conclusions.

2 Brief Description of TS and SACCS

In this section, we briefly describe a stateless cache consistency scheme TS [1] and a hybrid scheme SACCS [9]. A single cell environment is assumed in the following statement as in the original papers.

2.1 Time Stamp

In TS, a BS broadcasts IRs which include all data object IDs updated during the past kL seconds (where k is a positive integer), every L seconds. An MU cannot answer queries until it receives a broadcast IR. After an MU receives an IR message, it deletes all invalid data objects from its cache. Then the MU answers the buffered queries using cached data objects if the queried data objects are in its cache, otherwise it sends missing queries to BS to retrieve the queried data objects.

If an MU has been disconnected longer than kL seconds, it must drop all cached data objects to maintain the cache consistency after its wakeup. Stateless schemes in [3]-[7] are similar to TS. These strategies do not consider the effects due to mobility of MUs. They either do not consider the mobility or simply assume that all BSs broadcast the same IRs. But it is impractical for all BSs to broadcast same IRs if the BSs are belong to different MSCs. Hence it is necessary to provide some solutions for roaming MUs to maintain cache consistency.

2.2 Scalable Asynchronous Cache Consistency Scheme

SACCS [9] is a highly scalable, efficient, and low complexity algorithm due to the following three simple features: (1) Use of flag bits at server (MSC) and MU's cache to maintain cache consistency; (2) Use of an ID for each entry in MU's cache after its invalidation in order to maximize the broadcast bandwidth efficiency; (3) Rendering of all valid entries of an MU's cache to *uncertain state* when it wakes up.

In SACCS each data object in a BS is associated with a flag bit. If the flag bit is set, the BS must broadcast the data object IR to MUs and reset the flag bit when the data object is updated. Otherwise, the BS does not need to broadcast an IR. The flag bit is set when an MU retrieves the data object from the BS. Hence, unlike synchronous periodic broadcast IR schemes (e.g., [1]), most of the unnecessary IRs can be avoided and consequently substantial bandwidth can be saved.

A data entry in an MU's cache is set to $ID - only\ state$, i.e., the data object is deleted from the entry but its ID is kept, after the data object becomes invalidated. All the valid cache data entries in MUs are set to *uncertain* state after an MU wakes up. This mechanism makes the handling of sleep-wakeup pattern and mobility very simple.

All entries with data objects in uncertain state or ID-only state can be used to identify useful broadcast messages for validation and triggering of data object downloading. Hence, all the MUs strongly cooperate in sharing the broadcast resource.

In the past, cache consistency maintenance strategies focused on single cell environments, without considering the effects of mobility. In the following section, we will present a detail description on the *three* extension strategies based on TS and SACCS to maintain cache consistency for roaming MUs.

3 Cache Consistency Maintenance Strategies for Roaming MUs

In this section, we propose three types of cache consistency maintenance strategies for roaming MUs. At first, we claim that a flag bit can be set for each cached data object in an MSC to reduce the IR traffic for TS scheme.

As in SACCS, when a data object is requested, its flag bit is set. Only data object with a set flag bit need broadcast its IR during its update. The flag bit is reset after its IR is added into periodical IR message (broadcast in the next kL seconds). The difference between TS and SACCS is that the IRs need to be broadcast k times in TS but only once in SACCS. Consistency in TS is maintained due to the following: (1) If the flag bit is not set, no MU requests the data object since its last update, and hence no valid data copy in any MU's cache; (2) If an MU was in sleep state and missed its IRs during its broadcast time (kL seconds), its cached data objects must be dropped after wakeup. This happens when the MU's sleep time is longer than kL. So for TS, we can use a flag bit for each cached data object in MSC to reduce the IR traffic. We will use flag bits for TS in this paper to reduce the IR traffic.

We give detailed descriptions of the three type of strategies in the following subsections.

3.1 Homogeneous IR Strategy

This type of strategy can only be used for intra-roaming MUs. Here, all BSs broadcast the same IRs. When an MU roams among these cells, there is no different between a roaming MU and a static MU for maintaining the cache consistency. Consequently, the impact of mobility of an MU is minimized.

However, the total IR traffic for each cell is increased as the number of cells increases. This is because IR of any retrieved data object must be broadcast to all cells even if it is only retrieved in one cell. In an MSC cache, only a *single* flag bit is needed for each cached data object in both HMTS and HMSACCS. When a data object is retrieved, the corresponding flag bit is set. The flag bit is reset after a data object is updated. The advantage for this scheme is that there is no significant work at MSCs due to MU mobility and the impact of MU mobility is minimized. But the total IR traffic is increased compared with that of the inhomogeneous case. The simulation results in Section 4 show that this approach is superior to others for fast roaming MUs.

3.2 Inhomogeneous IR without Roaming Check Strategy

In this set of strategies, the IR messages vary in different cells. A data object in an MSC cache has multiple flag bits, each flag bit corresponds to a cell, if the data object is retrieved by a cell, the corresponding flag bit is set. When the data object is updated, a BS broadcasts its IR only if the corresponding flag bit is set for that cell. All flag bits are reset after the data object is updated.

In IMSACCS, a sleep-wakeup event is forced when an MU roams. In other words, the MU sets all its valid data objects into uncertain state after its roaming. There is no cache consistency conflict in such an MU, because an uncertain entry must be refreshed or checked before its usage. In IMTS, all cached data objects are simply dropped because there is no cache consistency guarantee for roaming MUs.

The advantage of this set of strategies is the reduction in IR traffic. However, the cache hit ratio is also reduced due to a forced sleep-wakeup event or dropped cache objects. The simulation results in Section 4 show that the approach is superior to others in a system with slow roaming MUs and has better scalability than homogeneous based strategies. The strategy can be used for both intra- and inter-roaming MUs.

3.3 Inhomogeneous IR with Roaming Check Strategy

In this case, the IR traffic treatment is the same way as in inhomogeneous IR without roaming check strategy. When an MU roams into a new cell, it immediately sends an ID list to an MSC through its new BS. The ID list includes all cached data object IDs for CKTS and all valid data object IDs for CKSACCS.

After an MSC receives the ID list, it sets the corresponding flag bit for each data object of that cell. Then any coming IRs for these data objects are broadcast to that cell. Therefore, after the ID list is sent, an MU can use its cache as the same way as in a single cell. The cost of roaming handling for this scheme is an extra uplink for sending the ID list for each roaming. But it can save some bandwidth used for downloading data objects for CKTS and reduce some confirmation uplinks for SACCS. Also the total IR traffic is reduced compared to homogeneous strategies. One disadvantage is the increased uplinks for sending the roaming check message resulting in more battery power consumption.

4 Performance Evaluation

Extensive simulations are done to evaluate the three types of schemes under various multi-cell environments. In the simulation, the query process for each MU is assumed to be Poisson distribution, and the access probability for each data object is followed by a Zipf-like distribution [10], in which the data object access probability p_x is proportional to its rank r, i.e., $p_x = c/r^z$, where c is a normalized constant and z is a parameter between 0 and 1.

The MU's roaming process is also assumed to follow a Poisson distribution with an average sojourn time T_r in a cell. When an MU roams from one cell to another, all requests that are sent to the former BS are dropped, i.e., the in-session hand off and data transaction are not considered in our simulation. The MSC cache is considered infinite and its cache consistency is maintained by wired cache consistency maintenance strategies [11]-[13]. The data object updates can only be done in the MSC and follow a Poisson process. The sleep-wakeup process is modeled as a two state Markov chain with MUs alternating between sleep and awake states.

For SACCS based strategies, in each BS, all downlink messages are buffered in a queue, in which the IR messages have the highest priority, all other messages are scheduled as a FIFO (first in first out) queue and can be scheduled only if there is no IR message in the queue. All uplink messages in a cell are scheduled as FIFO queue policy for both TS and SACCS based strategies. Each cell has an uplink channel with W_u (bps) and a downlink channel W_d (bps). The following parameters are defined in our simulation:

- N: number of data objects
- M: number of MUs in the whole system
- C: MU's cache size ($bytes$)
- B: number of BSs
- λ: average query arrival interval for an MU (sec)
- T_u: average update time interval for a data object (sec)
- T_s: period time of a sleep-wakeup cycle for an MU (sec)
- T_r: average sojourn time in a cell for an MU (sec)
- s: ratio of sleep time to the sleep-wakeup period for a MU
- W_u: uplink channel bandwidth (bps)
- W_d: downlink channel bandwidth (bps)
- b_p: size of a data object ($bytes$)
- b_u: uplink message size ($bytes$)
- b_d: downlink control message size ($bytes$)
- D: average query delay
- UPQ: average uplink per query (i.e., the total number of uplinks divided by the total number of queries)

In all studied cases, we set $z = 0.95$, $b_u = 10$ $bytes$, $b_d = 8$ $bytes$, and IR broadcast period as 10 sec and broadcast window size as 20 for TS based strategies. In case 1, the access rank for all MUs are start from data object 1, i.e., the data object 1 has the highest access rank, the data object 2 is the second highest and so on. In all other cases, the access rank for an MU is shifted by a random number in the range 1 to 100. For example, an MU randomly picks an access rank shift number 60, this means that the data object 61 is the highest access rank, the data object 62 is the second highest and so on up to N, then back to data object 60, which is the lowest access rank.

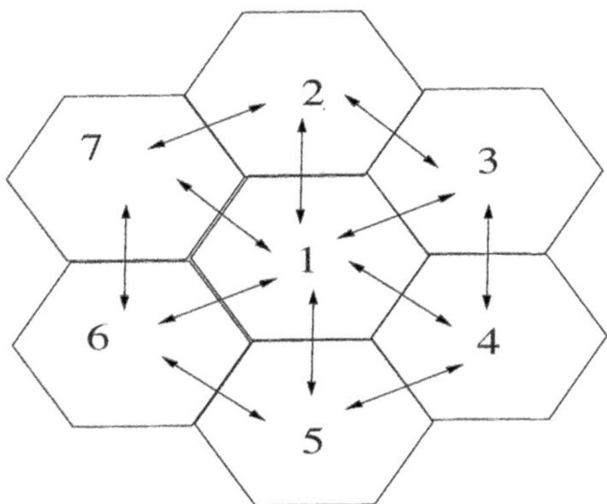

Fig. 2. Seven Cell configuration, all cells are controlled by the same MSC

Two important performance metrics: average query delay (D) and average uplink per query (UPQ) are studied in our simulations. UPQ is calculated as the total number of MU uplinks divided by total number of MU queries. For CKSACCS and CKTS, an extra uplink for sending a roaming ID check list to an MSC is also counted as an uplink when an MU roams, so the UPQ may be larger than 1 for CKSACCS and CKTS in some cases.

Now, we consider an MSC with *seven* cells. Figures 2 shows the configuration of the system. Initially, each cell has the same number of MUs (i.e., $M/7$ MUs). An MU from cell 1 has equal probability (1/6) of moving to any of 6 neighbor cells. An MU in any other cell has 1/6 probability of moving to cell 1 and 5/12 probability of roaming to another neighboring cell. This balances the number of MUs in each cell, and avoids to MUs centered into a single cell.

In the first case, we evaluate all 6 schemes. In the last three cases, we study only the SACCS based schemes because the communication channels are congested by TS based schemes.

4.1 Effects on Average MU Sojourn Time

In this case, we study the effects on average sojourn time of an MU to evaluate the performance of 6 different strategies. The parameters in this case are set as: $N = 4000$, $M = 350$, $C = 100$ $Kbytes$, $B = 7$, $T_s = 1800$ sec, $T_u = 300$ sec, $W_d = 10000$ bps, $W_u = 1000$ bps, $b_p = 1$ $Kbytes$, $\lambda = 25$ sec and $s = 0.5$.

From Figures 3 and 4, we can see that HMSACCS has the best performance both on D and UPQ in short average sojourn time (less than 1600 sec). When the average sojourn time is longer than 1600 sec, the IMSACCS has the best performance on D but a little bit more UPQ than HMSACCS. The performance for CKSACCS is very close to IMSACCS for slow roaming MUs. But very poor performance for fast roaming MUs due to the extra uplinks for sending an ID list to the MSC. In the short average sojourn time case, the cost of extra uplink for IMSACCS and CKSACCS is much more than the cost of extra IR traffic for HMSACCS, so a poorer performance of IMSACCS and CKSACCS is observed. When the average sojourn time increases, the cost of extra uplink for IMSACCS and IMTS decreases to lower than the cost of extra IR messages in HMSACCS. Then the performance of IMSACCS and CKSACCS is better than that of HMSACCS.

From Figures 5 and 6, it is clear that HMTS always has the longest D but the least UPQ. This indicates that the bandwidth cost of extra IR traffic in HMTS is more than the cost of extra data object downloaded in IMTS and CKTS. CKTS has the shortest average delay almost throughout the sojourn time because it reduces IR traffic and also saves all valid data objects during the roaming. But CKTS requires more uplink bytes to send the cached ID list, and hence incurs more MU battery power consumption.

From Figures 3 to 6, we can see that SACCS based strategies perform much better than TS based strategies. Ds for SACCS based strategies are a magnitude lower than those in TS based strategies. UPQs for SACCS based strategies are 10% less than that of TS based strategies. This shows that SACCS based

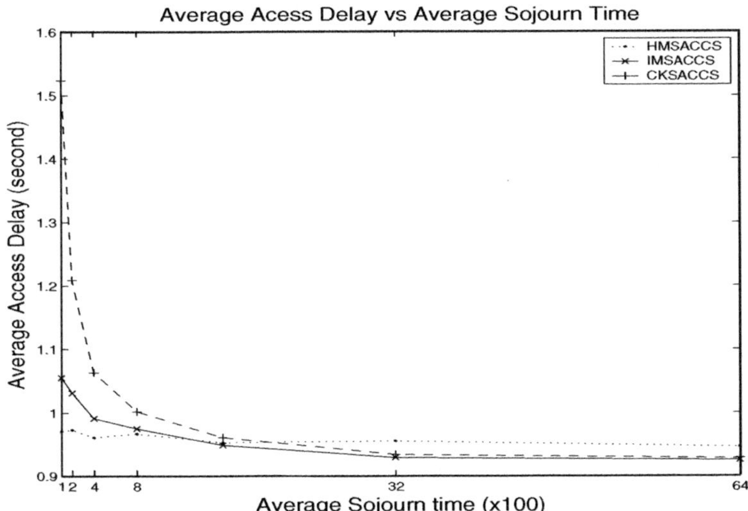

Fig. 3. Average Access Delay vs Average MU Sojourn Time for SACCS Based Schemes

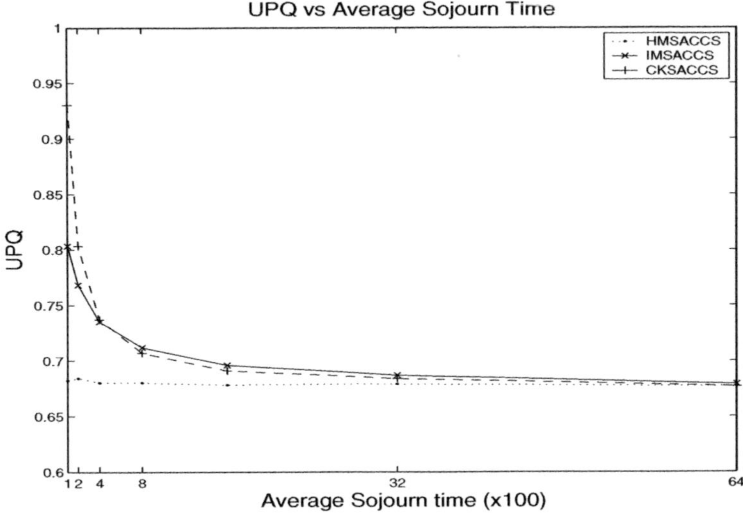

Fig. 4. UPQ vs Average MU Sojourn Time for SACCS Based Schemes

strategies have much better performance than TS based strategies in multi-cell environments, similar to *single* cell environments [9].

The results show that homogeneous IR for SACCS based strategy is more efficient for fast roaming MUs, and inhomogeneous IR strategy is more efficient for slow roaming MUs. For TS based strategies, inhomogeneous IR strategy has

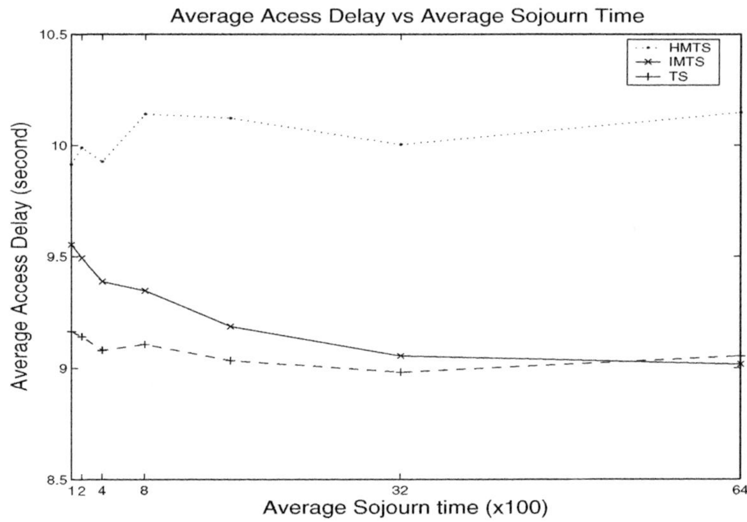

Fig. 5. Average Access Delay vs Average MU Sojourn Time for TS Based Schemes

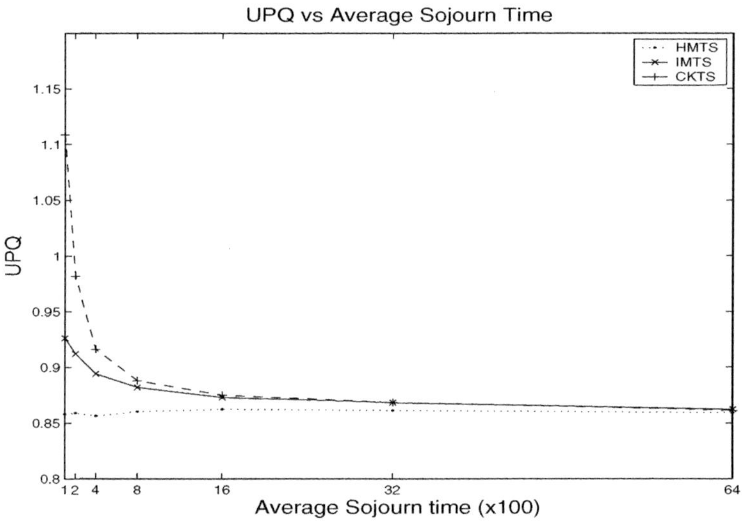

Fig. 6. UPQ vs Average MU Sojourn Time for TS Based Schemes

shorter D for both fast and slow roaming MUs. This is because the each IR is broadcast multiple times in TS, and results in much more bandwidth cost of IR traffic in TS based strategies.

4.2 Effect of the Number of MUs

In this case, we have ten types of data objects, each type of data object has different size, update time and the fraction of the whole database. Table 1 summarizes the ten types of data objects. Here, we set a small size for faster update data objects and large size for slower update data objects. This is reasonable for real web environments.

The λ of an MU has equal probability from a set values of (20, 40, 60, 90, 120) *sec*. Similarly, the T_s of an MH has equal probability from a set values of (600, 1200, 1800, 2400, 3600) *sec*, s from a set values of (0.2, 0.35, 0.5, 0.65, 0.8) and T_r from a set values of (100, 600, 1800, 3600, 18000) *sec*. The other parameters are set as: $C = 1\ Mbytes$, $W_d = 100\ Kbps$ and $W_u = 2\ Kbps$.

Table 1. Ten Types of Data Object for Subsection 4.2

Data Type	1	2	3	4	5	6	7	8	9	10
Size(Bytes)	400	800	1.2*K*	1.6*K*	2*K*	2.4*K*	2.8*K*	3.2*K*	3.6*K*	4*K*
T_u (*sec*)	10	60	120	300	600	1.2*K*	1.8*K*	3.6*K*	7.2*K*	18*K*
percentage(%)	5	5	10	10	20	20	10	10	5	5

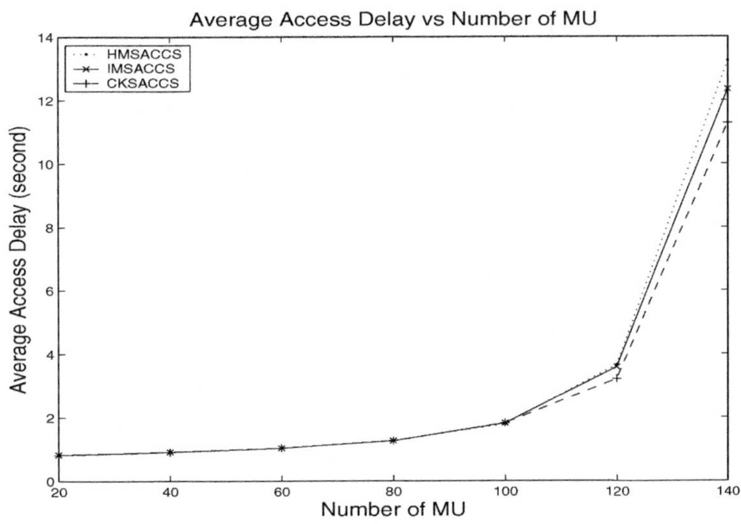

Fig. 7. Average Access Delay vs Number of MU for SACCS Based Schemes

From Figures 7 and 8, we can see that in a system with small number of MUs, there is no much difference on D for three strategies. When the number of MUs is over 120, CKSACCS has the shortest D, and HMSACCS has the longest. For example at $M = 140$, Ds for HMSACCS, IMSACCS and CKSACCS are 13.2 *sec*, 12.4 *sec* and 11.3 *sec*, respectively. This indicates that the total IR traffic

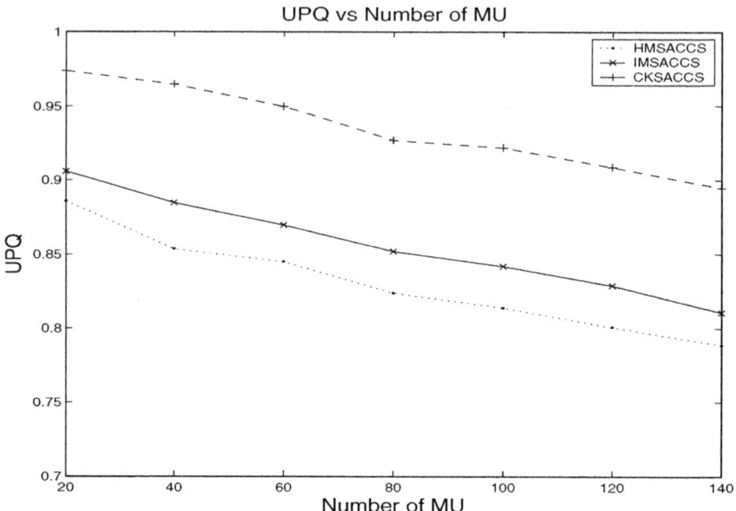

Fig. 8. UPQ vs Number of MUs for SACCS Based Schemes

is increased as the number of MUs increases and it makes the cost of IR traffic more important on the whole system performance. Hence, for a system with a large number of MUs, the *inhomogeneous* schemes are better than *homogeneous* schemes on the D metric. The UPQ for HMSACCS is always the least, and UPQ for IMSACCS is always less than that of CKSACCS. This means that CKSACCS outperforms others on D but incurs more MU battery power due to sending of roaming check ID list. Among these strategies, IMSACCS is in the middle on both performance metrics. The result shows that inhomogeneous IR strategy is more scalable than homogeneous IR strategy.

4.3 Effect of Database Size

In this case, we study the effects of the database size. Similar to previous subsection, we have ten types of data objects shown in Table 2. Each MU has a λ randomly choosing from a set value of (20, 40, 60, 90, 120) sec, T_s from a set values of (500, 1000, 1500, 2000, 3000) sec, s from a set values of (0.2, 0.35, 0.5, 0.65, 0.8) and T_r from a set values of (300, 1000, 2000, 3600, 86400) sec. The other parameters are set as: $N = 10K$, $M = 700$, $W_d = 100\ Kbps$, $W_u = 2\ Kbps$ and $C = 1\ Mbytes$.

From Figures 9 and 10, we can see a similar pattern for all three schemes as in Subsection 4.2. In small database system, HMSACCS has the best performance, but it has the longest D as the database size is over 8000. The D for IMSACCS is a little bit longer than that of CKSACCS, while the UPQ for IMSACCS is less than that of CKSACCS. The results show that inhomogeneous scheme is more scalable than homogeneous based scheme in term of database size.

Table 2. Ten Types of Data Object for Subsection 4.3

Data Type	1	2	3	4	5	6	7	8	9	10
$Size(Bytes)$	500	$1K$	$2K$	$4K$	$5K$	$8K$	$10K$	$20K$	$40K$	$50K$
$T_u(sec)$	10	60	120	300	600	$1.8K$	$3.6K$	$7.2K$	$18K$	$86.4K$
$percentage(\%)$	5	5	10	10	20	20	10	10	5	5

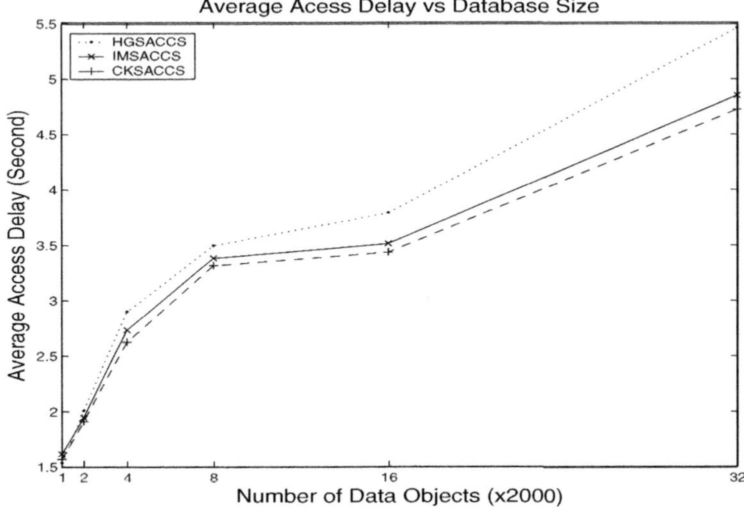

Fig. 9. Average Access Delay vs Database Size for SACCS Based Schemes

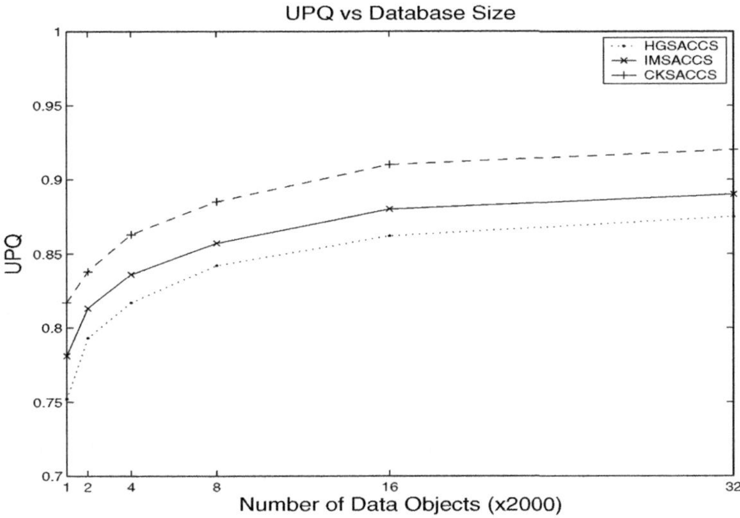

Fig. 10. UPQ vs Database Size for SACCS Based Schemes

4.4 Effect of MU Cache Size

In this subsection, we study the effects of MU cache size. Again, we have ten types of data objects shown in Table 3. Each MU has a λ randomly choosing from a set values of (20, 40, 60, 80, 100) *sec*, T_s from a set values from a set values of (500, 1000, 1500, 2000, 3000) *sec*, s from a set values of (0.3, 0.4, 0.5, 0.6, 0.7) and T_r from a set values of (100, 500, 1000, 1500, 2000) *sec*. The other parameters are set as: $N = 10k$, $M = 400$, $W_d = 200\ Kbps$, $W_u = 2\ Kbps$.

Table 3. Ten Types of Data Object for Subsection 4.4

Data Type	1	2	3	4	5	6	7	8	9	10
$Size(Bytes)$	500	$1K$	$2K$	$4K$	$8K$	$10K$	$15K$	$20K$	$40K$	$50K$
$T_u(sec)$	60	120	300	$1.2K$	$1.8K$	$3.6K$	$7.2K$	$18K$	$86.4K$	$259.2K$
$percentage(\%)$	5	5	10	10	20	20	10	10	5	5

From Figures 11 and 12, we can see that the performance of the three strategies becomes better as the MU's cache size increases. This effect is more significant in the small cache size situation. When the cache size increases from $0.5M$ to $1M$, D decreases from about $6sec$ to about $3.5sec$ for the three strategies. When cache size is over $1M$, the improvement of system performance slows down. HMSACCS has the least UPQ in all cache sizes. The D for CKSACCS is a little bit shorter than that of IMSACCS but the UPQ is larger than IMSACCS for the entire range of cache sizes from 0.5 M to 3 M. This means that CKSACCS saves some downlink bandwidth at the cost of more uplink bandwidth and MU battery power.

Through the above four cases, we have the following conclusions on the SACCS based strategies: (1) For intra-roaming MUs, homogeneous IR strategy is more efficient for fast roaming MUs, and inhomogeneous IR strategy is more scalable; (2) for inter-roaming MUs, IMSACCS have a little bit longer D but less UPQ, and CKSACCS has a shorter D but cost more UPQ.

Based on the above conclusions, we have the two guideline for cache system design: (1) For fast roaming MUs and small systems, homogeneous IR strategy should be used for intra-roaming MUs. (2) An inter-roaming MU should use CKSACCS if it has enough battery power, and should use IMSACCS if the power supply is a consideration, the two strategies can coexist in the same system.

5 Conclusion

In this paper, we propose three types of cache consistency maintenance strategies for roaming MUs: homogeneous IR, inhomogeneous IR without roaming check, and inhomogeneous IR with roaming check. All strategies are applied to TS and SACCS to evaluate the system performance under various multi-cell environments. The simulation results show that SACCS based schemes perform better than TS based strategies on all average sojourn time ranges. But no single SACCS based strategy is superior to others in all cases.

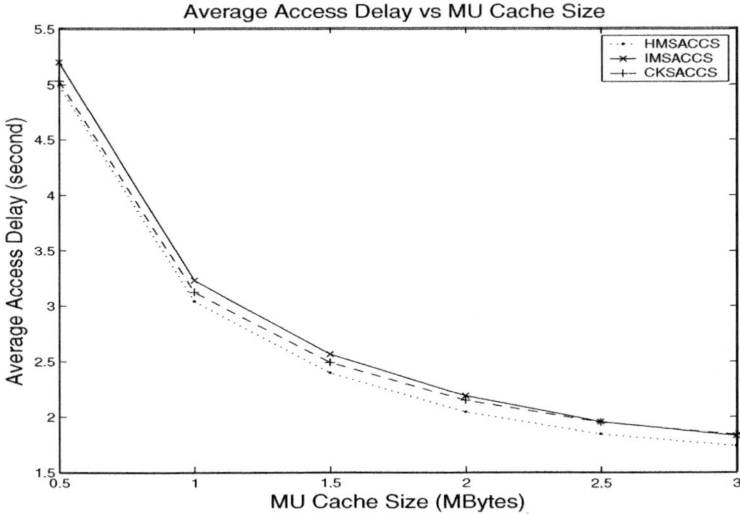

Fig. 11. Average Access Delay vs MU Cache Size for SACCS Based Strategies

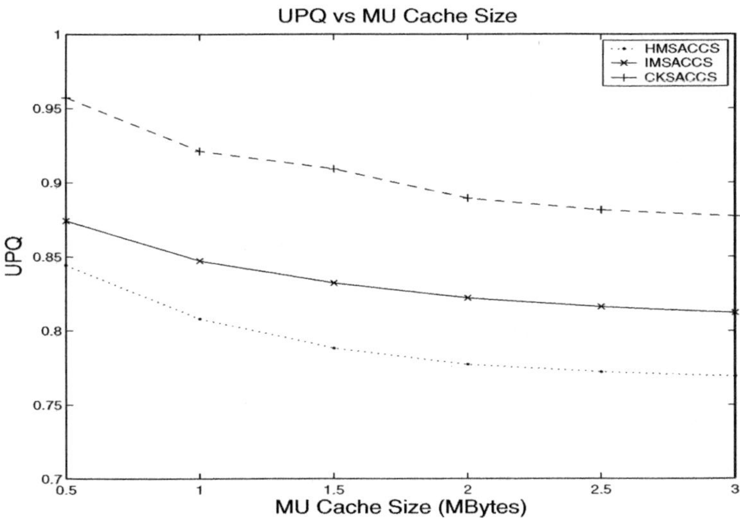

Fig. 12. Average Uplink per Query vs MU Cache Size for SACCS Based Strategies

After evaluating the performance of the strategies, we introduce two cache system design guidelines : (1) For fast intra-roaming MUs and small systems, homogeneous IR strategy should be used. (2) An inter-roaming MU should use CKSACCS if it has enough battery power, and use IMSACCS if the power supply is a consideration.

How to dynamically apply the two design guidelines to different multi-cell environments to achieve overall optimal system performance is our next work. Also cache replacement is also a consideration in our further works.

References

1. D. Barbara and T. Imielinksi," Sleeper and Workaholics: caching strategy in mobile environments ", *In Proceedings of the ACM SIGMOD Conference on Management of Data*, pp1-12, 1994.
2. J. Jing, A. Elmagarmid, A. Heal, and R. Alonso. "Bit-sequences: an adaptive cache invalidation method in mobile client/server environments", *Mobile Networks and Applications*, pp 115-127, 1997.
3. Q. Hu and D.L. Lee, "Cache algorithms based on adaptive invalidation reports for mobile environments", *Cluster Computing*, pp 39-50, 1998.
4. G.Y. Liu and G.Q. McGuire Jr, "A mobility-aware dynamic database caching scheme for wireless mobile computing and communications," *Distributed and Parallel Databases*, pp 271-288, 1996.
5. K.L. Wu, P.S. Yu and M.S. Chen, "Energy-efficient caching for wireless mobile computing", *In 20th International Conference on Data Engineering*, pp 336-345, 1996.
6. G. Cao, "A scalable low-latency cache invalidation strategy for mobile environments", *ACM Intl. Conf. on Computing and Networking (Mobicom)*, August, pp200-209, 2001
7. K. Tan, J. Cai and B. Ooi, "An evaluation of cache invalidation strategies in wireless environments", *IEEE Trans. on Parallel and Distributed System*, pp789-897, 12(8), 2001
8. A. Kahol, S. Khurana, S.K.S. Gupta and P.K. Srimani," A strategy to manage cache consistency in a distributed mobile wireless environment", *IEEE Trans. on Parallel and Distributed System*, 12(7), pp 686-700, 2001.
9. Z. Wang, S. Das, H. Che and M. Kumar,"SACCS: Scalable Asynchronous Cache Consistency Scheme for Mobile Environments", *Technical Report UTA-CSE-Cache#01*, Dept of Computer Science and Engineering, The University of Texas of Arlington, July 2002.
10. L. Breslau, P. Cao, J. Fan, G. Phillips and S. Shenker, "Web caching and Zipf-like distributions: evidence and implications, ", *Proceedings of IEEE INFOCOM'99*,pp126-134, 1999
11. H. Yu, L. Breslau and S. Shenker, "A scalable web cache consistency architecture", *In Proceedings of the ACM SIGCOMM*, August, 1999.
12. H. Yu, L. Breslau and S. Shenker, "A scalable web cache consistency architecture", *In Proceedings of the ACM SIGCOMM*, August, 1999.
13. P. Cao and C. Liu, "Maintaining strong cache consistency in the World-Wide Web", *In Proceedings of the International Conference on Distributed Computing Systems*, pp 12-21, 1997.

Resilient Data-Centric Storage in Wireless Ad-Hoc Sensor Networks

Abhishek Ghose, Jens Grossklags, and John Chuang

University of California at Berkeley
aghose@eecs.berkeley.edu, {jensg,chuang}@sims.berkeley.edu

Abstract. Wireless sensor networks will be used in a wide range of challenging applications where numerous sensor nodes are linked to monitor and report distributed event occurrences. In contrast to traditional communication networks, the single major resource constraint in sensor networks is power, due to the limited battery life of sensor devices. It has been shown that data-centric methodologies can be used to solve this problem efficiently. In data-centric storage, a recently proposed data dissemination framework, all event data is stored by type at designated nodes in the network and can later be retrieved by distributed mobile access points in the network. In this paper we propose Resilient Data-Centric Storage (R-DCS) as a method to achieve scalability and resilience by replicating data at strategic locations in the sensor network. Through analytical results and simulations, we show that this scheme leads to significant energy savings in reasonably large-sized networks and scales well with increasing node-density and query rate. We also show that R-DCS realizes graceful performance degradation in the presence of clustered as well as isolated node failures, hence making the sensornet data robust.

1 Introduction

Wireless sensor networks have emerged as a promising solution for a large number of monitoring applications, in fields as diverse as climatic monitoring, tactical surveillance, and earthquake detection. With improvements in sensor technology, it has become possible to build small sensor devices with relatively high computational power at low costs.

In most communication networks, naming of nodes for low-level communication leverages topological information. An example of this is the Internet (*point-to-point communication model*) where IP addresses are assigned to each node, and these serve as unique node identifiers in IP routing. Such a naming scheme is not very efficient in a sensor network scenario, since the identity of individual sensor nodes is not as important as the data associated with them. *Data-centric* models have been proposed for sensor networks, in which the sensor data itself (as contrasted to sensor nodes) is *named*, based on attributes such as event-type or geographic location. In particular, data-centric routing [5] and data-centric storage [9] have been shown to be energy-efficient in a sensor network scenario.

M.-S. Chen et al. (Eds.): MDM 2003, LNCS 2574, pp. 45–62, 2003.

Ratnasamy *et al.* have proposed data-centric storage (DCS) [7,9] as a data-dissemination paradigm for sensor networks. In DCS, data is stored, according to event type, at corresponding sensornet nodes. All data of a certain event type (*e.g.*, Temperature measurements) is stored at the same node. A significant benefit of DCS is that queries for data of a certain type can be sent directly to the node storing data of that type, rather than flooding the queries throughout the network (unlike data-centric routing proposals [4,5]). DCS is based on the low-level routing functionality provided by the GPSR geographic routing algorithm [6], and on distributed hash-table functionality provided by peer-to-peer lookup algorithms such as CAN [8] and Chord [10]. It is shown in [7] and [9] that DCS offers reduced total network load and peak (hotspot) network usage.

In this paper, we propose replication of control and data information in a DCS framework as the primary mechanism for reducing the data retrieval traffic and increasing resilience to node failures. We propose the storage of data of a particular type at one of several *replica nodes* in the network assigned to this type, and storage of control and summary information pertaining to this type at geographically distributed *monitor nodes* in the network. By increasing the number of nodes where data can be stored for each event-type, as well as maintaining summary and control information at several nodes, we decrease both (1) the average cost of storing data and (2) the average cost of querying data. Our preliminary results suggest that this scheme, which we call Resilient Data-Centric Storage (R-DCS), outperforms existing schemes in terms of scaling to a large number of nodes and a large number of queries. We also show that, in the case where nodes in the sensor network are unreliable and experience random failures, our scheme does not experience a dramatic increase in the number of messages sent as the node failure rate is increased. R-DCS also maintains a high query success rate in this scenario. Hence our scheme realizes graceful performance degradation in the presence of node failures.

The rest of the paper is organized as follows. In Section 2, we describe some basic concepts of sensor networks and give a brief overview of DCS. In Section 3, we propose intelligent replication in sensor networks for resilience and scalability, building upon the DCS framework. We analyze the cost structure of R-DCS analytically in Section 4. Section 5 presents simulation results showing that R-DCS can lead to significant performance improvements in certain scenarios. Finally, we present our conclusions in Section 6.

2 Background and Related Work

In this section we describe basic concepts in the domain of sensor networks, describe data organization in a typical sensor network, and give a brief overview of DCS [9].

2.1 Overview

Wireless sensor networks have certain unique features which must be accounted for in any data dissemination methodology designed for such networks. Sensor devices built using state-of-the-art technology have significantly higher processing capabilities and storage capabilities than available bandwidth [4] (this is in contrast to wired networks, where an explosion of available bandwidth has led to a drastic reduction in its relative cost). The reason for this difference is that sensors typically have limited battery-life. Hence they must use low-power (and consequently, low-bandwidth) wireless communication techniques to conserve battery power. This encourages the use of computational techniques to reduce the total communication overhead in the network.

The gateway through which sensor networks communicate with the external world (*e.g.*, a monitoring terminal or the Internet) is called an *access point*. We use the term *access path* to refer to the set of data paths from the sensor nodes to the access point. In a typical scenario, we can expect these access points to have higher communication load than other sensornet nodes. In a high-traffic scenario, such access points can become a bottleneck point in the sensornet (*hotspot*). An essential design requirement, then, is to minimize the peak amount of traffic flowing through these access points. This issue is addressed in Section 3.4.

2.2 Data Organization in Sensor Networks

Prior research [1,4] has shown that the tight energy constraints of wireless sensor networks can be more efficiently achieved by using an *attribute-based* naming system than by commonly used topological naming schemes (*e.g.*, IP). Such attributes could be pre-defined to reduce the overhead during actual communication. For example, we could classify all sensor data in an environmental sensing network as being of types "temperature", "pressure" and "humidity" and name all such data by including these pre-defined *event attributes* in the data itself.

At the lowest level, when an event occurs the sensors record and store the event data locally, and *name* this data based on its attributes. The low-level output from sensors (*observations*) is *named* based on the attributes of the associated data. This data can be handled in a number of different ways. Three canonical approaches [9] are considered: *External Storage (ES)* in which all event data is stored at an external storage point for processing, *Local Storage (LS)* in which all event information is stored locally (at the detecting node) and *Data-Centric Storage (DCS)* in which all event data is stored by event-type within the sensornet at designated nodes. These three methods involve substantially different assumptions and cost-benefit tradeoffs, which we analyze in Section 4.

Queries are used to retrieve event information from the sensornet. It is important to consider the ratio of query traffic to event-detection traffic while designing a sensor network. Each of the canonical approaches described above (as well as our approach: R-DCS) has different relative costs for query and event traffic. Hence, depending on the function of a sensor network, one of the above *canonical approaches* might be more useful than the others.

2.3 Data-Centric Storage

In this section we describe the concept of DCS [9] and the mechanisms required to support it. DCS uses a distributed hash-table (DHT) and offers the following interface: (1) the **Put(dataName, dataValue)** primitive to store the value of the data corresponding to a certain event at the sensornet node corresponding to the dataName (which serves as the *key* in the DHT and is typically based on the *event-type*). The name of the data is typically based on the relevant *event-type*. (2) the **Get(dataName)** primitive to retrieve the value of the data stored at the node corresponding to the given dataName. DCS uses the GPSR [6] geographic routing algorithm for low-level routing. It then builds a DHT [8,10] on top of GPSR.

DHT over GPSR The central idea in using a DHT is to hash the *name* of a certain event to a key (*dataName*) which is a location somewhere within the boundaries of the sensornet. The *put(dataName, dataValue)* primitive sends a packet with the given payload into the sensornet which is routed towards the location *dataName*. The *get(dataName)* primitive is routed to the node closest to the *dataName* location, which then transmits a packet to the node originating the query with the corresponding data. In a sensornet with completely stationary and reliable nodes, this approach is sufficient.

DCS Extensions In order to make DCS resilient to node failures and mobility, the authors [7,9] have proposed certain extensions to their basic scheme. The *storage node* for an event type periodically routes a *refresh* message to all nodes which had transmitted event-data to this node. Regular GPSR routing returns these refresh messages to the storage node along the network perimeter. In the intermittent time interval, the nodes in the sensor network could be displaced from their original locations. If a new node is closer to the *location of the original storage node* than the original storage node itself, this new node will become the storage node. Timer based algorithms ensure that in case a storage node dies in this fashion, the new storage node automatically starts generating refresh messages. This process is called the Perimeter Refresh Protocol [7] and is used to accomplish replication of (key,value) pairs and their consistent placement at the appropriate home nodes when the network topology changes.

In [7] Ratnasamy *et al.* have proposed a scheme called Structured Replication in DCS (SR-DCS) to achieve load-balancing in the network. SR-DCS uses a hierarchical decomposition of the key space and associates each event-type e with a hierarchy depth d. It hashes each event-type to a *root* location. It then computes $4^d - 1$ *images* of *root*. When an event occurs, it is stored at the closest image node. Queries are routed to all image nodes, starting at the *root* and continuing through the hierarchy. It has been shown through simulations that SR-DCS significantly improves the scalability of DCS, and is useful for frequently detected events. It must be noted, however, that SR-DCS does not involve actual *replication* of data. It only stores *one* copy of any event-data at the closest image node. If all nodes in a certain location fail simultaneously (clustered failures), SR-DCS might not be able to recover the data stored at these nodes. The *root* node is a single point-of-failure in the sense that if the *root* for event-type e fails, one might not be able to issue any queries for event-data corresponding to this type. These extensions are described in detail in [7].

3 Resilient Data-Centric Storage

In this section we describe our extensions to DCS which achieve the following: (a) Minimize query-retrieval traffic (hence saving energy consumption) (b) Increase data availability, ensuring that event information is not lost even with multiple node failures. The original version of DCS [9] has all events of the same type stored in one sensornet node. It is evident that if there are too many events of a particular type, then this storage node will become a bottleneck point (*hotspot*) in the network. Our data-dissemination scheme, Resilient Data-Centric Storage (R-DCS), overcomes these issues by a two-level replication strategy (*control* and *data*). We first describe the architecture of R-DCS and then outline its operational details.

3.1 Architecture

In R-DCS, we partition the coordinate space of the sensornet field into Z zones. We denote the set of available zones as $z_j : j = 1, \cdots, Z$. This zoning could be done on the basis of geographical boundaries, as shown in Figure 1. These zones can contain sensor nodes operating in three possible modes:

- **Monitor Mode**: Each zone has one monitor node for each event-type. The monitor node stores and exchanges information in the form of a *monitoring map* for each event-type. The monitoring map includes control and summary information in the following fields:
  - *List of zones containing replica nodes* (for forwarding event data and queries).
  - *List of zones containing monitor nodes* (for facilitating map exchange).
  - *Event summaries* (for facilitating summary-mode queries). The exact nature of event summaries depends on the event-type. For example, in a sensornet designed for temperature monitoring, the summary information could contain the number of events detected and the average temperature reading for each zone.
  - *Bloom filters* (for enabling attribute-based queries). Event-data is organized in the form of a set of attributes and their values. In the temperature monitoring case, for example, these attributes could be Event-time and Temperature. A user might want to access all temperature readings between 30 and 40. Bloom filters [2,3] offer an efficient way to support attribute-based queries. These are described in detail in Section 3.2. Note that this field is required only if we want to support attribute-based queries.
- **Replica Mode**: Each zone has at most one replica node for each event-type. The replica node, if present, is always the same as the monitor node. In addition to performing the functions of a monitor node, the replica node actually stores event-data for the given event-type.

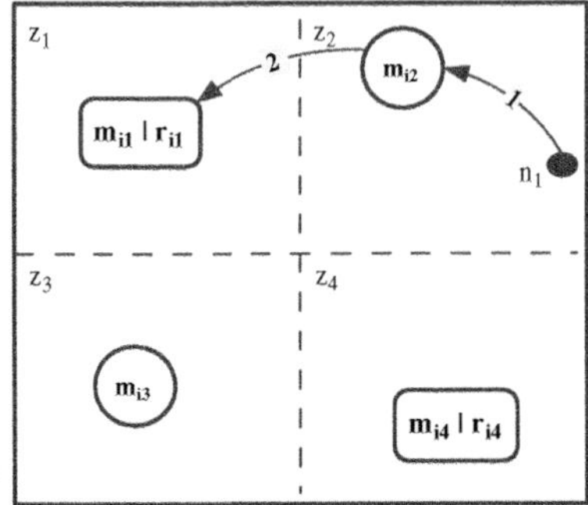

Fig. 1. Event Storage in RDCS

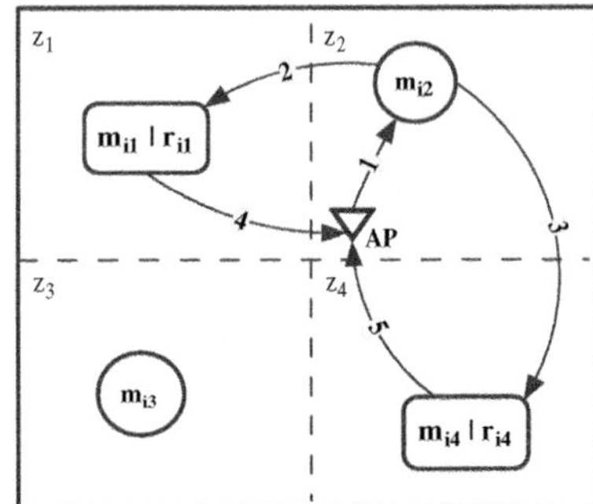

Fig. 2. Querying in R-DCS (*list*)

- **Normal Mode**: All nodes which are not *monitor* or *replica* operate in this mode. A normal node may originate or forward (*i.e. route*) event-data, but is not involved in storing any event data or control information.

Let E denote the number of event-types in the sensornet. Let M_i be the total number of monitor nodes for each event type e_i. Let R_i be the number of replica nodes for each event-type e_i. The following system constraints must be satisfied for each $i = 1, \cdots, E$.

1. $R_i \leq M_i \leq Z$ - This holds because each zone may have at most one replica node and one monitor node, and all replica nodes are monitor nodes as well. Under normal operations without clustered node failures (a majority of nodes failing in one zone), there will be one monitor node per zone: $M_i = Z$.
2. $R_i \geq 1$ - There must be at least one replica node in the network, since all event-data for each event-type is stored at the respective replica nodes.

For our DHT, we use a hash function H which is a function of the event-type e_i and the zone z_j. If event-type e_i in zone z_j hashes to a location $(x_{ij}, y_{ij}) \equiv H(e_i, z_j)$, then a sensor node m_{ij} geographically closest to (x_{ij}, y_{ij}) is the monitor node for event-type i within zone j. Depending on local decision rules (described in Section 3.2), this monitor node may also serve as a replica node r_{ij}. For load balancing, it is desirable that the function $H(e_i, z_j)$ be chosen such that for each zone z_j, different event-types e_i hash to distinct nodes.

3.2 Operational Procedures

In this section, we describe the methodology for performing common sensornet operations such as data storage and queries.

Event Storage A sensing node (situated in zone j) sends an event of type e_i to the monitor node in the same zone m_{ij}. If this monitor node is also the replica node r_{ij}, then the event-data is stored at r_{ij}. If not, the data is forwarded to the *closest* replica node for this event-type. The closest replica node can be determined from the information in the *list of replica nodes* field of the local monitoring map. The target replica node stores the event-data and updates its local copy of the monitoring map. This operation is illustrated in Figure 1.

Event Query We envision three types of queries in an R-DCS system - *summary*, *list* or *attribute-based*. These are now described in detail, along with mechanisms to support these in R-DCS.

- **List**: A list query for an event-type is a request for *all* stored data for events of this type. A querying node in zone z_j sends the query for event-type e_i to the local monitor (and possibly replica) node m_{ij}. The monitor node then duplicates the query and forwards it to all other active replica nodes $r_{ix} : x = 1, \cdots, Z$ in the sensornet. All active replica nodes reply directly to the querying node with event-data. This operation is illustrated in Figure 2.
- **Summary**: As the name suggests, the querying node requests a summary of event information for an event-type. A querying node in zone z_j sends the query for event-type e_i to the local monitor node m_{ij}. The monitor node responds with the *event summary* information from the local monitoring map. This is illustrated in Figure 3.
- **Attribute-based**: An attribute-based query requests data for all events which match certain constraints on their attribute values. A querying node in zone z_j sends the query for event-type e_i to the local monitor (and possibly replica) node m_{ij}. The monitor node then duplicates the query and forwards it to all other active replica nodes in the sensornet with Bloom filter *matches*. Bloom filters are explained in some detail in Section 3.2. All active replica nodes reply directly to the querying node with event-data.

Periodic Update and Exchange of Monitoring Map When an event of type e_i occurs in zone j, it updates the local monitoring map in m_{ij}. However for

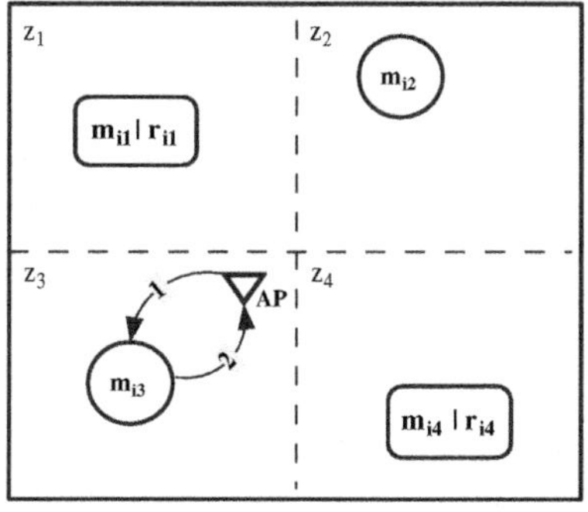

Fig. 3. Querying in R-DCS (*summary*)

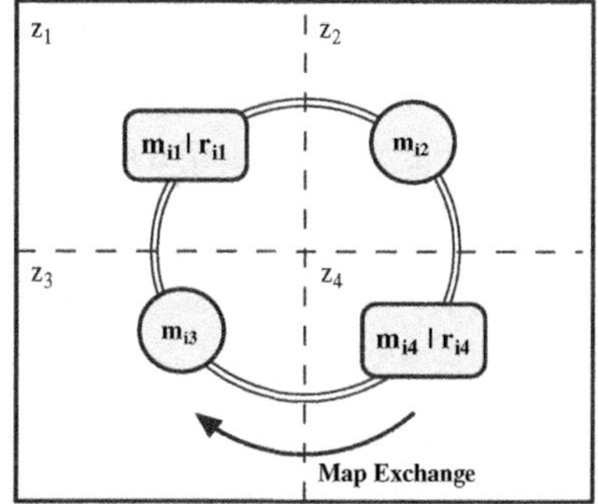

Fig. 4. Logical Ring in R-DCS

global consistency of information such as the number of replica nodes for type e_i and event-summary information, these monitoring maps must be exchanged between the respective monitor nodes at periodic intervals. For this purpose, all active monitor nodes for a type e_i form a logical ring as shown in Figure 4. Each zone has two adjacent zones. When a monitor node receives a new map, it adds its own local updates (based on events received since the last map update) and forwards it to the next monitor node.

Switching between Modes A node in zone j switches from *normal* mode to *monitor* mode (for event-type i) when it becomes the node closest to the location $(x_{ij}, y_{ij}) \equiv H(e_i, z_j)$. If due to node mobility, a new node $m_{ij'}$ becomes closer to this location, then the monitoring map is handed over from m_{ij} to $m_{ij'}$, and $m_{ij'}$ becomes the new monitoring node in zone j. $m_{ij'}$ also becomes a replica node $r_{ij'}$ if $m_{ij} \equiv r_{ij}$ was a replica node, and the relevant stored data is handed over from r_{ij} to $r_{ij'}$.

A node switches from *monitor* mode to *replica* mode and vice-versa based on certain local criteria such as event storage and query traffic loads, as well as residual energy in the node. In R-DCS, every monitor/replica node logs all its storage and query traffic loads for a certain window of time $(t - \tau, t)$, as well as its residual energy at time t, and uses a composite of this information to determine the current *activity coefficient* $A_{ij}(t)$ of a monitor/replica node for event-type e_i in zone j. The residual energy term accounts for the fact that sensornet nodes have limited energy, and hence replica nodes try to conserve energy (by becoming a monitor node) when their residual energy is running *low*.

We define two threshold values for the activity coefficient for event-type e_i: a lower threshold $\lambda_{i(r\to m)}$ and an upper threshold $\lambda_{i(m\to r)}$ (with a buffer zone in between to account for hysteresis effects). A monitor-node m_{ij} switches to replica mode when it's current activity coefficient exceeds the upper threshold, *i.e.*, $A_{ij}(t) > \lambda_{i(m\to r)}$. Conversely, a replica-node r_{ij} switches to monitor-mode when it's current activity coefficient goes below the lower threshold. The only constraint is that there must be at least one replica node in the network. Further, any replica node switching to monitor mode must handoff its stored data to one of the other replica nodes. The values of $\lambda_{i(r\to m)}$ and $\lambda_{i(m\to r)}$ are determined based on the relative costs of storage, queries and of conserving residual energy.

Handling Node Failure In the above discussion we assumed that all nodes are stable and able to continuously route, monitor or store data. Consider the more realistic case, when nodes fail with a certain failure probability f. We will discuss the effect of node failure and mechanisms to solve this problem in the following sections.

- **Failure of Monitor and Replica Nodes** When a monitor node in a certain zone fails, we can use PRP described in [7] and in Section 2.3 to find/elect an alternate monitor node for this zone. PRP also ensures that the new monitor node has a copy of the monitoring map through local refresh mechanisms. If this failed node is also a replica node, the event data stored in it might be lost. However using PRP, it is possible to recover at least part of this data through local refreshes.

- **Clustered Node Failures** Under normal operating conditions, there is one monitor node per zone *i.e.*, $M_i = Z$ for each event-type e_i. These nodes form a logical ring for the purpose of updating their maps. However, if all nodes in a particular zone fail, it results in a break of the logical ring of monitor nodes. Note that in every map-update cycle, the monitor nodes *mark* their presence by updating the *List of monitoring nodes* field of the monitoring map. By checking this field, it is possible to discover which zone(s) have experienced clustered node failures, and to route around these zones to reconstruct a logical ring of $M_i < Z$ monitor nodes. We can employ different techniques (*e.g.*, beaconing mechanisms) to recover from such failures.

Bloom Filter-Based Attribute Matching Bloom filters [2,3] can be used to support attribute-based queries of the event-list stored at each of the replica nodes for an event type. It is a method to represent a set of elements as a compact summary supporting membership queries. Consider a set of v elements to be represented using an u-bit long vector with all entries initially set to zero. A set of k independent hash functions $h_1, \cdots, h_k$ is chosen where each function has a range of 1 to u. For each element x to be represented, the bits at positions $h_1(x), \cdots, h_k(x)$ in the bit vector are set to 1. A bit may be set to 1 multiple times. This bit vector serves as a summary. To find out if an element y is in the event-list, we check the positions $h_1(y), \cdots, h_k(y)$ in the bit vector. If they are all set to 1, then we can infer that y is in the event-list, though there is a (small) probability of being wrong (false positive). The critical feature of a Bloom filter is that the probability of false positives decreases exponentially with u, if the number of hash functions k is chosen optimally. It is shown in [3] that the probability of a false positive is given by $(1-(1-\frac{1}{u})^{kv})^k$, which gives the optimal value of k as $\frac{u}{v}\ln 2$.

We now explain how this concept can be applied to the problem of attribute-based queries in a sensor network. Consider the temperature monitoring example. The Bloom filter can be used to represent compactly which replica nodes contain temperature readings in certain temperature ranges. Suppose a query node sends the query "Get all event data for temperature readings between 30 and 40" to the local monitor node. By Bloom filter-based matching, we can obtain a list of replica nodes containing relevant temperature readings. The local monitor node can then forward the query to this restricted set of replica nodes. Note that the usage of Bloom filters is optional. It is appropriate only when there are a large number of replica nodes, making flooding of all replica nodes expensive.

3.3 Performance Measures

We need to specify relevant performance measures to quantify the benefits of our DCS mechanisms. As mentioned before, data dissemination algorithms should seek to minimize communication in order to extend overall system lifetime. For consistency and for the sake of easy comparison, we use the same metrics to quantify communication overhead as Ratnasamy *et al.* [7,9]. These metrics (assuming that all packets have the same size) are:

- **Total Usage:** Total Number of Packets sent in the sensornet.
- **Hotspot Usage:** Maximal Number of Packets sent by any particular sensornet node.

As previously mentioned, when an event of a particular type occurs at any sensornet node, it transmits this event data to the closest monitor node, which then forwards it to the closest replica node for storage of events of that type. If the data storage operation fails (this could occur if the destination node is unreachable or malfunctioning), this data is retransmitted to one of the other R_i replica nodes. This process is iterated until the data is successfully stored. If all the R_i replica nodes for event type i fail simultaneously, then an attempt to store or retrieve data of this type will *fail.* To measure the success rate of queries for event data, we use the metric **Query Success Rate**, which is the mean percentage of queries (averaged over all Event Types) which return a successful response. *Total Usage*, as a function of node failure rate f, can also be used as a resilience metric. Total Usage is expected to increase as the node failure rate increases (due to more retransmissions). However, a drastic increase in Total Usage at higher node failure rates would imply that our scheme is not robust in the presence of node failures. Hence R-DCS aims to achieve graceful degradation (with respect to Total Usage) in system performance, as f is increased.

4 Analytical Results

We now derive analytical expressions for energy costs and savings obtainable with R-DCS. We consider a sensornet with N nodes equipped to detect E event types. Let D_{total} denote the total number of events detected, Q the number of queries issued, D_q the number of events detected for these queries, R be the number of replica nodes, M the number of monitor nodes and ω the frequency of monitoring map updates.

4.1 Energy Savings

The communication costs of the three canonical methods [9] and of R-DCS can be estimated using the fact that the asymptotic cost of a flood is $O(N)$ and that of direct routing from one random node to another (using GPSR) is $O(\sqrt{N})$. For R-DCS, we estimate the total message count (*total usage* T) and the number of messages at the busiest point (*hotspot usage* H). We reproduce from [9] the corresponding expressions for each of the aforementioned *canonical methods* [9] for completeness. The formulae presented give the *expected values* of these quantities. We consider a scenario in which we have one access point. The results can be easily generalized for multiple access points. We assume that the message counts at the access point is a good estimate of hotspot usage, since it is likely to be the busiest area of the network.

1. **External Storage**:
   The cost of storing each event (sending to external store) is $O(\sqrt{N})$. There is no cost for queries since event information is already external.

$$T = D_{total}\sqrt{N} \qquad H = D_{total} \tag{1}$$

2. **Local Storage**:
   Storing event information locally incurs no cost. Queries are flooded to all nodes at a cost of $O(N)$. Responses are routed back to the query-source at a cost of $O(\sqrt{N})$.

$$T = QN + D_q\sqrt{N} \qquad H = Q + D_q \tag{2}$$

3. **Data-Centric Storage**:
   Storing event information at a sensornet node depending on the event-type incurs a cost of $O(\sqrt{N})$. Queries for a particular event-type are routed to the storage node, which returns a response, both at a cost of $O(\sqrt{N})$. Total usage depends on whether a full-*listing* of events (l) is required (uses one packet for every instance of an event for each event-type) or whether a *summary* (s) is sufficient (uses only one packet for each event-type).

$$\begin{array}{ll} T = D_{total}\sqrt{N} + Q\sqrt{N} + D_q\sqrt{N} & H = Q + D_q \text{ (l)} \\ T = D_{total}\sqrt{N} + Q\sqrt{N} + Q\sqrt{N} & H = 2Q \quad\;\; \text{(s)} \end{array} \tag{3}$$

   In the case of Structured Replication with DCS (SR-DCS), the costs of storage is decreased whereas the cost of queries is increased (compared to plain DCS). For a hierarchy-depth d, the storage cost for a single event decreases from $O(\sqrt{N})$ to $O(\sqrt{N})/2^d$. The cost of routing a single query through the complete image-node hierarchy increases from $O(\sqrt{N})$ (for DCS) to $O(2^d\sqrt{N})$. Total usage is given by:

$$\begin{array}{ll} T = D_{total}\sqrt{N}/2^d + Q2^d\sqrt{N} + D_q2^d\sqrt{N} & H = Q + D_q \text{ (l)} \\ T = D_{total}\sqrt{N}/2^d + Q2^d\sqrt{N} + D_q2^d\sqrt{N} & H = 2Q \quad\;\; \text{(s)} \end{array} \tag{4}$$

4. **Resilient Data-Centric Storage**:
   Since event information is stored at the node closest to the event-location, storage costs are reduced from $O(\sqrt{N})$ to $O(\sqrt{N/R})$. Queries of a particular event-type are routed to the closest monitor node at a cost of $O(\sqrt{N/M})$. In the *summary* case described above, this monitor node can directly return a response (based on the monitor-maps it receives from the other nodes) at a cost of $O(\sqrt{N/M})$. In the *list* case, the query must be forwarded to all R replica nodes (at a cost of $O(R\sqrt{N})$), and all of them return responses, at a cost of $O(\sqrt{N})$. Let the frequency of monitoring map updates be ω. The cost of exchanging monitor maps is $O(2\omega\sqrt{NM})$ for each event-type.

$$
\begin{array}{lr}
T = D_{total}\sqrt{N/R} + Q(\sqrt{N/M} + R\sqrt{N}) + D_q\sqrt{N} + 2\omega E\sqrt{NM} & (l) \\
T = D_{total}\sqrt{N/R} + Q\sqrt{N/M} + Q\sqrt{N/M} + 2\omega E\sqrt{NM} & (s) \\
H = Q + D_q \quad (l) \qquad\qquad H = 2Q \quad (s) &
\end{array} \quad (5)
$$

From the above formulae, we can draw a number of conclusions. If N is increased, other parameters constant, the local storage method incurs the highest total message count. If $D_q \gg Q$ and events are summarized, DCS/R-DCS have the lowest hotspot usage. If $D_{total} \gg D_q$, ES has significantly higher hotspot usage than the other schemes. In general, DCS/R-DCS are preferable in cases where (1) N is large. (2) $D_{total} \gg D_q \gg Q$ (Many detected events and not all event types queried). However, LS may be preferable if the number of events is large compared to system size ($D_{total} > Q\sqrt{N}$) and event *lists* are used.

Comparing the total message counts for R-DCS and DCS/SR-DCS (R in R-DCS is equivalent to 4^d in SR-DCS), we see that R-DCS will outperform DCS in the *summarized* case, for typical values of the scenario parameters. Considering $N = 10000$, $E = 100$, $D_{total} = 10000$, $Q = 50$, $D_q = 5000$, $R = 4$, $M = 16$ and $\omega = 0.1$ we get $T = 1020000$ for DCS, $T = 540000$ for SR-DCS and $T = 513000$ for R-DCS. In the *list* case, the $T = 1505000$ for DCS, $T = 1510000$ for SR-DCS and $T = 1029250$ for R-DCS. Hence we see that R-DCS could provide significant energy savings for both the *summarized* and the *list* case. R-DCS presents an intermediate solution between LS (free storage, expensive queries) and DCS (both at moderate cost). It can be expected to give good performance when $D_q \gg Q$.

4.2 Increased Resilience

An obvious benefit of R-DCS over DCS is the increased resilience to node failures. Having M monitor nodes and R replica nodes for each event-type ensures that all information corresponding to an instance of a particular event-type is not lost when one node fails. Further, since these nodes exchange and replicate *monitoring maps*, it is extremely unlikely that the control and event-summary information is lost.

Let us consider a situation in which nodes in the sensor network are unreliable, so that they are *on* (*i.e.*, functioning correctly) with probability $1 - f$, and *off* (*i.e.*, malfunctioning) with probability f. Here $0 \leq f \leq 1$. $f = 0$ corresponds to the case of perfectly reliable nodes. As in the previous section, we approximate the communication costs (T) for R-DCS. Local refresh mechanisms like PRP [7] are used to ensure that there is one monitor node in every zone, except in the case of clustered node failures. *Normal* nodes in any zone j communicate with the local monitor (and replica if it exists) nodes m_{ij} (r_{ij}) for data of event-type e_i. In any message, the monitor (replica) node includes a list of *viable* destinations in the header. For example, when a source monitor node m_{ij} needs to store its event data at one of the R_i replica nodes for type i event data, it constructs the destination list by computing its distances to all the type i replica nodes listed in the local monitoring map and then ordering the R_i replica nodes according to their distances from the source. Hence the destination of first choice is the closest replica node, the first *alternate* destination is the second-closest replica node, and so on. m_{ij} then routes the message to each one of the R_i possible destinations in its list, in order of preference. If all R_i destinations are exhausted without success, the message is dropped.

In Resilient Data-Centric Storage, a single message has a cost $O(\sqrt{N})$, when the destination is *on* (with probability $1-f$). The probability of the destination being *off* is f, in which case the message is sent to the first *alternate* destination. In the second transmission, with probability $1-f$ the message is successfully sent at a total cost of $2O(\sqrt{N})$ [which consists of a cost $O(\sqrt{N})$ for the first failed message and a cost $O(\sqrt{N})$ for the second successful message] or it fails again with probability f and so on. This iterative process is represented by Equation 6.

$$\begin{aligned} C &= (1-f)\sqrt{N} + f((1-f)2\sqrt{N} + f((1-f)3\sqrt{N} + \ldots)) \quad R \text{ terms}, R \geq 2 \\ &= \sqrt{N} C_{Rf} \end{aligned} \tag{6}$$

$$C_{Rf} = \frac{1-(R+1)f^R + Rf^{R+1}}{1-f} \tag{7}$$

We can now give the expressions for T in this case.

$$\begin{aligned} T &= D_{total}\sqrt{N/R}C_{Rf} + Q(\sqrt{N/M} + R\sqrt{N})C_{Rf} + D_q\sqrt{N} + 2\omega E\sqrt{NM} \quad (1) \\ T &= D_{total}\sqrt{N/R}C_{Rf} + Q\sqrt{N/M}C_{Rf} + Q\sqrt{N/M}C_{Rf} + 2\omega E\sqrt{NM} \quad (s) \end{aligned} \tag{8}$$

Let us denote the probability of successful storage of data corresponding to a particular event instance as P_s (here the subscript i is omitted, but it must be noted that we are considering a particular event-type e_i). Further let us denote the probability of *clustered* node failure as f_c, in which a majority of the nodes in a particular zone malfunction. As described earlier, this operation succeeds if all R replica nodes for the corresponding event type do not fail simultaneously. The probability of simultaneous failure of R replica nodes (assuming these failures are independent) is f^R. Clearly, $P_s = 1 - f^R$. For a *summary* query to be successful, two conditions have to be satisfied: (1) The event data being queried must have been successfully stored (which occurs with probability P_s). (2) The local monitor node must be alive storing this summary data. Since we assume that local refresh mechanisms like PRP [7] ensure the existence of a monitor node except in the case of clustered node failures, this monitor nodes is alive with probability $1 - f_c$. Hence the mean query success rate Q_s is given by

$$Q_s = (1-f^R)(1-f_c) \qquad \text{(s)} \tag{9}$$

For a *list* query to be successful, the second condition is modified to all monitor nodes being alive. This occurs with probability $(1-f_c)^M$. Hence the mean query success rate is given by

$$Q_s = (1-f^R)(1-f_c)^M \qquad \text{(l)} \tag{10}$$

From equations 7 and 8, we can see that if f is increased, other parameters constant, then T increases. This should be obvious intuitively, since a greater number of unreliable nodes imply more retransmissions. Also from equations 9 and 10, it is clear that if R is increased keeping f constant, Q_s increases. Similarly if f is increased, Q_s decreases (for the same value of R). These variations are further investigated in Section 5.4.

Comparing the total message counts and query success rates for R-DCS for different values of R and f, we see that R-DCS offers significant resilience to node failures with $R \geq 4$ for different values of f. In list mode, considering $N = 10000$, $E = 100$, $D_{total} = 10000$, $Q = 50$, $D_q = 5000$, $M = 16$, $f_c = 0.02$, $R = 4$ and $\omega = 0.1$, we get $T = 1029250$, $Q_s = 1$ for $f = 0$ and $T = 1355032$, $Q_s = 0.92$ for $f = 0.5$. With the other parameters constant but $R = 16$, we get $T = 839250$, $Q_s = 1$ for $f = 0$ and $T = 1170410$, $Q_s = 0.98$ for $f = 0.5$. Hence we see that as R is increased beyond 4, the query success rate Q_s approaches $1 - f_c$ as f is increased up to 0.5. The tradeoff here is the overhead of maintaining a large number of monitor/replica nodes.

Table 1. Simulation Parameters

Parameter	Value
N, Number of Nodes	100 - 100000
R, Number of Replicas	2-16
Z, Number of Zones	16
E, Number of Event Types	100
Q, Number of Event Types Queried	50
D_i, Instances of each Event-type i	100
f_c, Probability of Clustered Node Failure	0.02
ω, Frequency of Monitor Map Updates	0.1

5 Performance Evaluation

DCS has been shown to be a viable and robust data dissemination scheme using detailed simulations in $ns-2$ extended with a GPSR-based DHT [7,9], which included a full 802.11 MAC and physical radio layer. These simulations verified the correct functioning of the low-level aspects of DCS. Since R-DCS builds upon DCS mechanisms, we expect that R-DCS will be viable in a bandwidth-limited, contention prone medium as well. However these $ns-2$ simulations do not scale to more than 200 nodes [9]. Since we envisage R-DCS to be most useful in very large-scale sensor networks, we used a lightweight (*i.e.*, without radio details) simulator built in C to compare the performance of R-DCS with DCS, LS and ES. This simulator assumes stationary nodes in the sensor network, as well as instantaneous error-free packet transmission. We examine *total usage* and *hotspot usage*, as described before, as metrics to compare the relative performance of these algorithms. The system parameters used in most of the simulations are summarized in Table 1.

We individually vary the parameters R, N, Q, and the node failure rate and investigate the effect on system performance. In our simulations, there is one randomly chosen access point, which represents the node where queries enter the network. At the start of the simulation, all events are inserted into the DHT once, by sensors chosen uniformly at random; these are the sensors that measured the inserted events. We present results averaged over multiple simulations (10 iterations with different random seeds for each simulation run).

5.1 Variation of R, the Number of Replicas

In this section, we show the results of varying the number of replicas R, while the other parameters are kept constant as in Table 1. The value of N is chosen to be 10000. Figure 5 shows the variation of total messages with R for the *list* and *summarized* cases. From Figure 5, we observe that ES and LS have higher total usages than our data-centric schemes. At low values of R, R-DCS and its extensions are found to have approximately similar performance as DCS. However, as R is increased, the performance of R-DCS based schemes is significantly better than DCS. These conclusions are also supported by our analytical results in Section 4. Note also that the total usage for R-DCS based schemes is significantly lower in *summarized* case than in the *list* case, as expected.

In our simulations, we first consider a *normal* scenario, where events occur uniformly spread out over the sensornet field. We have also conducted simulations with a *modified* event density, wherein 80% of the events occur in one quadrant of the field, and 20% of events occur in the remaining three quadrants taken together. We expect that this will lead to congestion in the "crowded" quadrant of the network. Our results (which we do not include here for lack of space) show that load-balancing mechanisms built into R-DCS help it outperform the other schemes in this case.

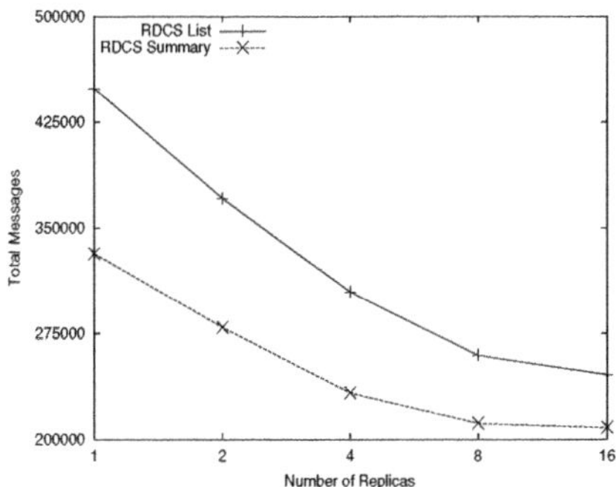

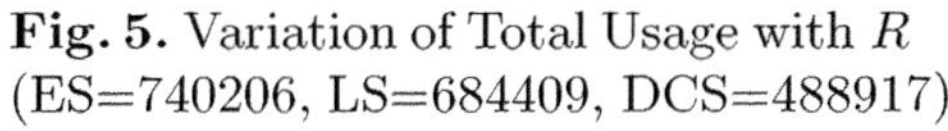
Fig. 5. Variation of Total Usage with R (ES=740206, LS=684409, DCS=488917)

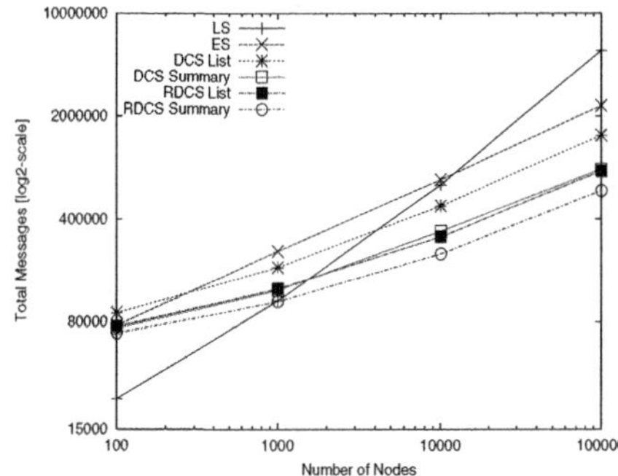

Fig. 6. Variation of Total Usage with Number of Nodes N

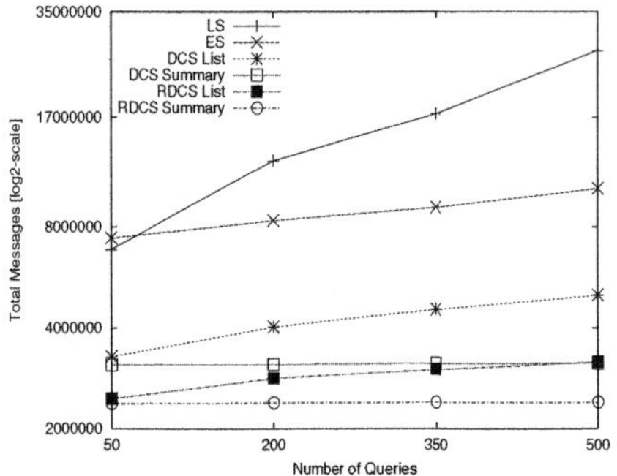

Fig. 7. Variation of Total Usage with Q

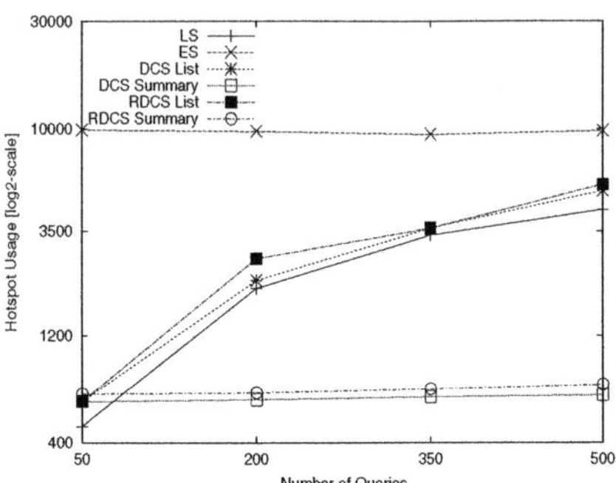

Fig. 8. Variation of Hotspot Usage with Q

5.2 Scaling to a Large Number of Nodes

In this section, we show the results of varying the number of nodes N, while the other parameters are kept constant as in Table 1. The value of R is chosen to be 4. Figure 6 shows the variation of total messages with n. From Figure 6, we see that all the methods have reasonably similar behavior for total usage, but LS has the lowest total usage at low N ($N = 100$) and ends up (at high N) with the highest value. Amongst the other schemes, our R-DCS schemes have the best performance, followed by DCS and then ES, for all values of N. These conclusions are supported by our analytical results in Section 4.

5.3 Scaling to a Large Number of Queries

In this section, we show the results of varying the number of queried event-types Q, while the other parameters are kept constant as in Table 1. The value of R is chosen to be 4 and $N = 10000$. Q is varied from 50 to 500. Since the number of instances of each event type is 100, this corresponds to a variation in the total number of queries from 5000 to 50000. Figure 7 shows the variation of total messages with Q, while Figure 8 shows the corresponding variation of hotspot usage. From Figures 7 and 8, we see that at low values of Q, LS offers good performance, but both total usage and hotspot usage increase linearly with increasing Q - hence LS does not scale well with more queries. ES has a medium total as well as hotspot usage, which is independent of Q. In both the *list* and *summarized* case, the graphs clearly show that R-DCS offers significant performance improvements over DCS which in turn does better than ES/LS with increase in Q. These experimental results are supported by our analytical expressions in Section 4, confirming the validity of our simulations.

5.4 Resilience to Node Failure

In the results presented so far, we assumed that nodes were stable. We now present simulation results for the case when nodes are *off* with probability f

and *on* with probability $1 - f$. We vary the value of f from 0 to 0.5 and observe the variation of T. The value of N is chosen to be 10000, and the other parameters are the same as in Table 1. Figure 9 shows the variation of total messages with f for the *summary* case in R-DCS. In fact for $R = 16$, T is almost constant for values of f ranging from 0 to 0.5. It must be noted that, as the node failure rate f is increased, the Query Success Rate Q_s is expected to decrease significantly as shown by Equation 9. For example, with $f = 0.4$ and $R = 2$, $Q_s = 70\%$ versus $Q_s = 100\%$ for $f = 0$. Our experimental results approximately agree with the analytical results derived in Section 4.2. These results shows that in a scenario where the probability of node failure is significantly high, R-DCS improves scalability by avoiding a dramatic increase in messages sent. In other words, R-DCS makes sensor data *resilient* to node failure.

6 Conclusions

In this paper, we have presented Resilient Data-Centric Storage as a means of reducing energy consumption and increasing resilience to node failures in a wireless ad-hoc sensor network. We present a methodology which stores event data at the closest of R replica nodes allocated (by the use of a DHT) for the event-type to which this data belongs. A set of monitor nodes exchange monitor maps to form a global image of all events of that event type which have occurred in the network. Since queries need to be routed to the closest monitor or replica node for an event-type, overall query traffic is reduced.

We have evaluated the viability and relative performance of our scheme vis-a-vis the original DCS scheme, Local storage and External storage. Through analytical results and simulations, we show that in a reasonably large sensor network with many detected events, R-DCS performs better than DCS, LS and ES. In particular R-DCS achieves significant energy-savings in the *summarized* mode of query responses. In all scenarios, R-DCS provides resilience to both clustered and isolated node failures.

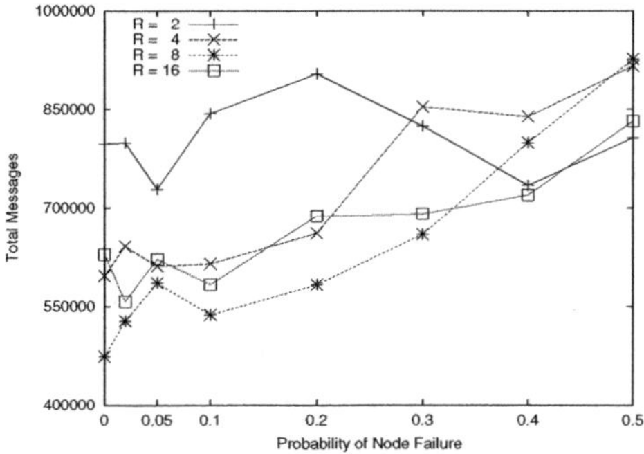

Fig. 9. Variation of Total Usage with node failure rate f

We believe that R-DCS will work well over a broad range of network scenarios. Robust and resilient storage of data within a sensor network could be useful for a large number of applications, such as monitoring meteorological data in turbulent conditions. Resilient DCS could also be used as a tool in providing service guarantees for applications running over sensor networks, analogous to how web-caching helps in providing service guarantees and improving data-availability over the Internet.

7 Acknowledgements

We would like to thank Sylvia Ratnasamy, Yin Li and Fang Yu for their help and guidance in the initial phases of the project and for making the source code of their DCS simulator available to us. We also thank the anonymous reviewers for their useful comments and feedback. This work is supported by the US National Science Foundation under Cooperative Agreement Number ITR-0085879.

References

1. W. Adjie-Winoto, E. Schwartz, H. Balakrishnan, and J. Lilley. The Design and Implementation of an Intentional Naming System. In *Proceedings of ACM SOSP-99*, pages 186–201, Dec. 1999.
2. B. H. Bloom. Space/Time Trade-offs in Hash Coding with Allowable Errors. *Communications of the ACM*, 13(7):422–426, July 1970.
3. L. Fan, P. Cao, J. Almeida, and A. Broder. Summary Cache: A Scalable Wide-area Web Cache Sharing Protocol. In *Proceedings of ACM SIGCOMM'98*, pages 254–265, Sept. 1998.
4. J. Heidemann, F. Silva, C. Intanagonwiwat, R. Govindan, D. Estrin, and D. Ganesan. Building Efficient Wireless Sensor Networks with Low-Level Naming. In *Proceedings of ACM SOSP-01*, 35(5):146–159, Oct. 2001.
5. C. Intanagonwiwat, R. Govindan, and D. Estrin. Directed diffusion: A Scalable and Robust Communication Paradigm for Sensor Networks. In *Proceedings of ACM MOBICOM-00*, pages 56–67, Aug. 2000.
6. B. Karp and H. T. Kung. GPSR: Greedy Perimeter Stateless Routing for Wireless Networks. In *Proceedings of ACM MOBICOM-00*, pages 243–254, Aug. 2000.
7. S. Ratnasamy, B. Karp, Y. Li, F. Yu, R. Govindan, S. Shenker and D. Estrin. GHT: A Geographic Hash Table for Data-Centric Storage. In *Proceedings of ACM WSNA-02*, Oct. 2002.
8. S. Ratnasamy, P. Francis, M. Handley, R. Karp, and S. Shenker. A Scalable Content-Addressable Network. In *Proceedings of ACM SIGCOMM 2001*, 31(4):161–172, Aug. 2001.
9. S. Shenker, S. Ratnasamy, B. Karp, R. Govindan, and D. Estrin. Data-Centric Storage in Sensornets. In *Proceedings of ACM HotNets-I*, Oct. 2002.
10. I. Stoica, R. Morris, D. Karger, F. Kaashoek, and H. Balakrishnan. Chord: A Scalable Peer-To-Peer Lookup Service for Internet Applications. In *Proceedings of ACM SIGCOMM 2001*, 31(4):149–160, Aug. 2001.

Shape-Based Similarity Query for Trajectory of Mobile Objects

Yutaka Yanagisawa, Jun-ichi Akahani, and Tetsuji Satoh

NTT Communication Science Laboratories,
NTT Corporation
{yutaka,akahani,satoh}@cslab.kecl.ntt.co.jp
http://www.kecl.ntt.co.jp/scl/sirg/

Abstract. In this paper, we describe an efficient indexing method for a shape-based similarity search of the trajectory of dynamically changing locations of people and mobile objects. In order to manage trajectories in database systems, we define a data model of trajectories as directed lines in a space, and the similarity between trajectories is defined as the Euclidean distance between directed discrete lines. Our proposed similarity query can be used to find interested patterns embedded into the trajectories, for example, the trajectories of mobile cars in a city may include patterns for expecting traffic jams. Furthermore, we propose an efficient indexing method to retrieve similar trajectories for a query by combining a spatial indexing technique (R^+-Tree) and a dimension reduction technique, which is called PAA (Piecewise Approximate Aggregate). The indexing method can efficiently retrieve trajectories whose *shape* in a space is similar to the shape of a candidate trajectory from the database.

1 Introduction

Recently, many location sensors such as GPS have been developed, and we can obtain the trajectory of users and moving objects using these sensors [12]. Trajectory data are widely used in location-aware systems [1], car navigation systems, and other location-based information systems, that can provide services according to a user's current location. These applications have stored in them a lot of trajectories, and these trajectories may include interesting individual patterns of each user. For example, by analyzing trajectories of users who work in a building, we can find important passages, rooms, stairs, and other facilities that are used frequently. The result of the analysis can be used for the management and maintenance of the buildings. In the case of a navigation system, a driver can check the route to a city by referring to the trajectories of other users who have driven to the city before. In another case, we can study characteristics to improve performance in a sport by analyzing the motion data measured by the sensors attached to the bodies of top sports players.

There have been many studies on managing mobile objects data (MOD) [2] [8] [11] [14] [16]. One of the most interesting of these is the development of efficient

M.-S. Chen et al. (Eds.): MDM 2003, LNCS 2574, pp. 63–77, 2003.

method to retrieve objects, which is indicated by either a spatiotemporal *range query* or a spatio-temporal *nearest neighbor query*. Both queries are defined as the distance between the trajectory of a mobile object and an indicated point in a space. For example, the range query is generally defined as the query for retrieving all objects which passed within a given distance of an indicated point, such as "retrieve all of the people who walked within one mile of the buildings at the time." The range query can also be defined as the query to retrieve all objects that passed within an indicated polygon.

In both cases, the query is defined using the distance between figures in a space. These distance-based queries are useful in location management of mobile objects [15], however, these queries do not have enough power to analyze the pattern of the objects' motion. As mentioned above, because we are interested in the extraction of the individual moving patterns of each object from the trajectories, it is necessary to develop more powerful tools to analyze the trajectories. Hence, we propose *shape-based* queries of trajectories in space for the analysis, for instance, "retrieve all objects that have a similar shape to the trajectory where a user walked in a shop." Using this query, we may classify the customers in the shop based on their shape patterns of the trajectories. In other words, our approach is based on the shape similarity between lines, while the existing approaches adopt the distance between points as the key to retrieve required objects.

It is difficult to define the similarity between lines in a space. However, we found this useful idea through research of time series databases [6] [7] [10]. The time series database systems can store time series data such as temperature, economic indicators, population, wave signals, and so on, in addition to supporting queries for extracting patterns from the time series data. Most of the time series database systems adopt the Euclidean distance between two time data sequences [7] for analysis; if two sequences, c, c', are given as $\langle w_1, w_2, \ldots, w_n \rangle$ and $\langle w'_1, w'_2, \ldots, w'_n \rangle$, the similarity can be defined as

$$D(c, c') = \sqrt{(w_1 - w'_1)^2 + \ldots + (w_n - w'_n)^2}.$$

(In Section 3, we describe the similarity in detail) Because trajectory is a type of time series data, the time series database is able to deal with trajectory. However, trajectory not only has a time series data feature, but also has a directed line in space feature. For example, it is difficult for the time series database to find data for a geographic and spatial query.

Therefore, in this paper, we present a data model for trajectories of mobile data, and a query based on the distance between two trajectories by extending the similarity used in the time series database systems. Moreover, we propose a new indexing method for retrieving required trajectories by queries based on our defined distance between trajectories. In Section 2, we describe our proposed data model for the trajectory. Section 3 describes the distance between two discrete directed lines for calculating similarities between two trajectories . In Section 4, we present both the processing method for our proposed query and an indexing

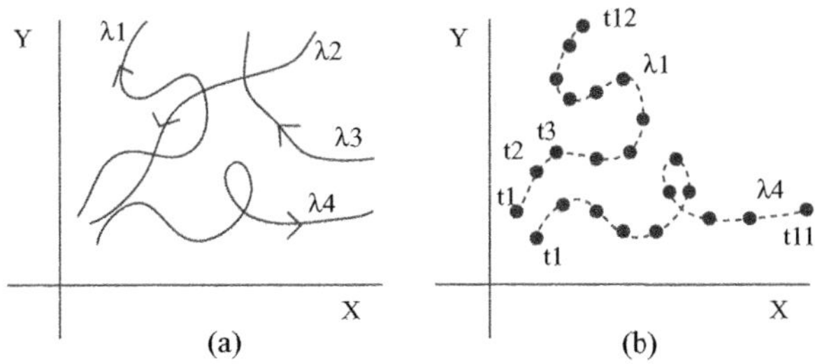

Fig. 1. Trajectory of Mobile Objects ((a) trajectory in the real world. (b) trajectory stored in a database)

technique that is an extension of Piecewise Aggregate Approximation (PAA) [7]. Finally the evaluation of our approach is shown in Section 5.

2 Trajectory of Mobile Objects

In order to effectively manage mobile objects, it is necessary to manage the location of each object at each time. Generally, a location management system can retrieve objects located in the indicated area at the indicated time [2]. However, we are interested in the similarity of the trajectory's shape in a space. In order to define the similarity between trajectories, it is necessary at first to define the trajectory as a figure drawn in space. Hence, we define the data model for the trajectory of mobile objects[1].

A real-world trajectory is a directed continuous line with a start and end point (Figure 1(a)). Given a two-dimensional space $\boldsymbol{R}^2$ and a closed time interval $I_\lambda = [t, t']$ with $t < t'$, a trajectory λ is defined as follows,

Definition 1 : *Trajectory*
A trajectory is the image of a continuous mapping of $\lambda : I_\lambda \rightarrow \boldsymbol{R}^2$.

This definition is a temporal extension of the definition of a simple line described in [3]. Next, we denote the length of trajectories in $\boldsymbol{R}^2$ as L_S and the interval of trajectories in temporal space as L_T:

Definition 2 : *Length of Trajectory in Space* $\boldsymbol{R}^2$
The length of trajectory λ during a period $[t_0, t_1]$ is denoted as $L_S(\lambda, [t_0, t_1])$ calculated as follows:

$$L_S(\lambda, [t_0, t_1]) = \int_{t_0}^{t_1} \sqrt{\left(\frac{dx}{dt}\right)^2 + \left(\frac{dy}{dt}\right)^2} dt, \text{ where } \lambda(t) = (x, y)$$

The length of the whole trajectory is denoted as $L_S(\lambda)(= L_S(\lambda, [t, t'])$.

[1] For simplification of the problem, we just focus on the trajectory of mobile objects; in other words, we do not discuss the data model of the other attributes of the objects, such as shape, name, and so on.

Definition 3 : *Temporal Interval of Trajectory*
The $\boldsymbol{x} = (x, y)$ is a vector in space $\boldsymbol{R}^2$. The temporal interval of trajectory λ between $\boldsymbol{x}_i$ and $\boldsymbol{x}_j$ on λ is defined as follows:

$$L_T(\lambda, [\boldsymbol{x}_i, \boldsymbol{x}_j]) = |t_j - t_i|, \text{ where } \lambda(t_i) = \boldsymbol{x}_i, \lambda(t_j) = \boldsymbol{x}_j, \text{ and } t_i, t_j \in I_\lambda[t, t']$$

$$L_T(\lambda) = |t' - t|$$

However, a location sensor device such as GPS does not continuously measure the coordinates of a mobile object, but samples such data. The measured data are thus a sequence of coordinates of positions shown in Figure 1(b). Hence, we define discrete trajectory $\dot{\lambda}$ as a discrete function. Each vector $\boldsymbol{x}_i$ represents a position of a mobile object at each time $\boldsymbol{T}_{\dot{\lambda}} = \{t_0, t_1, \ldots, t_m\}$ in the space.

Definition 4 :*Discrete Trajectory*
A discrete trajectory is the image of a discrete mapping: $\dot{\lambda} : \boldsymbol{T}_{\dot{\lambda}} \to \boldsymbol{R}^2$.

A discrete trajectory can be represented as a vector sequence $\langle \boldsymbol{x}_{t1}, \ldots, \boldsymbol{x}_{tm} \rangle$, also. If $\boldsymbol{T}_{\dot{\lambda}} = \{1, 2, \ldots, m\}$, we denote the discrite trajectory $\dot{\lambda}$ as just a simple vector sequence $\langle \boldsymbol{x}_1, \ldots, \boldsymbol{x}_m \rangle$. Additionally, where $\dot{\lambda}(t_i) = \boldsymbol{x}_i$, we introduce several notations; $\boldsymbol{T}_{\dot{\lambda}}(i) = t_i$, $\boldsymbol{X}_{\dot{\lambda}}(i) = \boldsymbol{x}_i$, and $|\dot{\lambda}|$ is the number of the vectors included in $\dot{\lambda}$ ($|\dot{\lambda}| = |\boldsymbol{T}_{\dot{\lambda}}|$). Next, we define the distance between two vectors $\boldsymbol{x}, \boldsymbol{x}'$ in $\boldsymbol{R}^2$.

Definition 5 :*Distance of Vectors*

$$D(\boldsymbol{x}, \boldsymbol{x}') = \sqrt{(x - x')^2 + (y - y')^2}$$

That is, this definition assumes that space $\boldsymbol{R}^2$ is Euclidean.

Although $\dot{\lambda}$ is a discrete line, it is necessary to deal with $\dot{\lambda}$ as a continuous line in a query. In order to satisfy this requirement, we define a function to convert the discrete line into a continuous line. There are various methods to calculate an approximate continuous line from a discrete line [4]. In our approach, we adopted the piecewise linear approximation because of its simplicity and popularity [15].

Definition 6 : *Piecewise Linear Approximation of* $\dot{\lambda}$
$\tilde{\lambda} : [t_0, t_m] \to \boldsymbol{R}^2$ is given as

$$\tilde{\lambda}(t) = \begin{cases} \dot{\lambda}(t) & \text{if } t \in \boldsymbol{T}_{\dot{\lambda}} \\ \frac{t - t_i}{t_{i+1} - t_i} \dot{\lambda}(t_i) + \frac{t_{i+1} - t}{t_{i+1} - t_i} \dot{\lambda}(t_{i+1}) & \text{if } t \notin \boldsymbol{T}_{\dot{\lambda}} \end{cases}$$

t_i must be selected under the condition: $t_i < t < t_{i+1}$

In the rest of this paper, we mainly discuss the features of both λ and $\dot{\lambda}$. Note that in this paper, we only mention the trajectory on $\boldsymbol{R}^2$, but our proposed model and techniques can obviously be adapted to the higher dimensional space $\boldsymbol{R}^n$.

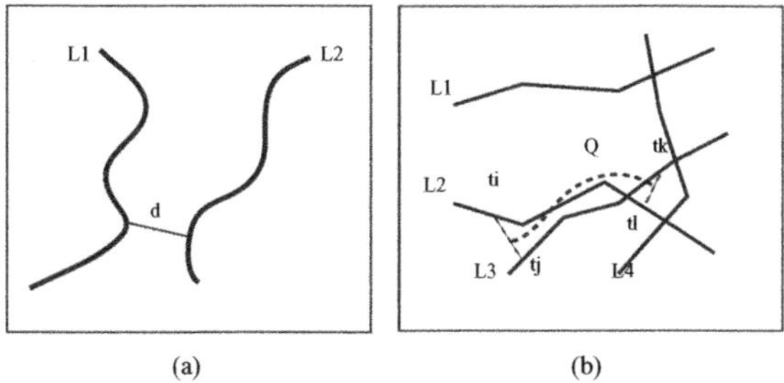

Fig. 2. Distance between trajectories

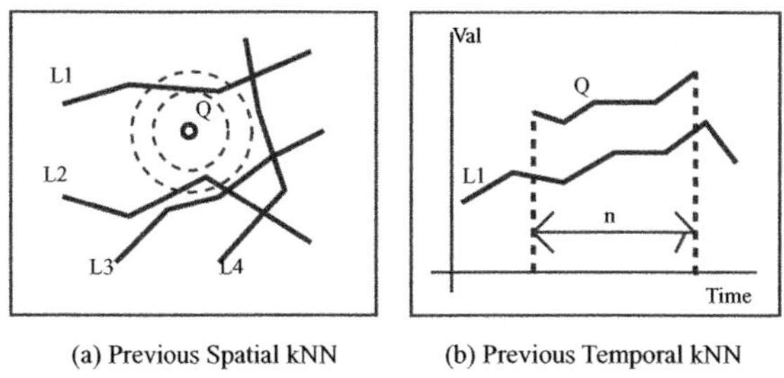

(a) Previous Spatial kNN (b) Previous Temporal kNN

Fig. 3. Existing kNN approaches

3 Similarity Query Based on Shapes of Lines

3.1 Shape-Based Approach

The similarity query is useful in its own right as a tool for exploratory data analysis[7], and it is a significant element in many data mining applications. For instance, we may find the optimum arrangement of items in a market by analyzing the trajectories of customers walking around in a shop. In addition to its usefulness in the trajectory database, the similarity query is one of the most interesting fields in time series databases. In time series databases, the similarity between two sets of time series data is typically measured by the Euclidean distance [6] [7], which can be calculated efficiently.

However, there have been few discussions on the similarity between two lines in space because the previous approaches for spatial queries have focused on the "distance" between a point and a line [2] [9] [15]. The interest of the previous approaches was mainly to find objects that pass a point near the indicated point, such as a car passing through a street. On the other hand, we are interested in the "shape" of the trajectory. In order to calculate shape-based similarities among trajectories, it is necessary to define a new similarity for the trajectories, as shown in Figure 2(b).

In general, the similarity query is represented as a k Nearest Neighbor Query (kNN) [2] [5] [9] . There are two types of existing approaches, one is based on spatial similarities, the other is based on similarity between two time series data. The example of the existing spatial kNN is illustrated in Figure 3(a). In this case, the answer is L_1, L_2 when K is 2. On the other hand, the similarity between two time series data is defined as the Euclidean distance between two time series, where the length of each is n. The distance is defined as the Euclidean distance between two n-dimensonal vector data [7] shown in Figure 3(b). While this distance of the time series data is based on shape, the distance is defined only in the case of $\boldsymbol{R}^1 \times T$ ($T = [0, \infty]$), but not in the case of $\boldsymbol{R}^n \times T$, shown in Figure 2(b). Since the trajectory has both spatial and temporal features, we consider three types of similarity queries for trajectories as follows:

Spatio-Temporal Similarity: based on a spatio-temporal feature in $\boldsymbol{R}^2 \times T$.
Spatial Similarity: based on a spatial only feature in $\boldsymbol{R}^2$ without temporal features.
Temporal Similarity: based on a temporal only feature in $\boldsymbol{R}^1 \times T$ without spatial features.

In the rest of this section, we define the similarity in the first two cases. We do not define the temporal similarity because this similarity is the same as the similarity defined for the time series databases.

3.2 Shape-Based Similarity Query

As mentioned above, the trajectory has a time series data feature. We define the similarity between two trajectories in the same manner as for the similarity defined in the time series query[7]. For the time series database, the similarity of the two time series data, where each has n value, is given by the Euclidean distance between vectors in $\boldsymbol{R}^n$. In [6] and [7], when there are two time series data, $c = \langle w_1, w_2, \ldots, w_n \rangle$, $c' = \langle w'_1, w'_2, \ldots, w'_n \rangle$, the distance $D(c, c')$ is defined as follows:

$$D(c, c') = \sqrt{(w_1 - w'_1)^2 + \ldots + (w_n - w'_n)^2}$$

This definition can be extended if each vector $\boldsymbol{x}$ is a vector in space $\boldsymbol{R}^2$, when the time series vectors are $\boldsymbol{X} = \langle \boldsymbol{x}_1, \boldsymbol{x}_2, \ldots, \boldsymbol{x}_n \rangle$, $\boldsymbol{X}' = \langle \boldsymbol{x}'_1, \boldsymbol{x}'_2, \ldots, \boldsymbol{x}'_n \rangle$, and the distance is $D(c, c')$. We define the distance between two time series vectors $D(\boldsymbol{X}, \boldsymbol{X}')$ by extending the definition of $D(c, c')$, as follows:

$$D(\boldsymbol{X}, \boldsymbol{X}') = \sqrt{D(\boldsymbol{x}_1, \boldsymbol{x}'_1)^2 + \ldots + D(\boldsymbol{x}_n, \boldsymbol{x}'_n)^2}$$

Based on this definition, we consider the shape-based similarity query for trajectories. Here, $\dot{\boldsymbol{\Lambda}}$ is the set of discrete trjectories stored in the database, and each $\dot{\lambda}_i$ ($\dot{\lambda}_i \in \dot{\boldsymbol{\Lambda}}$) is a discrete trajectory, such as $\dot{\lambda}_i = \langle \boldsymbol{x}_1, \ldots, \boldsymbol{x}_m \rangle$. The query trajectory $\dot{\lambda}_q$ is given as $\dot{\lambda}_q = \langle \boldsymbol{x}_1, \ldots, \boldsymbol{x}_n \rangle$. The shape-based range query can then be defined using $\dot{\boldsymbol{\Lambda}}$, $\dot{\lambda}_q$, and the previous defined distance between two time series vectors, as follows:

Input : $\dot{\boldsymbol{\Lambda}}$, $\dot{\lambda}_q$ and θ (θ is a natural number).
Output : $\dot{\boldsymbol{\Lambda}}_a$, $\{\dot{\lambda}_{a1}, \ldots, \dot{\lambda}_{ak}\} \in \dot{\boldsymbol{\Lambda}}_a$.

function Q_{range}(θ: integer, $\dot{\lambda}_q$, $\dot{\boldsymbol{\Lambda}}$) : $\dot{\boldsymbol{\Lambda}}_a$
begin
  var j : integer, $l := |\dot{\lambda}_q|$, $\dot{\boldsymbol{\Lambda}}_a := \phi$;
  for each $\dot{\lambda}_i$ in $\dot{\boldsymbol{\Lambda}}$ **do**
    for $j := 1$ to $|\dot{\lambda}_i| - l + 1$ **do**
    begin
      $\dot{\lambda}_{ij}$ = subsequence($\dot{\lambda}_i$, j, l);
        { *This function will return a subsequence of the original sequence* $\dot{\lambda}_i$,
          such as $\langle \boldsymbol{x}_j, \boldsymbol{x}_{j+1}, \ldots, \boldsymbol{x}_{j+l-1} \rangle$, *each* $\boldsymbol{x} \in \dot{\lambda}_i$}
      if $D(\dot{\lambda}_q, \dot{\lambda}_{ij}) < \theta$ **then**
        Add $\dot{\lambda}_{ij}$ to $\dot{\boldsymbol{\Lambda}}_a$;
    end;
  return $\dot{\boldsymbol{\Lambda}}_a$;
end.

Fig. 4. The process of the shape-based range query of trajectories

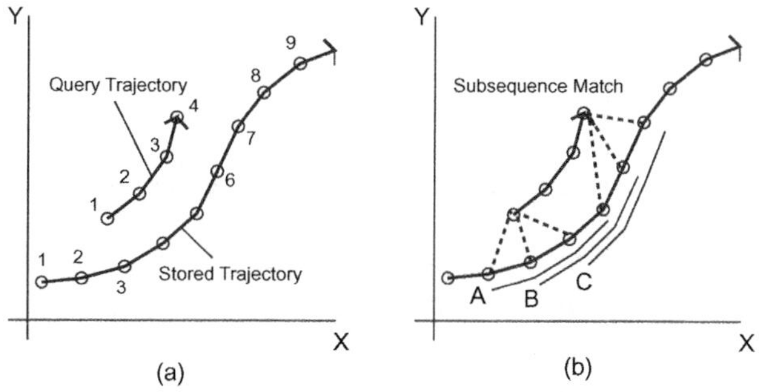

Fig. 5. Similarity query for trajectories

Definition 7 : *Shape-based Range Query*

The process for calculation of the shape-based range query $Q_{range}(\theta, \dot{\lambda}_q, \dot{\boldsymbol{\Lambda}})$ is given in Figure 4. The range query is defined as a subsequence match of trajectories as shown in Figure 5.

In addition, the nearest neighbor query can be defined using our distance between trajectories. In our definition, the temporal features are not indicated in the query, however, we consider that the temporal features can be indicated independently from the range query. For example, a query "Q_{range}(θ, $\dot{\lambda}_q$, $\dot{\boldsymbol{\Lambda}}$) $\wedge$ 11:00 $< \boldsymbol{T}_{\dot{\lambda}_{ai}}(1) <$ 12:00" means retrieving subsequences $\dot{\lambda}_{ai}$ where the distance between $\dot{\lambda}_q$ and $\dot{\lambda}_{ai}$ is less than θ. Moreover, the first vector in $\dot{\lambda}_{ai}$ is measured within the interval [11:00, 12:00].

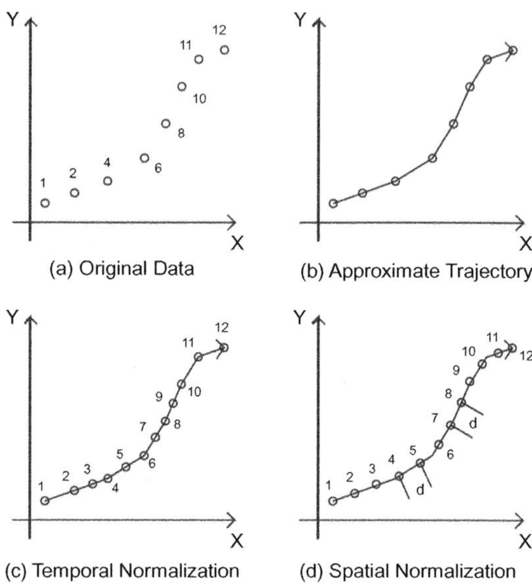

Fig. 6. Normalization of trajectories

3.3 Spatio-temporal Distance between Two Trajectories

Our defined distance $D(\boldsymbol{X}, \boldsymbol{X}')$ can be used only in the case where each vector $\boldsymbol{x} \in \boldsymbol{X}$ is measured by the same interval, that is $\Delta t = t_{i+1} - t_i$ $(i = 1, \ldots, n-1)$, where t_i is an interval from the time when $\boldsymbol{x}_i$ is measured. However, each vector in the trajectory is not always measured by the same interval Δt because sensor devices often lose the data. For example, a discrete trajectory illustrated in Figure 6(a) has no measured vectors at $t = 5, 7, 9.$. Therefore, to calculate the similarity using our definition, we define a temporal normalized discrete trajectory $\dot{\lambda}_{\Delta t}$ for trajectory λ, as follows:

Definition 8 : *Temporal Normalized Discrete Trajectory*
Given a trajectory λ defined for time interval $[t_S, t_E]$, and a natural number m, the temporal normalized discrete trajectory $\dot{\lambda}_{\Delta t}$ is defined as follows:

$$\dot{\lambda}_{\Delta t} = \langle \lambda(t_S), \lambda(t_S + \Delta t), \ldots, \lambda(t_S + m\Delta t) \rangle \text{ , where } t_S + m\Delta t = t_E$$

Intuitively, this discrete trajectory $\dot{\lambda}_{\Delta t}$ is the re-sampled trajectory per fixed interval Δt from λ, shown in Figure 6(c). In other words, $\dot{\lambda}_{\Delta t}$ is generated by dividing λ into equal interval Δt. For discrete trajectory $\dot{\lambda}$, we can use the piece-wise linear approximation $\tilde{\lambda}$ instead of λ. In the case of Figure 6, the temporal normalized discrete trajectory (Figure 6(c)) is generated from the approximate trajectory (Figure 6(b)).

Definition 9 : *Spatial-Temporal Similarity between two Trajectories*

Given two trajectories λ and λ' with the same temporal length (i.e. $L_T(\lambda) = L_T(\lambda')$) and a natural number m, the spatio-temporal distance (similarity) $D_{TS}(\lambda, \lambda')$ between λ and λ' is defined as follows:

$$D_{TS}(\lambda, \lambda') = \frac{1}{m+1}\sqrt{\sum_{i=0}^{m} D(\boldsymbol{X}_{\dot{\lambda}_{\Delta t}}(i), \boldsymbol{X}_{\dot{\lambda}'_{\Delta t}}(i))^2}, \text{ where } \Delta t = \frac{L_T(\lambda)}{m} = \frac{L_T(\lambda')}{m}$$

Note that $D_{TS}(\dot{\lambda}, \dot{\lambda}')$ can be defined as $D_{TS}(\tilde{\lambda}, \tilde{\lambda}')$. In this definition, the similarity is the Euclidean distance between trajectories represented as $m+1$ dimensional vectors, and the interval of each trajectory is normalized. Using this definition, it is possible to find trajectories whose shape is more similar to the query trajectory than can be found using previous methods.

3.4 Spatial Distance between Two Trajectories

In definition 8, we focused on the shapes of the trajectories in space $\boldsymbol{R}^2 \times T$. However, there are cases where the shapes in $\boldsymbol{R}^2$ (without temporal features) are just as important, such as for example in the case of finding similar trajectories to those of a specified user when focusing on the "spatial" shape. Hence, we also define the spatial similarity between two trajectories.

Definition 10 : *Spatial Normalized Discrete Trajectory*

Given a trajectory λ and a natural number m, the spatial normalized discrete trajectory $\dot{\lambda}_\delta$ is defined as follows;

$$\dot{\lambda}_\delta = \langle \lambda(t_0), \ldots, \lambda(t_m) \rangle, \text{ where } L_S(\lambda, [t_{i-1}, t_i]) = \delta, (i = 1, \ldots, m)$$

Similar to $\dot{\lambda}_{\Delta t}$, $\dot{\lambda}_\delta$ is generated by dividing λ into equal spatial length δ. In the case of Figure 6, the spatial normalized discrete trajectory (Figure 6(d)) is generated from the approximate trajectory (Figure 6(b)).

Definition 11 : *Spatial Similarity between Trajectories*

Given two trajectories λ and λ' with the same spatial length (i.e. $L_S(\lambda) = L_S(\lambda')$) and a natural number m, the spatial distance (similarity) $D_S(\lambda, \lambda')$ between λ and λ' is defined as follows:

$$D_S(\lambda, \lambda') = \frac{1}{m+1}\sqrt{\sum_{i=0}^{m} D(\boldsymbol{X}_{\dot{\lambda}_\delta}(i), \boldsymbol{X}_{\dot{\lambda}'_\delta}(i))^2} \text{ where } \delta = \frac{L_S(\lambda)}{m} = \frac{L_S(\lambda')}{m}$$

Using this definition, it is possible to find the trajectories whose spatial shape is similar to that of the query trajectory without temporal features.

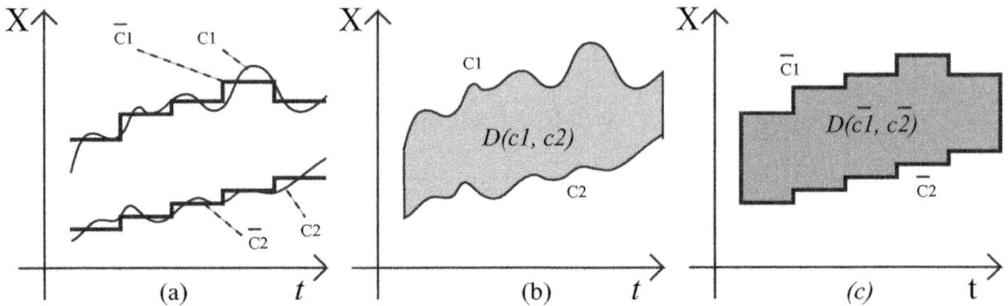

Fig. 7. Distance between two sequences c_1, c_2, and Distance between approximate sequences $\bar{c}_1$, $\bar{c}_2$

4 Indexing

With our proposed method for calculating similarity between trajectories, the database system can find the trajectories that have similar shapes to the shape of the query trajectory. However, the cost of calculating our defined similarity is very high, because it is necessary to calculate Euclidean distances between each point on the query trajectory and each point on all trajectories stored in the database. In general, database systems store a lot of trajectories, and the amount of data is increasing rapidly. Therefore, it is important to reduce the cost of calculating similarities. In this section, we present an indexing method to reduce the cost of calculating similarities, which is based on techniques for reducing the dimensions of vector data.

4.1 Piecewise Aggregate Approximation

Piecewise Aggregate Approximation (PAA) [7] is a technique for reducing the cost of comparing two sets of time series data. The essential idea of this technique is the reduction of the number of compared data using the lower limit values of time series data. Here, we describe only an outline of this technique because it was fully presented in [7].

As mentioned in Section 3, the similarity between two time series data sets can be defined as Euclidean distance between two sequences represented as multi-dimensional vectors. Even if the query sequence has a shoter length m than the candidate sequence, the similarity can be defined as the distance between the query sequence and each subsequence of the candidate sequence, as illustrated in Figure 3(b). According to the definition of the similarity between two sequences (mentioned in Section 3.3), when the length of the query sequence is m and the maximum length of a candidate sequence is n, the order of calculating the similarity is obviously $O(mn)$ for each stored sequence. In order to reduce cost of comparison, it is necessary to reduce the number of compared values in sequence.

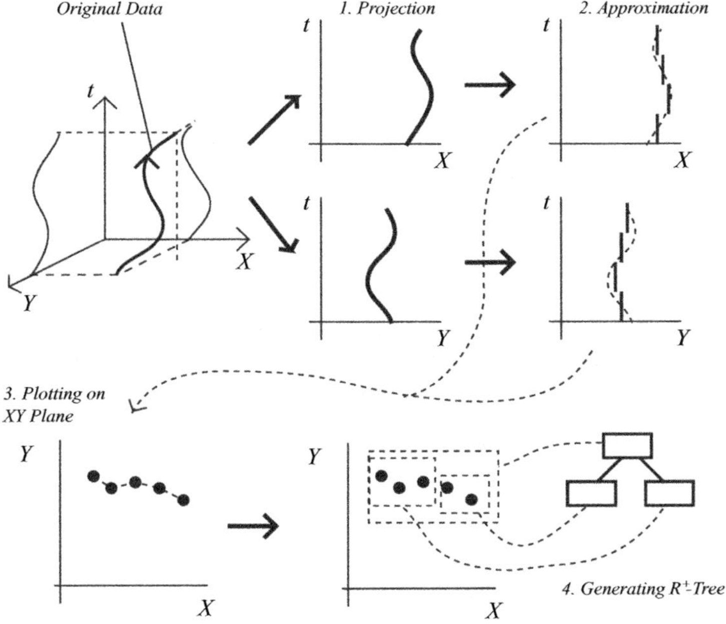

Fig. 8. The process for generating indexes to trajectories

The PAA is a technique for generating approximate sequences to efficiently calculate similarity. If the original sequence has n values, the approximate sequence has only k values (k is a factor of n and k is much less than n). Each member of the approximate sqeuence $\bar{c} = \langle \bar{w}, \ldots, \bar{w}_k \rangle$ is given as follows:

$$\bar{w}_i = \frac{1}{k} \sum_{j=k(i-1)+1}^{ki} w_j$$

In short, each $\bar{w}_i$ is calculated as the average of $\langle w_{k(i-1)+1}, \ldots, w_{ki} \rangle$. Moreover, it was proved in [7] that the approximate sequences $\bar{c}, \bar{c}'$ have a special relationship with the original sequences c, c' ($|c| = |c'| = k$):

$$D(\bar{c}, \bar{c}') < D(c, c')$$

This relationship means the distance between the approximate sequences is the lower limit of the distance between the original sequences. For example, in the case of Figure 7(a), the distance between c_1 and c_2 (Figure 7(b)) is always greater than the distance between approximate sequences $\bar{c}_1$ and $\bar{c}_2$, shown in Figure 7(c). Using this result, the database system can reduce the number of compared sequences with the query sequence.

4.2 Extended Indexing Method for Shape-Based Similarity Query

The PAA is a simple and efficient technique for reducing the number of compared time series data; however, this technique has only been adapted to the data in space $\boldsymbol{R}^1 \times T$. Hence, we extend this technique to trajectory data in space $\boldsymbol{R}^2 \times T$. Moreover, we present an efficient indexing method for trajectories by combining two techniques: PAA and a spatial indexing technique (R$^+$-Tree [13]) [2] We only describe the case of the spatial similarity query, but the essential idea can also be adapted to the spatio-temporal similarity query.

First, we define the extension of PAA for the 2-dimensional space:

Definition 12 : *2-Dimensional PAA (2D-PAA)*

Given a normalized spatial trajectory $\dot{\lambda}_\delta$ ($L_S(\dot{\lambda}_\delta) = n\delta$) and k, which is a factor of n, 2-Dimensional PAA of $\dot{\lambda}_\delta$ is:

$$\bar{\boldsymbol{X}} = \langle \bar{\boldsymbol{x}}_1, \bar{\boldsymbol{x}}_2, \ldots, \bar{\boldsymbol{x}}_k \rangle \text{ such that } \bar{\boldsymbol{x}}_i = (\bar{x}_i, \bar{y}_i)$$

$$\bar{x}_i = \frac{1}{k} \sum_{j=k(i-1)+1}^{ki} x_j, \; \bar{y}_i = \frac{1}{k} \sum_{j=k(i-1)+1}^{ki} y_j \text{ where } \boldsymbol{x}_j = (x_j, y_j)$$

Intuitively, 2D-PAA can approximate the trajectory's shape by calculating the center of the points contained in the trajectory. Using 2D-PAA and R$^+$-Tree, the indexes to trajectories can be generated as shown in Figure 8.

In order to retrieve trajectories whose distance to the query trajectory is less than θ, the database system processes the following steps;

1. Calculating $\bar{\boldsymbol{X}}_q = \langle \bar{\boldsymbol{x}}_{q1}, \ldots, \bar{\boldsymbol{x}}_{qk} \rangle$ for the query trajectory λ_q ($\lambda_{\delta q}$).
2. Searching sequences $\bar{\boldsymbol{X}}_1, \bar{\boldsymbol{X}}_2, \ldots$ (the length of each sequence is k) on R$^+$-Tree, such as $D_S(\bar{\boldsymbol{X}}_i, \bar{\boldsymbol{X}}_q) < \theta$.
3. Finding answer trajectories $\lambda_1, \lambda_2, \ldots$ such as $D_S(\lambda_i, \lambda_q) < \theta$.

In our approach, the number of compared data is reduced in two steps. The combination of both reduction techniques enables databases to retrieve answer trajectories efficiently. Therefore the trajectory databases can support the shape-based similarity query without a heavy load. Our indexing method's performance is evaluated in Section 5.

5 Performance Study

5.1 Experimental Settings

There are basically three variables that could affect the similarity between trajectories. The first is the length of a query, because as the length of query increases, the similarity decreases, but the cost of calculating similarities increases.

[2] APCA [6], which is an extension of PAA, uses R-Tree based indexing techniques. However, the indexing method only uses R-Tree techniques to retrieve time series data (in space $\boldsymbol{R}^1 \times T$) efficiently, not to retrieve sequences of vectors such as trajectories in space $\boldsymbol{R}^2 \times T$.

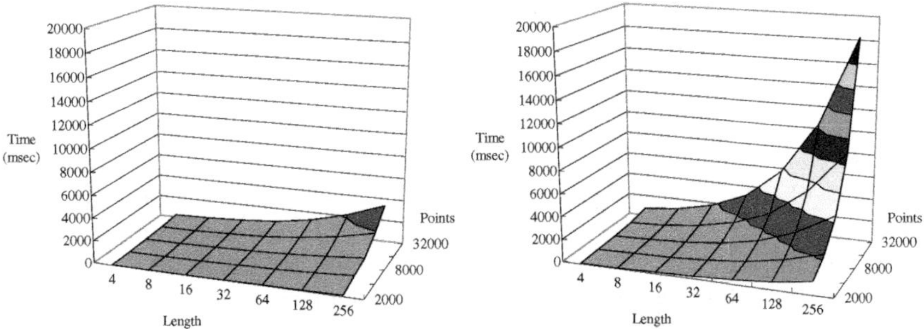

Fig. 9. The comparison of our proposed indexing method with the existing method

The second variable is the density of the points in a space, since the similarity increases and the number of compared trajectories increases with the density of the points. The final variable is the complexity of the trajectory's shape. Generally, the irregularity of an object, motion causes greater complexity of shapes; in other words, people walking randomly generate the most complex shapes.

Therefore, we generate sample trajectories by changing these variables. The number of trajectories is 2-32, the length of trajectorics is just 1000 (i.e., the maximum number of points is 32000), and the sampling interval Δt is fixed[3]. Each trajectory has shapes that represent people walking freely on a plane while changing speed and direction, and the frequency of the change in these values is altered in order to generate various complex shapes. In addition, the trajectories are embedded into a fixed area (size is 500×500) without any tendencies, in other words, the density of points can be controled by the number of embedded trajectories.

5.2 Efficiency

We have an experiment to assess the performance of our indexing method. For the experiment, we implemented two types of engine to find the trajectory which is the "nearest" to the query trajectory. The first engine checks every point on all trajectories (without any index), while the second engine checks only points filtered by our proposed index. We generated random trajectories for a query, and measured the calculation time required to find the nearest trajectory to the generated query trajectory. Since this is a simple performance evaluation of our approach, we gave a very simple query for each engine; the length of the query is fixed to k. In other word, the query trajectory $\dot{\lambda}_q$ has just k items,

[3] Even in the case where Δt is not fixed, we can take similar results from the case where Δt is fixed, because fixed trajectories can be mechanically generated from original trajectories with the piecewise linear approximation defined in Section 2.

such as $\dot{\lambda}_q = \langle \boldsymbol{x}_1, \ldots, \boldsymbol{x}_k \rangle$, and the approximate trajectory $\bar{\lambda}_q$ has only one item $\bar{\boldsymbol{x}} = (1/k \sum_{i=1}^{k} x_i, 1/k \sum_{i=1}^{k} y_i)$ where $\boldsymbol{x}_i = (x_i, y_i)$.

Figure 9 illustrates the experimental result in the case where the shape of trajectories is simple. While the left graph in the figure shows the calculation time with our indexing methods, the right graph shows without any index. With our proposed index, the calculation time is 60%–75% less than without the index in each situation. For example, in the case where the query length is 256 and the number of points is 32000 (it is the worst case), the calculation time is 3,325 msec with the index. On the other hand the calculation time without the index is 18,206 msec. Although the time taken to generate indexes (overhead) is 3,377 msec when the number of points is 32,000, it is less than the time required without the index. As a result of our experiment, the advantage of our indexing method is clear.

In addition, we measured the calculation time in cases where the shape of trajectories is very complex, such as trajectories where each object moves almost randomly. In this case, the calculation time is 15%–20% greater than the time in the simple case; however, the rate of increase is the same in cases with the index and without the index.

6 Conclusion

The main contribution of this paper is the presentation of an efficient indexing method for processing shape-based similarity queries for trajectory databases. In order to calculate similarity between trajectories, we defined discrete trajectories that were re-sampled per fixed interval. Furthermore, we described the performance of our proposed indexing method, and we show the advantage of our method over the existing methods.

As future work, we will implement the trajectory database system to evaluate our proposed model, Furthermore, we will develop several real application programs such as a car navigation system, a personal navigation system, and other location-wares.

References

[1] G. Chen and D. Kotz. Categorizing binary topological relations between regions, lines, and points in geographic databases. Technical Report TR2000-381, A Survey of Context-Aware Mobile Computing Research, Dept. of Computer Science, Dartmouth College, 2000.

[2] H. Chon, D. Agrawal, and A. E. Abbadi. Query processing for moving objects with space-time grid storage model. In *MDM2002 Conference Proceedings*, pages 121–129, 2002.

[3] E. Clementini and P. D. Felice. Topological invariants for lines. *IEEE Transaction on Knowledge and Data Engineering*, 10(1):38–54, 1998.

[4] L. E. Elsgolc. *Calculus of Variations*. Pergamon Press LTD, 1961.

[5] E. G. Hoel and H. Samet. Efficient processing of spatial queries in line segment databases. In O. Gunther and H. J. Schek, editors, *SSD'91 Proceedings*, volume 525, pages 237–256. Springer-Verlag, 1991.

[6] E. Keogh, K. Chakrabarti, S. Mehrotra, and M. Pazzan. Locally adaptive dimensionality reduction for indexing large time series databases. In *SIGMOD2001 Conference Proceedings*, pages 151–162, 2001.

[7] E. Keogh, K. Chakrabarti, M. Pazzani, and S. Mehrotra. Dimensionality reduction for fast similarity search in large time series databases. *Knowledge and Information Systems*, 3(3):263–286, 2001.

[8] G. Kollios, D. Gunopulos, and V. J. Tsotras. On indexing mobile objects. In *SIGMOD'99 Conference Proceedings*, pages 261–272, 1999.

[9] G. Kollios, V. J. Tsotras, D. Gunopulos, A. Delis, and M. Hadjieleftheriou. Indexing animated objects using spatiotemporal access methods. *IEEE Transactions on Knowledge and Data Engineering*, 13(5):758–777, 2001.

[10] Y.-S. Moon, K.-Y. Whang, and W.-S. Han. General match: A subsequence matching method in time-series databases based on generalized windows. In *SIGMOD 2002 Conference Proceedings*, pages 382–393, 2002.

[11] K. Porkaew, I. Lazaridis, and S. Mehrotra. Querying mobile objects in spatiotemporal databases. In C. S. Jensen, M. Schneider, B. Seeger, and V. J. Tsotras, editors, *SSTD 2001*, volume 2121 of *Lecture Notes in Computer Science*, pages 59–78. Springer-Verlag, 2001.

[12] N. Priyantha, A. Miu, H. Balakrishnan, and S. Teller. The cricket compass for context-aware mobile applications. In *MOBICOM2001 Conference Proceedings*, pages 1–14, 2001.

[13] T. Sellis, N. Roussopoulos, and C. Faloutsos. The R^+-tree: A dynamic index for multidimensional objects. In *VLDB'87 Conference Proceedings*, pages 3–11, 1987.

[14] A. P. Sistla, O. Wolfson, S. Chamberlain, and S. Dao. Modeling and querying moving objects. In *ICDE'97 Proceedings*, pages 422–432, 1997.

[15] M. Vazirgiannis and O. Wolfson. A spatiotemporal model and language for moving objects on road networks. In C. S. Jensen, M. Schneider, B. Seeger, and V. J. Tsotras, editors, *SSTD 2001*, volume 2121 of *Lecture Notes in Computer Science*, pages 20–35. Springer-Verlag, 2001.

[16] O. Wolfson, B. Xu, S. Chamberlain, and L. Jiang. Moving objects databases: Issues and solutions. In *Statistical and Scientific Database Management (SSDM'98) Conference Proceedings*, pages 111–122, 1998.

An Efficient Spatiotemporal Indexing Method for Moving Objects in Mobile Communication Environments

Hyun Kyoo Park[1], Jin Hyun Son[2], and Myoung Ho Kim[1]

[1] Div. of Computer Science, Korea Advanced Institute of Science and Technology, 373-1 Guseong-dong, Yuseong-gu, Daejeon, South Korea, 305-701
{hkpark, mhkim}@dbserver.kaist.ac.kr

[2] Dept. of Computer Science and Engineering, Hanyang University, 1271 Sa-1 dong Ansan, Kyunggi-Do, South Korea, 425-791
jhson@cse.hanyang.ac.kr

Abstract. The spatiotemporal databases concern about the time-varying spatial attributes. And one of the important research areas is tracking and managing moving objects for the location-based services. Many location-aware applications have arisen in various areas including mobile communications, traffic control and military command and control (C2) systems. However, managing exact geometric location information is difficult to be achieved due to continual change of moving objects' locations.

In this paper we propose the B^{st}-tree that utilizes the concept of multiversion B-trees. It provides an indexing method for future location queries based on the dual transformation. This approach can be applied for the range query on moving object's trajectories specifically in the mobile communication systems. Also we present a dynamic management algorithm that determines the appropriate update interval probabilistically induced by various mobility patterns to guarantee the query performance.

1 Introduction

The recent advances in mobile computing and sensing technology such as *GPS* have made it possible to perform location-based services via wireless links. One of the important problems for location-based services is to provide fast answers about range queries that retrieve the moving objects in a certain area. The spatiotemporal indexing method for the location management is one of the important issues for this purpose.

However, in the conventional database systems, the data remain unchanged unless it is explicitly modified. It means that the accuracy of the location information is difficult to be maintained in the database systems, since a large number of continual updates may not be properly processed. To reduce the number of updates required while keeping the reasonable data accuracy, functions of time

M.-S. Chen et al. (Eds.): MDM 2003, LNCS 2574, pp. 78–91, 2003.

that express the object's positions may be used. Then updates are necessary only when the parameters of the functions change "significantly" [6, 10].

In this paper we propose an indexing approach that can answer the queries about current or anticipated future positions of moving objects. We use one linear function per object's trajectory that is composed of the initial position and a function of velocity for a certain period of time. Then we index the location information by the ideas of *Space Filling Curve* and duality. Also we provide a dynamic management algorithm of the index structure to maintain the performance of range queries about future position.

2 Problem Statement and Motivation of Research

The database management system (DBMS) technology can provide a foundation for spatiotemporal applications in spite of some problems such as frequent updates. So, the spatiotemporal capabilities need to be integrated, adapted, and built on top of existing database management systems.

Previously most related work focused on historical queries for the spatiotemporal databases. However, the location-based services require future queries as well as the conventional historical queries. In this paper we concentrate on the range queries of moving objects in the future. An example future query may be "retrieve the patrol cars that are or will be in a given region within next 30 minutes". More generally, the location-based range queries considered in this paper have the following forms:

Type-1(Time-sliced) Query: "Find all objects that are(or will be) in a certain area at time *t*" □

Type-2(Time-window) Query: "Find all objects that are(or will be) in a certain area during time [*ts, te*]" □

In the location-based services, the positional information of moving objects is obtained via wireless communication. Then the real trajectories can be approximated as linear functions through the *Dead-Reckoning* policy that can predict the future position of the object. Then the trajectories can be reduced to lines starting from an initial location during a certain period of time. Since several mobile communication protocols have been developed as a solution of location management for next-generation communication system (e.g., *IS-41, GSM*), this assumption is practical [10].

One of the indexing methods for the lines is a *Spatial Access Method* (SAM) such as the R-tree [9]. However, there are several problems when handling the trajectories for the future position. Some of them are: (i) The *Minimum Bounding Rectangle* (MBR) assigned to a trajectory often has too large area, and (ii) implicitly all lines may extend to '*infinity*' until being updated. It may incur many overlaps of MBRs.

Hence we need an appropriate methodology to support indexing of trajectories rather than previous spatial access methods. In addition, the index structure should afford the frequent updates after storing the trajectory information.

3 Preliminaries

3.1 Related Work

Wolfson et al. [6] proposed a data model to represent moving objects. They proposed a framework called *Moving Object Spatio-Temporal* (MOST) for tracking objects and a temporal query language called *Future Temporal Logic* (FTL). Several spatiotemporal indexing methods were proposed for the concept of the paper [6].

In the past, research in spatial and temporal database has mostly been made separately. And for the theoretical basis, the community of computational geometry developed early work on moving points focused on bounding the number of changes in various geometric structures as the points move [6]. One of the result is the *Kinetic Data Structure* (KDS) by Basch et al. proposed in [2].

A survey of previous work on spatiotemporal databases can be found in the papers [4, 7, 9]. An important result of these is the TPR-tree by Saltenis et al. [9]. They proposed the R-tree based approach but we think that the R-tree based approach has some restrictions as a remedy of our problems. Another result by Kollios et al. [4] proposed a sophisticated approach for the future queries that can outperform the kd-B-tree and the R-tree. In their work, they used the *Hough-X* transformation for indexing moving points, however they assumed that the space is linear or quadratic and the velocities are fixed.

To support the future queries those we are concerning, we use multi-version structure concept [3, 11] and extend the study of [4]. Also an appropriate prediction of mobility patterns can reduce the difficulties in studying the continuous updates in spatiotemporal indexing and the performance degradation in the multi-version structure. Yet much of the literature to date has not considered it. So we refer the information theory and the stochastic process [1] to predict update intervals for our dynamic update algorithm in the mobile communication environment [10].

In the practical environment, the location information is acquired from mobile communication systems. And various location management scheme were suggested [5, 10] and the dynamic location management scheme in the next generation communication network [10] is used as a framework of our method.

3.2 Duality and Space Filling Curves

In the mobile communication environment, moving objects send their location information and we can translate them as a set of tuples $\{ t, x, y, Fx(t), Fy(t)\}$ where t is the time when the update message is sent. x, y denotes the coordinates of the location. $Fx(t)$ and $Fy(t)$ that are functions of time denote velocity vectors

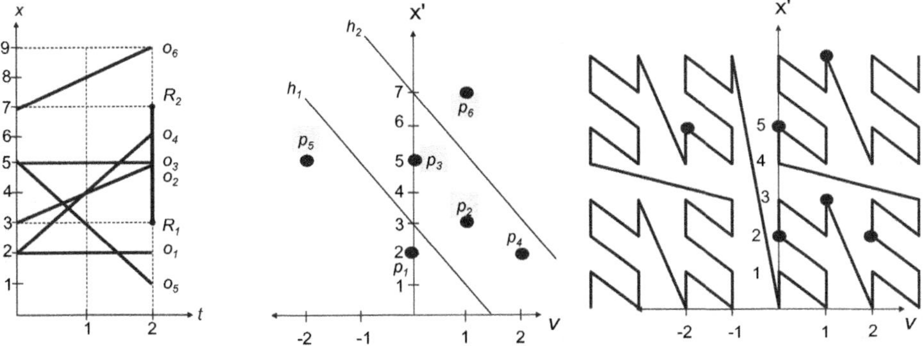

Fig. 1. Trajectories of moving points in the primal and the dual space

at x and y, respectively. This location information is transformed to a point in the dual space as follows:

In d-dimensional space R^d, the hyperplane represented as a function of $x_d = \nu_1 x_1 + \ldots + \nu_{d-1} x_{d-1} + \nu_d$ in the primal space is transformed to a point $(\nu_1, \ldots, \nu_d)$ in the dual space and the point $(\nu_1, \ldots, \nu_d)$ in the primal space can be transformed to a line of $x_d = -\nu_1 x_1 - \ldots - \nu_{d-1} x_{d-1} - \nu_d$ in the dual space [7].

Figure 1 shows the geometric representation of moving objects in the primal and the dual spaces. Generally, a trajectory in the primal xt plane induces a static point p in the dual $x'\nu$ plane. Similarly a query range R induces a query strip σ that is bounded by two hyperplanes. For example, the trajectories of objects $(o_1, \ldots, o_6)$ in the primal space can be transformed to points $(p_1, \ldots, p_6)$ in the dual space. Also extreme points R_1 and R_2 of range R in the primal space can be represented as a strip between the hyperplanes h_1 and h_2 in the dual space. Hence the query of retrieving trajectories intersecting a range R parallel to the x axis can be rewritten to a query of retrieving set of points in the query strip σ.

In this paper we consider *Space Filling Curve* for an indexing in the dual space. In Figure 1, the *Peano* curve provides a linear ordering in multidimensional spaces starting from a lower left corner. Let $SFC(x, v)$ be the function that assigns a linear address for two-dimensional point (x, v). Then the dots in the dual space have unique addresses that can be mapped as key values in one-dimensional index structure.

4 The Proposed Index Structure for Moving Objects

In this section, we describe the B^{st}-tree; namely the Spatiotemporal B-tree. The B^{st}-tree incorporates features of B-tree and the multi-version tree [3, 11]. Since the B^{st}-tree stores the whole trajectory of a moving object as pieces, we need a multi-version access method that stores the records with their own life-time to handle the updated records.

Although moving object traces out a certain trajectory in three-dimensional spatiotemporal space, we describe the two-dimensional case for simplicity. Then the three dimensional queries can be achieved by conjunctions of each trees.

In Figure 2(a), object O_2 changes its trajectory at time t_2 and t_4, and O_3 changes at every timestamp while O_1 keeps its movement function all the time. The moving object checks its location periodically. We define unit time Δt and normalize to 1 for convenience. Since it is impractical to monitor the location at every unit time, the object monitors its location every time interval $\tau = n\Delta t$. For example, Figure 2(b) shows the state of object O_2. Then there are two updates at t_2 and t_4. Here we define two different updates.

Definition 1. ***Adaptive Update***
The Adaptive Update(AU) is an update operation to make an index of current location information for all moving objects.

Definition 2. ***Instantaneous Update***
The Instantaneous Update(IU) is a demanded update by individual moving object when a movement function of an object is changed.

The time interval between two consecutive *AU* is represented as T_u in Figure 2(b) and determined by a dynamic method described in Section 5. *IU* is occurred when the deviation from the expected trajectory exceeds the distance threshold that is predefined by the *Distance-based update* protocol [10].

In our scheme, we discard the notion of *Location Area* borders and communication cells in most *PCS* or *Cellular* systems. Dynamic mobility management in wireless communication services is known to provide better cost-effectiveness and our method can be dynamically adapted to this scheme [5, 10].

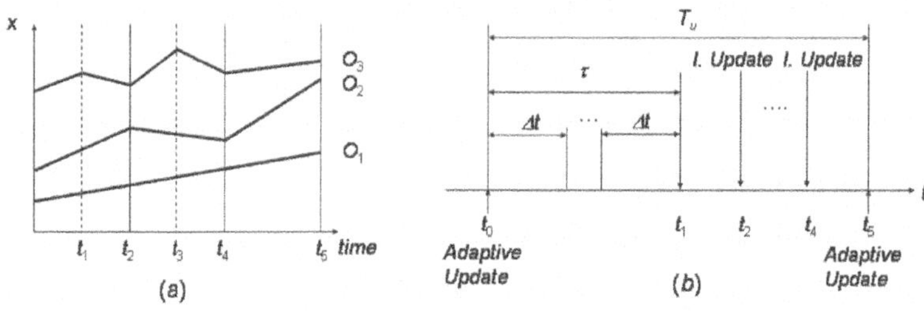

Fig. 2. Representation of trajectories and the update parameters

4.1 Overview of the B^{st}-tree

Although a multi-version structure keeps both the current and old data, it may experience a significant overhead when a large number of updates. Thus we may need an appropriate management strategy to handle this case. The B^{st}-tree has two types of nodes, i.e., *Branch nodes* and *Version nodes*, as shown in Figure 3. When AU occurs, a new *Branch* is made with the update timestamp. Then the *Branch node* has a timestamp as a key value and a link to the *Version node*. For example, an AU occurred at time 0 and 5. Then we have two *Branches* of $b_1(0)$, $b_2(5)$ as shown in Figure 3.

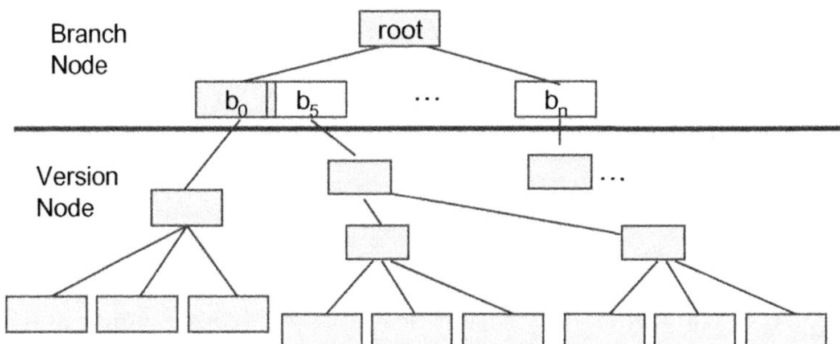

Fig. 3. The illustration of the B^{st}-tree structure

Definition 3. ***B^{st}-tree Record Structure***
The record in a Version node of the B^{st}-tree has the form (Key, TS, Pointer) where Key, TS and Pointer denote a key, life-time of a record and a link to a node at the next level (intermediary) or of the actual record (leaf) respectively.

In the *Version node*, the index key value is acquired from the method described in Section 3.2. The life-time TS is set of (t_s, t_e): the start and end time of the record respectively. The currently valid record has the TS value of $(t_s, *)$ and the updated old record has (t_s, t_e) value.

4.2 Fundamental Operations

In this section we describe the insert and update operations. The B^{st}-tree is partially persistent; therefore insertions and updates only are applied to current data pages. To insert a new record or update an existing record, the B^{st}-tree starts searching the current *Branch* node until finding the appropriate leaf node which the record must be added.

To differentiate from the inserted record, we call the updated record a *variant*. In the update operation, we do not perform the delete operation but invalidate the currently existing record. The invalidation changes the t_e value of the record to current timestamp. Hence the update operation inserts a *variant* into a proper node that is logically the same as the insert operation followed by invalidation. This process is the same as a conventional temporal database. However, if there is modification of the structure by the split or merge operation, the operations are not trivial.

Definition 4. ***Node Splitting***
The P_{max} *and* P_{min} *are node split parameters to control the node occupancy. The values of parameters are* $P_{min} \leq P_{max} \leq B$*, where* B *is the node capacity.*

When an update operation triggers the modification of structure, the valid records of overflowed node are copied to a new assigned node with the current timestamps. Then we invalidate all the records in the previous node and insert the *variant* into the new node. The node split is divided into two cases where N_v is the number of valid records in a node.

Case 1. $Nv > P_{max}$; Create two nodes. And copy all the valid records in the overflowed node to the new nodes by their key values. □

Case 2. $P_{min} \leq Nv \leq P_{max}$; Create one node. And copy all the valid records in the overflowed node to the new node. □

In Figure 4, the upper illustration describes the Case 1 and the lower illustration describes the Case 2 updates. In the Case 1, the update operation copies the valid records in overflowed node A (e.g., the records have key value 5, 6, 8, 21) into nodes B and C at time 5. Then we change the life-time value '*' of the TS in node A to 5. After all we insert a *variant* of key value 25 in C. In the Case 2, there are two valid records (e.g., the records have key values 5, 7) in overflowed node. Then we assign one node and perform the similar update procedure in the Case 1. In the case of the number of valid records being less than the value of P_{min}, the underflow condition is occurred. When the underflow condition is occurred, we merge valid records in underflowed node with sibling node as the same way of the update procedure. Therefore, when a node is created, it is ensured that it contains less than the value of P_{max} valid records.

4.3 Range Searching in the Dual Space

The range searching in the primal space induces the infinite range problem in the dual space as shown in Figure 1. For example, query range of [R1, R2] in the primal space in Figure 1 derives infinate query strip in the dual space. So, we need a "*Pruning*" method to limit the query strip to avoid the infinite range problem in the dual space.

The pruning phase is processed as follows. Since the possible positive and negative maximum velocity of objects can bound the query strip in the dual space, we can set the query strip to two rectangular regions as shown in Figure 5. The maximum velocity trajectories T_1 and T_3 in the primal space are represented by points T_1^* and T_3^* in the dual space. The staying objects' trajectories that may intersect the query range are depicted points T_2^* and T_4^* in the dual space in the same way.

This restriction of range query is intuitive since the moving objects have maximum velocities in real world. Then those four points make a super set of

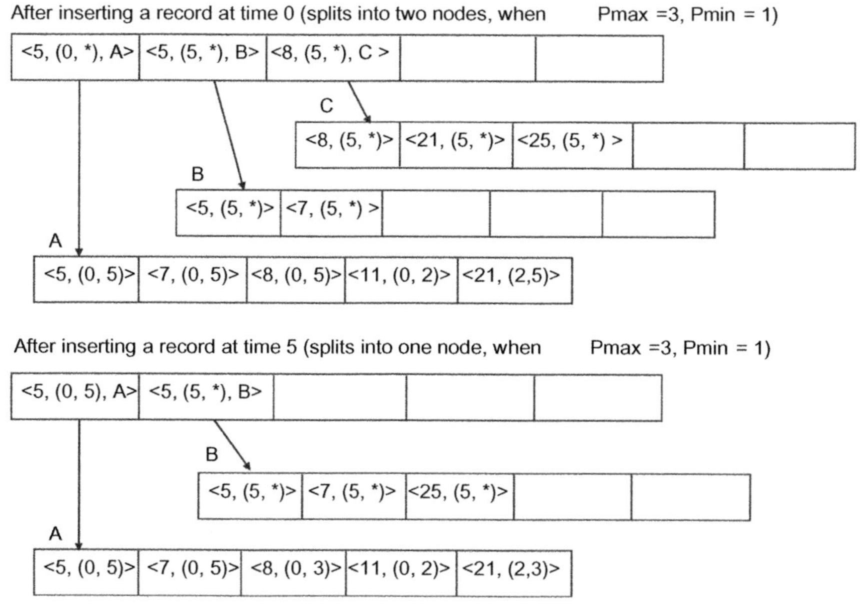

Fig. 4. Illustration of updates with structure modification

range query result. For time-window range queries, we only consider the range at time t_e during the pruning phase since it contains all possible trajectories between t_s and t_e. Then the query strip in the dual space is bounded by rectangles the same as the previous time-slice queries. Based on this observation, we can define corresponding $Lower\_Left(LL)$ for the lowest key value and analogously $Upper\_Right(UR)$ value in a query range region. The result set is composed of several data pages based on the *Peano* value in an ascending order. However all the pages in the key ranges $[LL, UR]$ may not be the result set. The result set corresponds to a set of leaves that are composed of several node region that is represented by I_n of the *Version node* in Figure 5. Each node has its key range $LL \leq l_i \leq u_i \leq UR$ where l_i and u_i are the lower and upper key values of each node i respectively.

To retrieve the exact pages rather than the all pages in range $[LL, UR]$, we first search for the leaf node that contains the lowest value larger than LL value. Then follow pointers to find subsequent leaves until the node that has the larger l_i value than UR value. The algorithm of excluding the unnecessary pages shown in Figure 5 was introduced in several works [8], so that we do not mention the details here.

5 Dynamic Management of the B^{st}-tree

In this paper we propose a dynamic algorithm to select the appropriate update interval (the value of T_u in Figure 2) in the mobile environments. The determi-

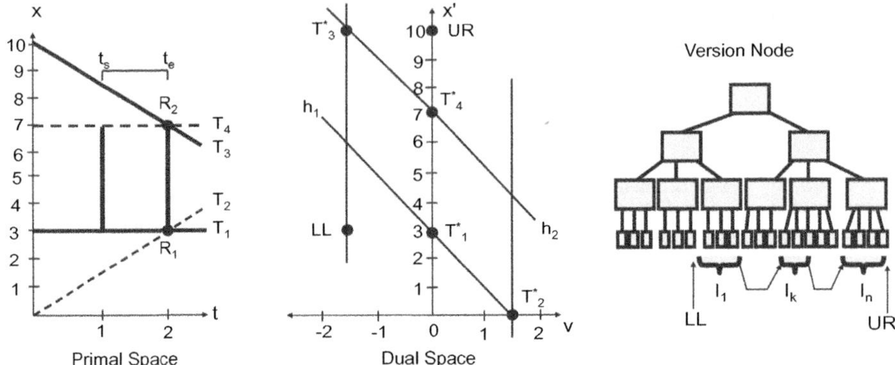

Fig. 5. Primal-Dual representation for range query

nation of the interval between two consecutive AU depends on the duration that objects keep their movement linear and the performance of *Version node*. Various mobility models found in the literature [10], however most index structures assume linear movement with constant speed [2, 12], that does not reflect reality. In our approach, we adopt the multi-version structure to reflect updates. So we model a moving object as a dynamic linear function and use the cost model that is suggested by Tao et al. [11] for the evaluation of the performance and suggest the parameter determination methodology.

5.1 Cost Model for Dynamic Management

The performance of B^{st}-tree depends on the number of node accesses that are affected by the mobility pattern that defines how random the movement is. The movement pattern is described as *Randomness* as;

Definition 5. ***Randomness***
The Randomness defines how random the mobility pattern is. If on average, v *new variants by IU will be occurred in a Branch, then the Randomness is* v/N_v *where* N_v *is the valid number of records at a certain time.*

At a timestamp, the expected number of N_v is $ln2(P_{max}/B)$ since the B^{st}-tree follows the average utilization of B-tree. Hence after a certain number of updates, the performance of *Version node* approaches the degradation point [3, 11]. At the degradation point, the *Version node* produces worse performance than a single ephemeral B-tree.

The value of *Affordable Randomness* is determined mostly by the value of P_{max} that is selected as a tuning parameter. Then we need the value of *Affordable Randomness* for query parameters (q_k, q_t). In Eq. (1), the value of *Affordable Randomness* for query parameters is shown.

$$AffordableRandomness(q_k, q_t) \cong \frac{\frac{N_v}{N_n} q_t - 1}{\frac{ln2(P_{max}/B)}{1 - ln2(P_{max}/B)}(q_t - 1)} \quad (1)$$

where q_k and q_t denote query range of key and time space respectively. We can allow *IU* in the interval of *AU* until *Affordable Randomness* by queries and the value of *Affordable Randomness* can determine the optimal *Branch* duration T_u in the B^{st}-tree.

5.2 Probability-Based Dynamic Update Algorithm

In continuous time, the movement of objects can be described as a *Brownian* motion and the location distribution of an object in the future is a stochastic process as time progresses [1]. Furthermore, the velocity of object is a stationary *Gauss-Markov* process since the initial location and the ensuing location estimation process have the same probability law. Hence, the predictive location of moving object in the future is correlated in time and modeled by some finite states with its velocity. In the linear motion assumption, we discretize the continuous movement into intervals to check the location information as shown in Figure 2.

Let define $\alpha = e^{-\lambda_p \tau}$ ($0 \leq \alpha \leq 1$, where $0 \leq \lambda_p$), the rate of *IU*. Then the velocity of moving object in discrete time is;

$$v_n = \alpha v_{n-1} + (1-\alpha)\mu + \sqrt{1-\alpha^n} x_{n-1} \tag{2}$$

In the sense of *Gauss-Markov* process, the general representation of velocity in terms of the initial velocity v_0 is

$$v_n = \alpha^n v_0 + (1-\alpha^n)\mu + \sqrt{1-\alpha^n} \sum_{i=0}^{n-1} \alpha^{n-i-1} \zeta_i \tag{3}$$

In the equation, ζ_n is an independent, uncorrelated, and stationary *Gaussian* process, with mean $\mu_\zeta = 0$ and $\sigma_\zeta = \sigma$, where μ and σ are the asymptotic mean and standard deviation of v_n when n approaches infinity [1, 5].

To expect the update interval, the moving object checks its location periodically. Let τ be the location information inspection period as shown in Figure 2, and $s_{k\tau-1}$ be the displacement from an initial state. Then each moving object triggers *IU* at the kth location inspection where the condition of $|s_{k\tau-1} - \mu_{s_k}|$ is greater than a distance threshold. Probabilistically the expected location information at k^{th} location inspection is x_k that we can derive from the Eq. (3)

$$x_k = \sum_{i=k\tau}^{i=k\tau+\tau-1} v_i \quad \text{and} \quad s_{k\tau-1} = \sum_{i=0}^{k-1} x_i \tag{4}$$

$s_{k\tau-1}$ is a *Gaussian* random variable with mean $\mu_{s_k} = \sum_{i=0}^{k-1} \mu_{x_i}$. The expected value of location x_k and displacement $s_{k\tau-1}$ is $hatx_k = x_k - \mu_{x_k}$ and $hats_k = s_{k\tau-1} - \mu_{s_k}$. Then the following recursive iteration predicts the update interval probabilistically.

$$\hat{s}_k = \hat{s}_{k-1} + \hat{x}_{k-1} \tag{5}$$

Table 1. Definition of fuctions and parameters for the update interval

Functions	*Contents*
$\overline{p}_k(\hat{s}, \hat{x})$	Probability there is no update up to time $k-1$ where $\hat{s}_k = \hat{s}$, $\hat{x}_{k-1} = \hat{x}$
$p_{k-1}(s, \hat{x})$	Probability there is no update up to time k where $s_k = s$, $x_{k-1} = x$
$p_k(s, x)$	Probability there is no update up to time k where $s_k = s$, $x_k = x$
$\overline{F}(k)$	Probability there is no update up to time k
$f(k)$	PDF of time k between two consecutive updates

From the previous analysis, our deterministic algorithm provides the predictive k^{th} interval of each *Branch*. In Table 1, we define some random variables and probability density functions used in our algorithm. Probabilistically the AU interval guarantees the query performance. So, the location distributions at time $k > 0$ can be found from $k = 1$ until $f(k) = \overline{F}(k-1) - \overline{F}(k)$ is sufficiently small, with initial condition of $\overline{p}_k(\hat{s}, \hat{k}) = f_{\hat{x}_0}(\hat{x})\delta(\hat{s} - \hat{x})$ where $k = 0$, distance threshold δ and $f_{\hat{x}}$ the PDF of $\hat{x}_k$ for the x direction. where . The $\overline{F}(k)$ is acquired from Eq. (6).

$$\overline{F}(k) = \int_{-\infty}^{\infty}\int_{-\infty}^{\infty} p_{k-1}(\hat{s}, \hat{x}) d\hat{x} d\hat{s} \tag{6}$$

In the algorithm, the iteration will stop when the value of k exceeds the *Affordable Randomness* iteratively for each object. Finally, we set the *AU* interval for the new *Branch* from the current time to the value of k.

```
Algorithm determine_interval_of_Adaptive_Update
   step 1: calculate k of all objects and enqueue in queue_k
   step 2: dequeue minimum k in queue_k
   step 3: while randomness > affordable_randomness do loop
     increase the number of variants and calculate randomness;
     dequeue the current k in queue_k;
     calculate new k of the current object;
     enqueue the new k in queue_k;
```

6 Experimental Results

We evaluated the B^{st}-tree in the *Sun Ultra-60* with 256MB main memory system. The experimental data set was generated with $GSTD$[1] software that simulates the trajectories of moving objects. The B^{st}-tree allows *variants* to be inserted until it reaches until the *Affordable Randomness*. In Figure 6, we investigated

[1] Synthetic moving object dataset generation tool, SIGMOD Record, 29(3).

the *Affordable Randomness* for the time-sliced and time-window queries where the moving objects are uniformly distributed. This is the expected result that is suggested in [11]. We investigated the performance of the B^{st}-tree with the uniformly distributed datasets here. As a result, for the time-slice query, it is preferable not to allow *IU*. Also for the time-window query, the practically *Affordable Randomness* converges almost 30%. In the *Affordable Randomness*, our method is efficient for the construction and reflection of updates.

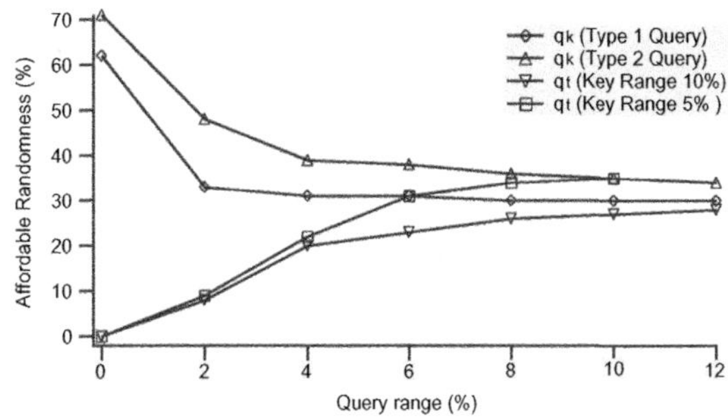

Fig. 6. *Affordable Randomness* for the various queries

And the R*-trees consume more construction time than the B^{st}-tree. We do not reflect the TPR-tree algorithm since it is not available to get the end time of the movement functions when we construct the trees. Hence we construct the R*-tree after one iteration of *GSTD* and use the trajectories. To evaluate *IU* performance, we use 500,000 objects to construct trees and update 1,000 objects up to 20,000 objects. The results of comparisons are shown in Figure 7.

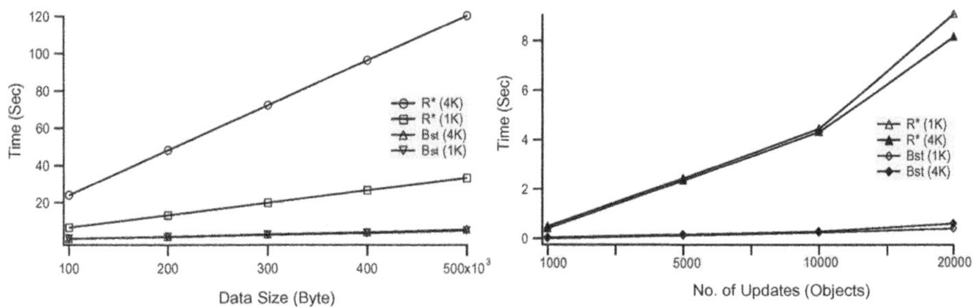

Fig. 7. Performance comparison of construction and update time

The size comparison of the R*-tree and the B^{st}-tree is shown in Figure 8. For future queries in our environment, the size of the R*-tree is not affected by the variance of the *Randomness* significantly but the B^{st}-tree is increased exponentially. Although in the 30% *Randomness*, the size of the B^{st}-tree is less than half of the R*-tree. Furthermore, the node accesses in answering the time-slice query of 5% key range in the 20,000 objects are shown in Figure 8. Since the future queries are affected by the *Randomness* directly, the 0% *Randomness* matches the optimal performance and the B^{st}-tree increases the number of node accesses almost linearly less than 25% *Randomness* range.

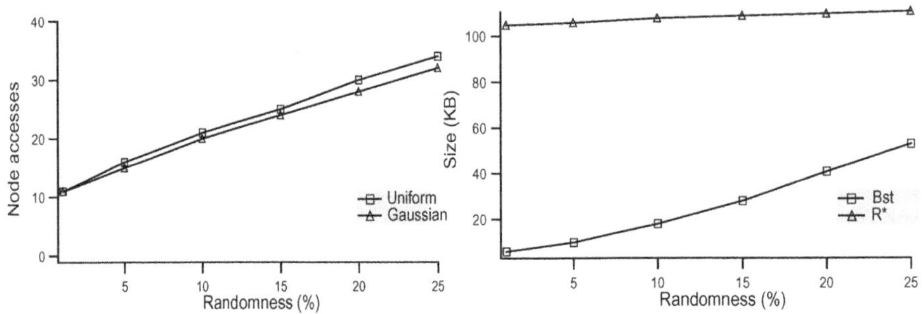

Fig. 8. Node access and size comparison by the *Randomness*

7 Conclusions and Future Work

The location-based service is expected being used widely in the near future, and it is an important application area of the spatiotemporal databases. In this paper we have proposed an index structure, named the B^{st}-tree as a spatiotemporal indexing method for the future location queries in mobile communication environments. The contribution of our work can be summarized as follows.

(i) An efficient indexing approach for moving objects to support the historical and future queries: While the R-tree based indexing is very time consuming for construction or updating, our index structure requires less time consuming and supports $O(log_B N + t)$range query performance.

(ii) A comprehensible probability-based determination of update interval: The proposed algorithm can predict the behavior of an index structure for guaranteeing the query performance. Our update policy uses the mobile communication framework that can be very useful for the next generation communication environments, specifically the *Distance-based* management protocol.

For further work, we are working on the effects of the mobility patterns that can be generated in various environments. We use the discrete time interval for the prediction, and we will analyze our method in continuous time.

Acknowledgement: This work was supported in part by grant No. R01-1999-00244 from the interdisciplinary Research program of the KOSEF.

References

1. Ross, S.: Stochastic Processes 2nd Ed., John Wiley & Sons (1996)
2. Basch, J., Guibas, L. Hershberger, J.: Data Structures for Mobile Data. Proc. of the 8th Annual ACM-SIAM Symposium on Discrete Algorithms.(1997) 747–756
3. Bertimas, D., Tsitsiklis, J.: Introduction to Linear Optimization. Athena Scientific.(1997)
4. Varman, P., Verma, R.: An Efficient Multiversion Access Structure. IEEE TKDE Vol.9(3) (1997) 391–409
5. Liang, B., Haas, Z.: Predictive Distance-Based Mobility Management for PCS Networks. Proc. of IEEE INFOCOM.(1999) 1377–1384
6. Wolfson, O., Sistla, P., Chamberlain, S., Yesha, Y.: Updating and Querying Databases that Track Mobile Units. J. of Distributed and Parallel Databases. (7). (1999) 257–287
7. Agarwal, P., et al.: Efficient Searching with Linear Constraints. J. of Computer and System Sciences. 61. (1999) 194–216
8. Ramsak, F., et al.: Integrating the UB-Tree into a Database System Kernel. Proc. of VLDB. (2000) 263–272
9. Saltenis, S., et al.: Indexing the Positions of Continuously Moving Objects. Proc. of SIGMOD. (2000) 331–342
10. Wong, V., Leung. V.: Location Management for Next-Generation Personal Communication Networks. IEEE Network 14(5). (2000) 18–24
11. Tao, Y., Papadias, D., Zhang, J.: Cost Models for Overlapping and Multi-Version Structures. Proc. of ICDE. (2002) 191–200
12. Kollios, G., Gunopulos, D., Tsotras, V., "On Indexing Mobile Objects", Proc. of PODS, (1999) 262–272

DynaMark A Benchmark for Dynamic Spatial Indexing

Jussi Myllymaki and James Kaufman

IBM Almaden Research Center
{jussi,kaufman}@almaden.ibm.com

Abstract. We propose a performance benchmark for dynamic spatial indexing that is directly geared towards Location-Based Services (LBS). A set of standard, realistic location trace files is used to measure the update and query performance of a spatial data management system. We define several query types relevant for LBS: proximity queries (range queries), k-nearest neighbor queries, and sorted-distance queries. Performance metrics are defined to quantify the cost (elapsed time) of location updates, spatial queries, and spatial index creation and maintenance.

1 Introduction

The DynaMark benchmark is designed to evaluate the performance of middleware and database servers supporting Location-Based Services (LBS). With mobile devices quickly penetrating many areas of business and consumer activities, new types of data and queries are becoming important. Managing the continuously changing data representing the current, past, or future locations of mobile users requires dedicated support from the underlying database system. LBS queries such as "where are my friends?" will be typical of new LBS applications. The high update load intrinsic to these applications places unique requirements on a spatial database system. Real-time alerts triggered by a spatial condition, such as "alert me when my child leaves from school," will require more processing power and more sophisticated spatial indexing methods and query evaluation strategies.

The DynaMark benchmark defines a method to evaluate the scalability of a standalone database server or an entire LBS system. It consists of synthetic but realistic datasets that simulate the movement of an arbitrary number of mobile users, and a standard set of spatial queries. The benchmark performance metrics include the cost of inserting and updating a user's location, the size of a spatial index, the cost of executing various types of spatial queries, and the cost of creating and maintaining a spatial index.

The performance metrics are evaluated as a function of the number of simulated mobile users. In this way the benchmark provides data on the scalability of the system under test. The metrics are also evaluated as a function of time (number of updates performed), which helps determine whether a system is capable of sustaining its update or query performance over extended periods of time.

M.-S. Chen et al. (Eds.): MDM 2003, LNCS 2574, pp. 92–105, 2003.

Our experiments have shown this not to be the case with some systems, as their spatial index efficiency deteriorates over time and overall system performance suffers.

The paper is organized as follows. In Section 2, we review existing work on spatial data management, moving object databases, and database benchmarking. In Section 3, we discuss spatial queries typical of Location-Based Services and define the scope of the benchmark. The details of the benchmark are described in Section 4. We present a summary and directions for future work in Section 5.

2 Related Work

The focus in spatial data management has traditionally been on Geographic Information Systems (GIS), VLSI, and mechanical CAD applications. Dozens of multidimensional access methods have been proposed for managing spatial data, including grid indices, quadtrees, R-trees, k-d-trees, and space-filling curves such as Z-order [5]. Most of this research has been done in the context of static data (e.g. cartographic data) and few albeit relatively expensive queries (e.g. spatial joins). Moreover, the data usually consists of complex spatial objects, e.g. polylines and polygons with hundreds of vertices or more [13].

In contrast, a spatial index for location-based services contains a very large number of simple spatial objects (e.g. points) that are frequently updated. These "moving object databases" pose new challenges to spatial data management [18]. The workload is characterized by high index update loads and relatively simple but frequent queries (e.g. range queries). A location update may merely contain the position of a user or it may include the user's trajectory [14, 10]. Supporting trajectories adds additional requirements to the index and query scheme [1]. A location update may also expire at a certain point in time [12]. Finally, location data is inherently imprecise because location determination yields only an estimate of a user's location.

Several benchmarks exist for transaction processing in relational databases (Wisconsin benchmark [2] and TPC-C), object-relational databases (BUCKY [4]), and object-oriented databases (OO7 [3]), but to our knowledge no standard benchmarks exist for LBS applications and dynamic spatial indexing. The SEQUOIA 2000 benchmark [15] was motivated by Earth Sciences and focused on retrieval performance over large datasets of complex spatial objects. Several spatial access methods were compared in [7], but apart from an initial insertion cost, the benchmark focused on query performance. A qualitative comparison of spatio-temporal indexing methods is presented in [16].

Use of realistic datasets is critical to gaining meaningful data from performance evaluation experiments. Most LBS experiments to date have used spatial data generated by random walks or algorithms using simple distribution functions. For instance, [7] looked at six random distributions for spatial data: diagonal, sinus, bit, x-parallel, clusters, and uniform distributions. A few generators have been proposed for creating spatio-temporal data. The GSTD generator [17] creates data for randomly moving points and rectangles. GSTD was later

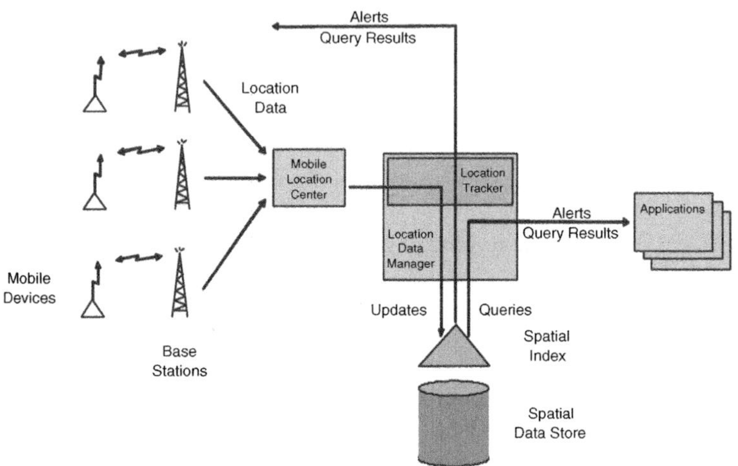

Fig. 1. Simplified view of data flow in a location tracking system.

extended to account for buildings and other "infrastructural" obstructions, but it still allowed objects to move free of influence from other moving objects. The Oporto toolkit [11] describes an elaborate scheme with moving point objects (fishing boats), moving extended objects (schools of fish), stationary extended objects that change size (areas with plenty of plankton that attract fish, and storms that boats try to avoid), and static extended objects (harbors).

Our work advances the state of the art in two ways: by proposing a standard benchmark for quantitatively measuring the scalability of spatial data management systems for location-based services, and by proposing a method of generating realistic, three-dimensional spatio-temporal datasets that accurately model the movement of people in a virtual city consisting of multi-lane streets and multi-floor buildings with elevators. A realistic traffic flow model creates traffic jams and other non-stochastic interaction between moving objects.

3 Benchmarking Location-Based Services

The DynaMark benchmark focuses on LBS applications that require continuous tracking of mobile users (Figure 1). Many of the query types required by continuous tracking, e.g. calculating the distance between two objects or finding objects that are within a certain distance from some other object, are very similar to spatial predicates and functions found in traditional GIS systems. The main differences between GIS systems and continuous tracking is that in GIS systems spatial data is assumed to be fairly static, spatial objects may have complex shapes (e.g. polygons with holes), and applications are dominated by expensive spatial queries. In contrast, continuous tracking involves managing simple point objects that are highly dynamic and supporting queries that are extremely fre-

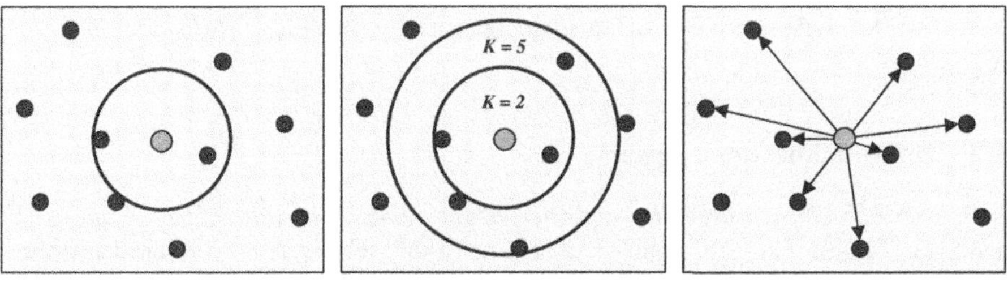

A) Proximity query. B) K-nearest neighbor. C) Sorted-distance query.

Fig. 2. Sample proximity, k-nearest neighbor, and sorted-distance queries.

quent, fairly simple, and have a dynamic reference point (current position of a user).

In the following we describe three types of spatial queries that form the basis of many LBS applications (see also illustrations in Figure 2). The queries are typically centered around the location of a user issuing the request.

3.1 Proximity Query

Find all objects that are within a certain range. This query is also known as the range query or window query and returns mobile users or stationary objects that fall within a specified range condition. The range condition may vary along different axes hence the query forms a 2D ellipse (Figure 2A) or 3D ellipsoid depending upon the topological dimension of the space in question. Alternatively, the range can be expressed by forming a 2D rectangle or 3D ortho-rhombus.

Whereas spatial indexing methods may have little difficulty in computing the precise answer to a rectangular range query, a radial range query may require processing in two steps: first filter out spatial objects that, based on their minimum bounding rectangle (MBR), cannot possibly satisfy the constraint, and then compare the precise extent of the objects with the constraint. Note that the filter step is identical to a rectangular range query and indeed a radial range query may be executed by first performing a rectangular range query with the MBR of the query ellipse or ellipsoid and then computing the distance from the matching objects to the reference point.

3.2 K-Nearest Neighbor Query (kNN)

Find the K nearest objects (Figure 2B). The query may search other mobile users or it may search stationary objects such as coffee shops, gas stations, hospitals, etc. Unless the spatial indexing method has built-in support for this query type (e.g. bottom-up evaluation in a tree index), it is very expensive to execute because it may require issuing several range queries to find the required number of neighbors. An alternative albeit expensive execution strategy is to compute the distance to all other objects, sort by distance, and select the closest K.

3.3 Sorted Distance Query

List each object in increasing distance order relative to the reference point (Figure 2C). This is like a K-nearest neighbor query where K is the total number of objects, but the difference is that during query evaluation the number of objects inspected by the application is not known. Also, the result must include distance information. This query could be used by an application to process arbitrary distance-based triggers, or by a client program to display distance information to points of interest.

3.4 More Advanced Queries

Next-generation LBS applications may require significantly more complex spatial queries than those described above. Consider an application that triggers a real-time alert to a parent when their child leaves from school. The trigger may further be parameterized by other profile information, e.g. "child is not being accompanied by another parent" or "child left school before school is over." More processing power and more sophisticated spatial indexing methods and query evaluation strategies are needed in such cases.

Ultimately, spatial queries could be supplanted by spatio-temporal queries that look at the historical location data or future predictions instead of current location data only. Consider queries such as *path queries* (find all objects that are staying or not staying on a predefined route), *directional queries* (find all objects approaching or leaving another object), *and traffic queries* (find locations where traffic jams have formed).

Support for a much larger query domain will enable a variety of spatial data services. For instance, one application of spatial data mining is to characterize the paths of mobile users and determine important hubs where their paths cross or where kiosks and retail stores should be set up to catch the attention of mobile users. These queries also require spatio-temporal indexing support. As new classes of queries are developed for various LBS applications, it will be necessary to compare the relative performance of spatial data managements systems and quantitative measurements to aid in performance optimization for each application domain.

3.5 Historical Location Traces Vs. Dynamic Spatial Data

The DynaMark benchmark, as defined today, does not incorporate predictive queries, spatial data mining, or spatio-temporal queries on historical (time series) data. We note that historical data may require indexing methods that are very different from those built for dynamic, continuously changing data. Indeed, we envision future LBS systems as consisting of separate stores for current data and historical data. This distinction is pervasive in data management, with the former being analogous to Online Transaction Processing (OLTP) and the latter to data warehousing and Online Analytical Processing (OLAP). Also note that historical data accumulates over time but is not continuously updated, which is a focus of the proposed benchmark.

Predictive queries and data mining are primarily concerned with accuracy and predictive power rather than pure update/query performance. We have therefore not included them in the performance benchmark.

Several research groups are investigating spatio-temporal queries that operate on user trajectories [14, 10]. An example of this category is a query that predicts a user's current location based on the user's last reported location, velocity, and direction. This type of trajectory query appears to be a natural extension to the benchmark and we intend to support it in a future version of the benchmark once more systems have the capability to run those types of queries.

The reason trajectories are needed is two-fold: they result in fewer updates to the index and they help produce a more accurate prediction of where users really are. We argue that extending the benchmark with trajectories would not convey much additional information about the update performance of the underlying spatial index, because updating an index with a user's trajectory requires essentially the same amount of processing as updating position data alone. A system that can sustain 500 position updates per second is likely to support about as many trajectory changes per second. Hence the spatial benchmark reflects the update performance of a spatio-temporal index reasonably well.

As far as accurately predicting where a user is, we note that this introduces an entirely new dimension to comparing spatial data management systems. Location data for mobile users is inherently out of date. Indeed the precision of any spatial query result depends as much on user behavior and path dimensionality as it does on the particular algorithms used to process a spatial query. Accurate determination of the query precision requires new metrics. Although the DynaMark benchmark could easily be extended to include trajectory queries, it is not clear a priori how to assign a general cost function that effectively combines performance and precision metrics. For example, how many milliseconds in query cost does a 10% improvement in accuracy correspond to? One system may perform very fast, while another is slow but more precise. The system chosen may depend on the LBS application used, the type and speed of mobile objects tracked (cars on roads vs. pedestrians walking randomly), reporting accuracy, and other factors. For this reason we suggest that "predictive" queries that apply algorithms to historical location traces be evaluated with a separate (perhaps application-specific) benchmark.

Even without path information or trajectory algorithms, any system designed to manage dynamic spatial data must also handle instantaneous velocity. The instantaneous velocity can be viewed as establishing a minimalist category of trajectory query, or viewed as a measurement that defines when spatial data should expire. Both views are supported by the DynaMark benchmark. Queries with a very short time window, e.g. a minute or two, are reasonably well imitated by purely spatial queries whose range is enlarged based on the known data collection time. In other words, a spatial buffer is created around the original query window that corresponds to the maximum distance covered by any of the objects in the time period. Knowledge of the maximal velocity establishes this maximal uncertainty. If we assume that mobile objects report their locations periodically and that the reports have a limited validity period (basically until the next report is scheduled), then the objects cannot have moved more than a limited distance from their last reported location. Note that the resulting accuracy naturally depends on the assumed speed of the moving objects and the type of LBS applications. For example, a query issued by a user in a car traveling at 60 miles per hour (one mile per minute) is likely to have wide range, say, a 10 mile radius, in which case extending the range by another mile or two does not significantly hamper query performance.

4 Benchmark Specification

4.1 Goal

The goal of the DynaMark benchmark is to provide quantitative measures of the scalability of middleware and database systems used to support LBS applications. As location data is typically stored in a spatial index, the benchmark ultimately measures the scalability of the underlying index. While some indexing methods appear naturally attuned to update-intensive workloads (e.g. grids), others appear to provide better selectivity for queries (e.g. R-trees). The implementation and maintenance of the spatial index also has a significant impact on the overall performance. A tree index, for example, may become too "loose" for applications requiring continuous updates and can result in degraded selectivity. Periodic rebuilding or tuning of the tree may be too expensive for some set of problems.

Given the variation in requirements for different location based applications, we define the following set of simple performance metrics to help discriminate between systems under various workloads. A *location update cost* (elapsed time per update) indicates the rate at which location updates can be applied to the index, in absence of queries. An *index size* metric measures the main memory and/or disk storage requirement of the index and is useful in determining how the index evolves over time under a continuous update load. A *query cost* (elapsed time per query) indicates the rate at which one of the aforementioned spatial queries can be performed. Together, the location update cost and query cost determine the maximum number of mobile users that can be tracked in real time.

For spatial data management systems that have a separate base table and spatial index, we also define metrics for measuring the cost of index creation and maintenance. An *index creation cost* metric measures the elapsed time in creating a new spatial index on an existing base table. An *index rebuild cost* measures the elapsed time in tuning or optimizing the index for future updates and queries.

4.2 Database Schema for Location Tracking

In any practical schema for location tracking, a unique global ID is needed to identify each moving object. For simplicity, we assume that the ID is an integer number. In practice, the ID could be a Mobile Identification Number such as a mobile phone number or the IP address of a tracking device.

We assume that each moving object reports its location with some average periodicity. The report includes the timestamp when the location was determined or report was generated. In real location traces, for example GPS logs, the timestamp may be an absolute time in the GMT timezone. In simulated location traces, the timestamp is a double-precision floating-point number that measures the elapsed, simulated time in seconds from the start of simulation.

The location of a moving object is modeled as a triplet X, Y, and Z where X is the longitude in degrees, Y is latitude in degrees, and Z is elevation in meters.

4.3 Persistence, Recovery, and Transactions

Assuming that moving objects report their locations with a relatively high frequency, it is reasonable to consider spatial indexing methods that operate entirely in main memory and do not store the index or base tables in persistent storage. If the index and base tables are lost due to a server crash or disk failure, the data will be recovered relatively quickly from freshly received location updates.

Benchmark results must clearly indicate whether the system under test stores location data persistently or not. Benchmarks conducted on systems that guarantee the persistence of location data should be identified as "persistent systems," whereas main memory-based systems should be identified as "non-persistent systems."

Many spatial data management systems are built on database systems that provide robustness and recovery through transactions and logging. However, it can be argued that for some LBS applications transactional properties are not required. For example, a spatial query that returns uncommitted data (dirty reads) may be acceptable because the objects being tracked move relatively slowly and location updates are rarely if ever rolled back. Database systems that provide "non-logging" databases, tables, or indices, and reduced transaction isolation levels, are allowed by the benchmark as long as these features are clearly indicated in the benchmark results.

4.4 Benchmark Data Sets

The benchmark requires running several location trace files through the spatial data management system under study. The number of mobile users contained in a location trace file determines the size of an individual benchmark run. We propose the following population sizes in the standard benchmark: 10,000; 50,000; 100,000; 500,000; and one million mobile users. The benchmarks are accordingly known as the P10K, P50K, P100K, P200K, and P1M benchmarks. As technology evolves and it becomes feasible to handle much larger populations on future systems, the population sizes listed here can be naturally extended with larger sizes.

Each record in the location trace is a location update that adheres to the schema described earlier: object ID, timestamp, and X, Y, and Z coordinates of the location. The X and Y coordinates represent the longitude and latitude values of the location, and Z indicates elevation in meters.

It is assumed that each user's location is reported with some average periodicity, say T_P seconds. In a location trace for N mobile users, the first N records (first iteration) contain the initial locations of the N users, and the next N records (second iteration) contain the next location of those N users. Therefore, the first iteration results in a batch of INSERTs to the spatial index, whereas the remaining iterations are UPDATEs. The records need not be in an identical order in every iteration, allowing for some randomness in the order of wireless network packets. Rather, the records are in temporal order, sorted by the timestamp field of the location trace record. This is in accordance with the fact that delays in wireless networks cause location updates to be received in an indeterministic order.

We propose using a standard set of location trace files generated with City Simulator, a toolkit we have made available on the IBM developer Web site [6]. City Simulator generates realistic three-dimensional spatio-temporal data for large populations of mobile users moving about in a city. The simulator is written in Java for maximum portability and provides a powerful graphical user interface that allows the user to control several aspects of the simulation at runtime, including traffic flow, traffic congestion, and blocked streets and routes. A visualization display shows the progress of the simulation in real time. New city models can be imported as XML documents which describe the layout of the city, including roads, buildings, and vacant space.

4.5 UPDATE Benchmark

An UPDATE benchmark involves running one of the standard location trace files through the spatial index and maintaining the index such that it reports each user's last known location accurately. The location update cost metric represents the average update cost over a number of consecutive updates and is measured periodically (Figure 3A) to help detect any degradation in system performance over time. The number of updates N_U included in each sample should be large enough to smooth out irregularities but still provide frequent enough sampling. We propose calculating the average update cost for every block of 500 consecutive updates.

Most indexing methods have a fixed INSERT cost and UPDATE cost, but others show a gradually increasing UPDATE cost because the spatial index becomes less efficient. This is usually accompanied by a corresponding increase in index size. Measuring the update cost and index size over time clearly points out deficiencies in the underlying spatial indexing method or its implementation.

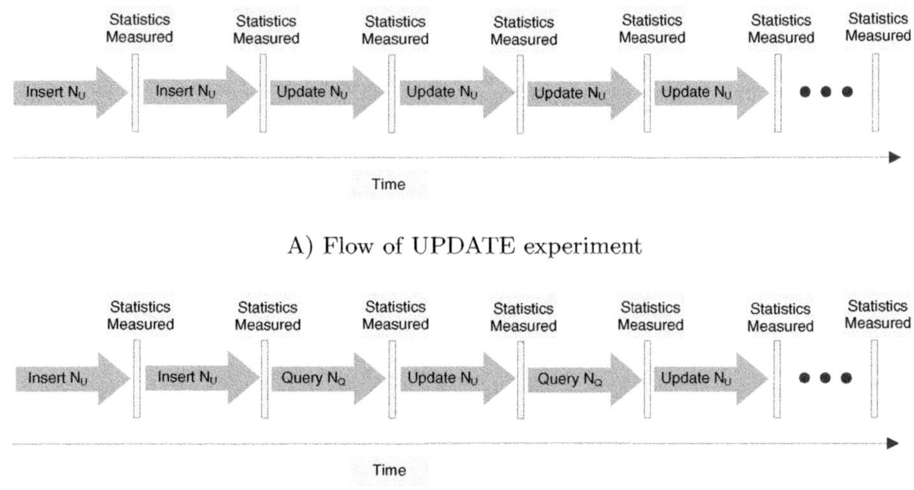

A) Flow of UPDATE experiment

B) Flow of QUERY experiment

Fig. 3. Execution plan for updates, queries, and statistics collection.

An UPDATE benchmark is identified by a combination of the data set size and the "U" suffix. For example, an UPDATE benchmark that uses the P100K data set is referred to as the "P100K-U" benchmark.

4.6 QUERY Benchmark

A QUERY benchmark runs a location trace through the spatial index as in the UPDATE benchmark, but in addition to maintaining the index, queries against the index are issued. Updates and queries are interleaved such that a batch of queries is executed after every so many updates (Figure 3B). The number of updates N_U should be fairly small to allow detailed performance metrics to be collected, and the number of queries N_Q should be large enough to smooth out irregularities. We propose executing 100 queries after every 500 updates.

We define the following standard set of queries.

Q1. Proximity query with a rectangular range corresponding to 0.1%, 0.5%, 1%, 5% and 10% of the area of the entire coordinate space. The size of the coordinate space is defined by the range of each of its dimensions. The range of a dimension is defined as the corresponding minimum and maximum coordinate values found in the location trace files. A query with areal coverage AC is referred to as the "Q1-AC" query. For example, Q1 with 0.1% areal coverage combined with the P100K location trace file is referred to as the "P100K-Q1-0.1" benchmark.

Q2. Proximity query with radial range corresponding to 0.1%, 0.5%, 1%, 5% and 10% of the area of the entire coordinate space. A query with areal coverage AC is referred to as the "Q2-AC" query. For example, Q2 with 0.1%

Table 1. Sample benchmarks and descriptions.

Benchmark	Description
P100K-U	UPDATE benchmark with location trace file containing 100,000 mobile users
P50K-Q1-0.1	QUERY benchmark with Q1 (proximity query with rectangular range), areal coverage 0.1%, and location trace file containing 50,000 mobile users
P200K-Q2-0.5	QUERY benchmark with Q2 (proximity query with radial range), areal coverage 0.5%, and location trace file containing 200,000 mobile users
P1M-Q3-100	QUERY benchmark with Q3 (K-nearest neighbor), K value 100, and location trace file containing 1 million mobile users
P10K-Q4	QUERY benchmark with Q4 (sorted-distance) and location trace file containing 10,000 mobile users
P200K-IC	INDEX benchmark that creates a new index for base table containing 200,000 mobile users
P1M-IR	INDEX benchmark that periodically rebuilds an existing index for location trace file containing 1 million mobile users

areal coverage combined with the P100K location trace file is referred to as the "P100K-Q2-0.1" benchmark.

Q3. K-nearest neighbor query with K value 1, 5, 10, 100, and 1000. A query with K nearest neighbors is referred to as the "Q3-K" query. For example, Q3 with K value 10 combined with the P100K location trace file is referred to as the "P100K-Q3-10" benchmark.

Q4. Sorted-distance query with a random number of objects inspected by the application (and hence returned by the index). The number is a random variable with uniform distribution between values 1 and 100. Note: the number of objects inspected is not part of the query since otherwise the query could be executed efficiently as a K-nearest neighbor query. It is important that during query execution the exact number of objects inspected is not known. The spatial index may execute the query incrementally, however. Query Q4 combined with the P100K location trace file is referred to as the "P100K-Q4" benchmark.

4.7 INDEX Benchmark

For spatial data management systems that have a separate base table and spatial index, we define an INDEX benchmark that measures the time it takes to create a new spatial index on an existing base table and the time it takes to rebuild an existing index. Rebuilding is defined loosely as the process of tuning the index so that future updates and queries perform better. This may mean rebuilding the index completely from scratch, or traversing the index, reorganizing data, tightening bounding boxes, and other actions deemed necessary for maintaining the index. A system may also choose not to do anything for index rebuild, in which case the index rebuild cost is zero.

The INDEX CREATE benchmark is conducted by first loading the first iteration of moving object data from the location trace. For example, for the P100K data set this means loading the first 100,000 records (100,000 moving objects) into the base table. The index is then created and the elapsed time measured. This benchmark is identified by a combination of the data set size and the "IC" suffix. In the case of the P100K data set, the benchmark would be referred to as the "P100K-IC" benchmark.

The INDEX REBUILD benchmark is conducted in conjunction with the UPDATE benchmark. As the base table is being updated according to the location trace file, the spatial index is rebuilt as deemed necessary by the database system. The elapsed time per index rebuild is measured and the cost displayed as a function of time during the benchmark run. The benchmark is identified by a combination of the data set size and the "IR" suffix. In the case of the P100K data set, the benchmark would be referred to as the "P100K-IR" benchmark.

4.8 Composite Performance Metrics

Given knowledge of the expected update vs. query frequencies in a particular LBS application, it is a simple matter to take a weighted average of the quantitative metrics provided by the benchmark to make a controlled comparison between LBS middleware candidates in support of that application. Fitting measured update and query cost data to an appropriate curve allows interpolation and extrapolation of that data. This in turn allows construction of an equation calculating the maximum population size supported by an LBS middleware candidate for a given application [9, 8].

It is also straightforward to calculate TPC-like composite performance metrics such as "queries per hour" (Qph) and "dollars per Qph" in system cost based on the benchmark metrics.

5 Conclusion

We have described a performance benchmark for Location-Based Services (LBS) and a method for conducting such experiments in a controlled manner. We presented an initial list of spatial queries whose performance should be evaluated in the benchmark, and discussed future extensions that exercise spatio-temporal indices. The dynamic spatial queries were: proximity query, K-nearest neighbor query, and sorted-distance query.

We defined a consistent set of quantifiable performance metrics to evaluate a spatial indexing method. A location update cost (elapsed time per update) indicates the rate at which location updates can be applied to the index, in the absence of queries. An index size measure tracks the memory requirement of the index and is useful in determining how the index evolves over time under a continuous update load. A query cost (elapsed time per query) indicates the rate at which one of the aforementioned spatial queries can be performed. We also defined index creation and maintenance metrics that measure the cost of creating a new spatial index on an existing base table and the cost of maintaining (rebuilding) that index. These metrics can be combined into composite metrics such as "queries per hour" and "maximum population size supported" in a relatively straight-forward manner.

References

[1] Pankaj K. Agarwal, Lars Arge, and Jeff Erickson. Indexing moving points. In *Proceedings of the ACM Symposium on Principles of Database Systems (PODS)*, pages 175–186, 2000.

[2] Dina Bitton, David DeWitt, and Carolyn Turbyfill. Benchmarking database systems: A systematic approach. In *Proceedings of the International Conference on Very Large Databases (VLDB)*, Firenze, Italy, October 1983.

[3] Michael J. Carey, David J. DeWitt, and Jeffrey F. Naughton. The OO7 benchmark. In *Proceedings of ACM International Conference on Management of Data (SIGMOD)*, pages 12–21, Washington, DC, June 1993.

[4] Michael J. Carey, David J. DeWitt, Jeffrey F. Naughton, Mohammad Asgarian, Paul Brown, Johannes Gehrke, and Dhaval Shah. The BUCKY object-relational benchmark. In *Proceedings of ACM International Conference on Management of Data (SIGMOD)*, pages 135–146, Tucson, AZ, May 1997.

[5] Volker Gaede and Oliver Günther. Multidimensional access methods. *ACM Computing Surveys*, 30(2):170–231, June 1998.

[6] James Kaufman, Jussi Myllymaki, and Jared Jackson. City Simulator spatial data generator, November 2001. http://alphaworks.ibm.com/tech/citysimulator.

[7] Hans-Peter Kriegel, Michael Schiwietz, Ralf Schneider, and Bernhard Seeger. Performance comparison of point and spatial access methods. In *Proceedings of the International Symposium on the Design and Implementation of Large Spatial Databases (SSD)*, pages 89–114, Santa Barbara, CA, 1989.

[8] Jussi Myllymaki and James Kaufman. High-performance spatial indexing for location-based services, November 2002. Submitted for publication.

[9] Jussi Myllymaki and James Kaufman. LOCUS: A testbed for dynamic spatial indexing. *IEEE Data Engineering Bulletin* (Special Issue on Indexing of Moving Objects), 25(2), June 2002.

[10] Dieter Pfoser and Christian S. Jensen. Querying the trajectories of on-line mobile objects. In *Proceedings of the ACM International Workshop on Data Engineering for Wireless and Mobile Access*, Santa Barbara, CA, May 2001.

[11] Jean-Marc Saglio and Jose Moreira. Oporto: A realistic scenario generator for moving objects. In *Proceedings of the International Workshop on Database and Expert Systems Applications (DEXA)*, pages 426–432, 1999.

[12] Simonas Saltenis and Christian S. Jensen. Indexing of moving objects for location-based services. In *Proceedings of the IEEE International Conference on Data Engineering (ICDE)*, San Jose, CA, February 2002.

[13] Shashi Shekhar, Sanjay Chawla, Siva Ravada, Andrew Fetterer, Xuan Liu, and Chang tien Lu. Spatial databases: Accomplishments and research needs. *IEEE Transactions on Knowledge and Data Engineering*, 11(1):45–55, January 1999.

[14] A. Prasad Sistla, Ouri Wolfson, Sam Chamberlain, and Son Dao. Modeling and querying moving objects. In *Proceedings of the IEEE International Conference on Data Engineering (ICDE)*, pages 422–432, Birmingham, UK, April 1997.

[15] Michael Stonebraker, James Frew, Kenn Gardels, and Jeff Meredith. The SEQUOIA 2000 storage benchmark. In *Proceedings of ACM International Conference on Management of Data (SIGMOD)*, pages 2–11, Washington, DC, June 1993.

[16] Yannis Theodoridis, Timos K. Sellis, Apostolos Papadopoulos, and Yannis Manolopoulos. Specifications for efficient indexing in spatiotemporal databases. In *Proceedings of the International Conference on Statistical and Scientific Database Management (SSDBM)*, pages 123–132, 1998.

[17] Yannis Theodoridis, Jefferson R. O. Silva, and Mario A. Nascimento. On the generation of spatiotemporal datasets. In *Proceedings of the International Symposium on Spatial Databases (SSD)*, pages 147–164, Hong Kong, China, July 1999.

[18] Ouri Wolfson, Bo Xu, Sam Chamberlain, and Liqin Jiang. Moving objects databases: Issues and solutions. In *Proceedings of the International Conference on Statistical and Scientific Database Management (SSDBM)*, pages 111–122, Capri, Italy, July 1998.

{ymchan, kwlam}@cs.cityu.edu.hk; csvlee@cityu.edu.hk

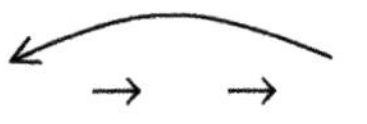

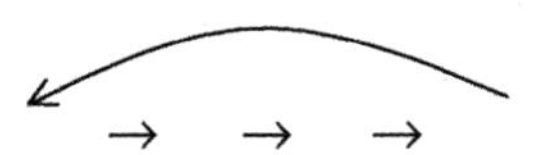

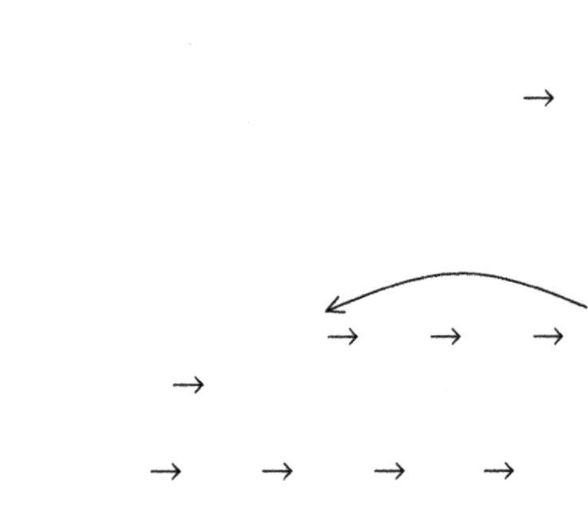

ϕ

$\rightarrow$

```
NRS(Qa) := φ ;
 do while Qa is active
 {
  Whenever an update transaction Uc commits, do
  {
   for each x ∈ WS(Uc)∩CRS(Qa)    do
    NRS(Qa) := NRS(Qa)∪{x};

   if RS(Uc)∩NRS(Qa)  ≠ φ  then
    for each y ∈[WS(Uc)\CRS(Qa)]    do
     NRS(Qa) := NRS(Qa)∪{y};
  }
 }
```

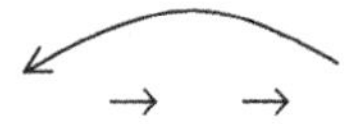

$\rightarrow$ $\rightarrow$

$\rightarrow$

	ϕ

ϕ

$\in$ $\cap$

$\notin$

$\notin$

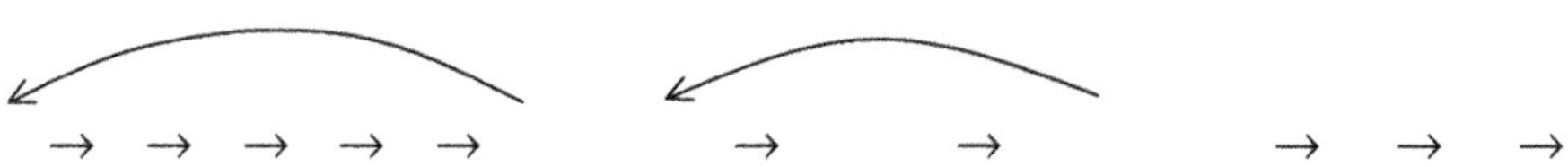

i.e. if $WS(U_c) \cap CRS(Q_a) \neq \phi$ or $RS(U_c) \cap NRS(Q_a) \neq \phi$ then
$CRS(Q_a) := CRS(Q_a) \cup RS(U_c)$;

i.e. Whenever a ROT Q of the same group as Q_a commits, do

```
if RS(Q)∩NRS(Qa) ≠ φ then {
  if NRS(Q) ∩ RS(Qa) ≠ φ then {
    abort Qa;
  }
  NRS(Qa):= NRS(Qa)∪NRS(Q);
  CRS(Qa):= CRS(Qa)∪RS(Q);
}
```

```
RS(Qa):= φ ;
do while Qa is active {
  Whenever an update transaction Uc commits, do
  {
    if WS(Uc)∩CRS(Qa)≠φ or RS(Uc)∩NRS(Qa)≠φ then
      CRS(Qa):= CRS(Qa)∪RS(Uc);
    for each x ∈ WS(Uc)∩CRS(Qa)    do
      NRS(Qa):= NRS(Qa)∪{x};
    if RS(Uc)∩NRS(Qa) ≠ φ then
      for each y ∈ [WS(Uc)\CRS(Qa)]    do
        NRS(Qa):= NRS(Qa)∪{y};
  }
```

```
Whenever a ROT Q of same group as Qa commits, do
  if RS(Q)∩NRS(Qa) ≠ φ  then {
    if NRS(Q) ∩ RS(Qa) ≠ φ  then {
         abort Qa;
    }
    NRS(Qa):= NRS(Qa)∪NRS(Q);
    CRS(Qa):= CRS(Qa)∪RS(Q);
  }
}
```

	ϕ	ϕ
	ϕ	ϕ
		ϕ
		ϕ

ϕ

ϕ

$\cap$ $\neq \phi$

$\in$ $\in$ $\cap$

$\in$ $\cap$

$\in$ $\cap$

$\notin$

$\in$ $\cap$

$\in$

$\in$

$\times$

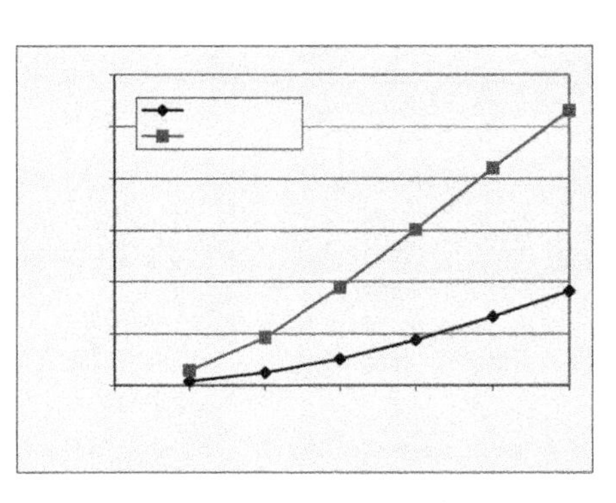

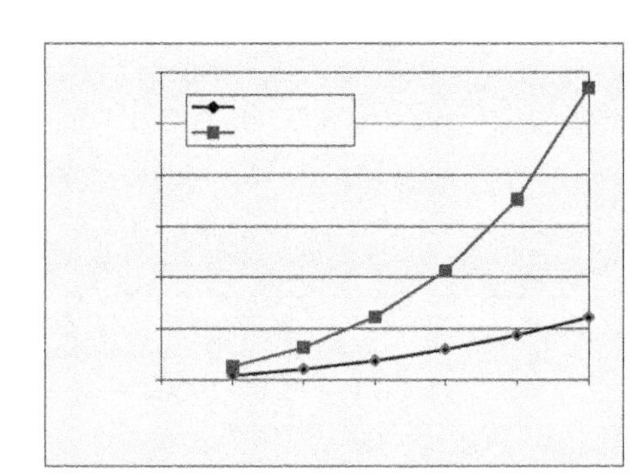

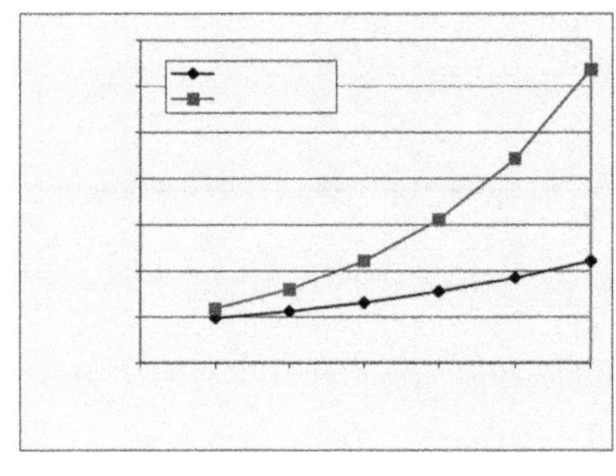

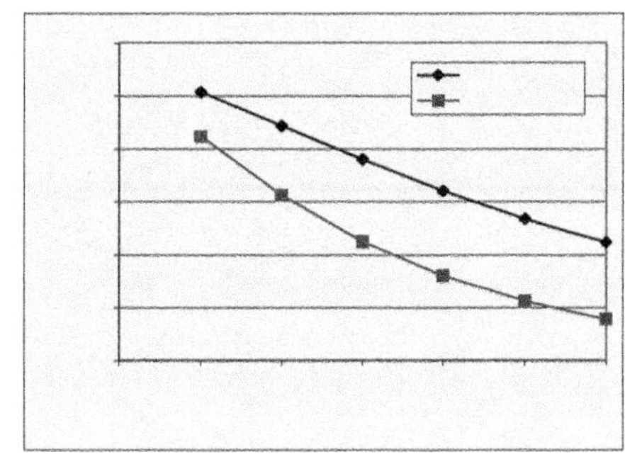

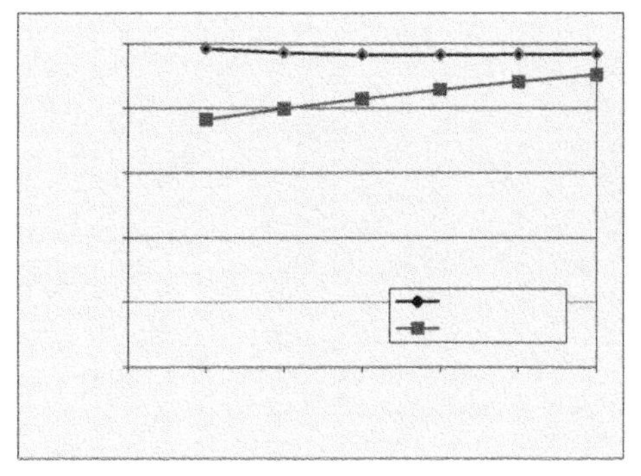
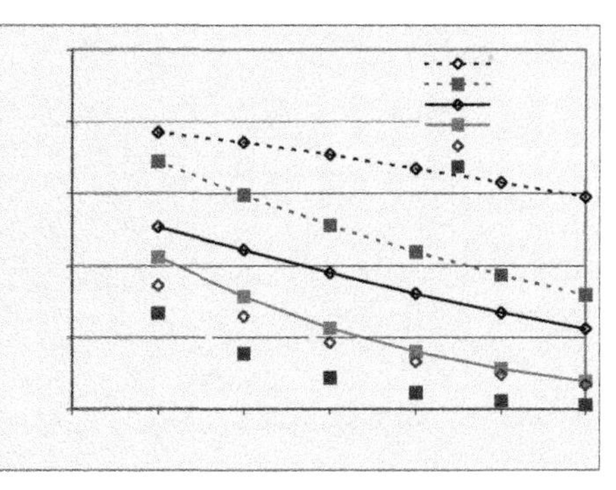

Publish/Subscribe Tree Construction in Wireless Ad-Hoc Networks

Yongqiang Huang and Hector Garcia-Molina

Stanford University, Stanford, CA 94305
{yhuang, hector}@cs.stanford.edu

Abstract. Wireless ad-hoc publish/subscribe systems combine a publish/subscribe mechanism with wireless ad-hoc networking. The combination, although very attractive, has not been studied extensively in the literature. This paper addresses an important problem of such systems: how to construct an optimal publish/subscribe tree for routing information from the source to all interested recipients. First we precisely define the optimality of a publish/subscribe tree by developing a metric to evaluate its "efficiency." The optimality metric takes into account both the goal of a publish/subscribe system (i.e., to route a set of events), and the characteristics of an ad-hoc network (for example, devices are resource limited). We propose a greedy algorithm, SHOPPARENT, which builds the publish/subscribe tree in a fully distributed fashion. A key feature is that this algorithm can be "subscription-aware", allowing it to use publication/subscription information in order to find a better outcome. Our simulations show that SHOPPARENT's performance is within 15% of optimal under normal configurations. We also study the effect of geographically localized subscriptions.

1 Introduction

A *publish/subscribe* system connects information providers with consumers by delivering *events* from sources to interested users. A user expresses interest in receiving certain types of events by submitting a predicate defined on the event contents. The predicate is called the user's *subscription*. When a new event is generated and *published* to the system, the publish/subscribe infrastructure is responsible for checking the event against all current subscriptions and delivering it efficiently and reliably to all users whose subscriptions match the event.

Many problems related to publish/subscribe have been tackled and solved. However, almost all of the research so far has concentrated on publish/subscribe systems in a fixed network. With increasing popularity of wireless handheld devices, there is a pressing need to extend publish/subscribe to a wireless environment [1, 2, 3, 4]. In this paper, we study publish/subscribe systems in a wireless ad-hoc network. Such a network is formed by wireless devices communicating without the benefit of a fixed network infrastructure, either because the infrastructure has been wiped out by a natural disaster, or because it is impractical to build one. As an example, in a military battlefield, thousands of wireless and mobile sensors such as satellites and equipment sensors report all kinds of information ranging from the location of enemy troops to whether the engine of a tank has overheated. There are also many parties interested in receiving certain

M.-S. Chen et al. (Eds.): MDM 2003, LNCS 2574, pp. 122–140, 2003.

types of information. An individual soldier may need to know the location of the nearest enemy troops, or whenever a missile has been fired. The above scenario requires the deployment of a highly scalable and efficient communication infrastructure, for which publish/subscribe is an ideal candidate.

In a wireless ad-hoc publish/subscribe system, nodes cooperate to deliver events from their publishers to interested subscribers. In this paper, we focus on constructing a *publish/subscribe tree* for such a system. A publish/subscribe tree is a node hierarchy along which new publications are sent in order to reach their subscribers. A tree construction algorithm produces such a tree, given information such as how the nodes are interconnected, and what each node has subscribed to.

A publish/subscribe tree is similar in function to a multicast tree in traditional multicast networks. Existing multicast protocols often aim to produce an "optimal" multicast tree which seeks to optimize one of a few possible metrics. For example, in order to minimize the total number of nodes participating in multicast, some protocols generate (or heuristically approximate) Steiner trees. In this paper we propose a new optimality criterion suitable for wireless publish/subscribe systems. Our metric measures how "efficiently" a publish/subscribe tree can transmit all publications to the interested nodes. Efficiency is important because wireless devices in ad-hoc networks typically have very limited resources, such as battery power. Hence it is desirable to accomplish the same goal with minimum work. Furthermore, our metric is tailored to a publish/subscribe system because it takes into account information such as user subscriptions and events, which are not applicable in regular multicast.

We present tree construction algorithms to produce efficient publish/subscribe trees based on the above metric. Our algorithms exploit knowledge of expected traffic on the publish/subscribe tree. For example, if a new node can connect to the tree in say two ways, it will select the connection that generates less new traffic for its parent, given the current subscriptions. As we will see, this "intelligent" tree construction can yield significant gains.

In addition to the algorithms, our other contribution is their evaluation. In particular, we devise models to simulate subscription and publication so that we can compare the performance of various algorithms.

We define the problem and our optimality metric in Section 2. We then give our algorithms for tree construction (Section 3). Finally, we discuss our evaluation model (Section 4) and present our results (Section 5).

2 Framework

Figure 1 illustrates a wireless ad-hoc publish/subscribe system. It consists of N wireless nodes, each identified by a globally unique id. The nodes communicate with each other wirelessly using radio, and cooperate to send, relay, and receive *events* (i.e., publications).

We define the *connectivity graph* G of a system as the graph whose vertices are the wireless nodes, and where an edge exists between two vertices if the two corresponding nodes are "neighbors." Two nodes are neighbors if they can talk to each other directly via radio, i.e., if they are a single hop away. See Figure 1 for an example G. Note that,

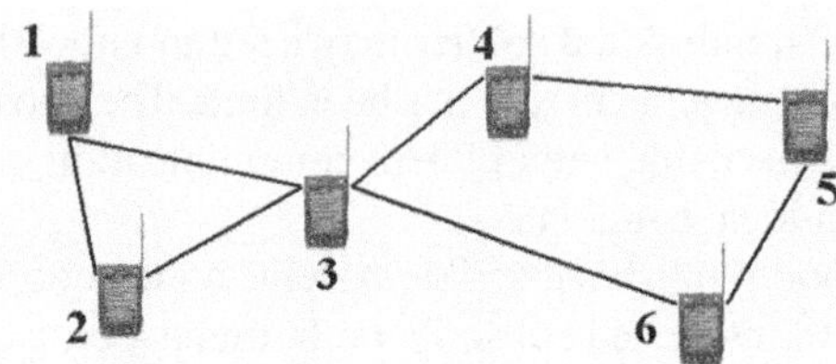

Fig. 1: A wireless ad-hoc publish/subscribe system. The lines form the connectivity graph G.

as in [5], the connectivity graph is assumed to be undirected, precluding the situation where one node can hear another, but not vice versa.

Each node i has a subscription s_i, called its *inherent subscription*, which is expressed as a predicate defined on the contents of an event. A node i is interested in an event e if and only if $s_i(e) = true$. Naturally, the inherent subscription is determined by the particular applications that are running on that node. Note that we do not make any assumptions about an event's content.

Of all the nodes in the system, one is designated as *root* of the publish/subscribe mechanism. Without loss of generality, we assume that the root is always labeled node 1. We assume that only the root node can publish new events. That is, a new event has to be generated at node 1, and then forwarded to other interested nodes. Although this seems like a limiting assumption, our algorithms can easily be modified when multiple nodes are allowed to publish. For example, we can construct one tree per potential publisher (per-source trees in multicast lingo). Alternatively, we can require any new event to be passed to the root first via unicast. However, a better method is to forward the event towards the root along the publish/subscribe tree itself, allowing shortcutting to tree branches along the way. The user is referred to [6] for a relevant discussion in the context of shared multicast trees.

Although neighbors of the root node can receive newly published events directly from the root, other nodes will have to rely on the help of intermediate nodes to relay events. For example, in Figure 1, node 3 needs to forward those events of interest to node 4. We use the term *publish/subscribe tree* (PST) to describe the tree rooted at node 1 that is formed by paths traveled by the events to reach their destinations.[1] We observe that the PST is a spanning tree of the connectivity graph G.[2]

We assume that every node in the system participates in the publish/subscribe protocol.[3] In particular, nodes that are not interested in any event will simply have an empty

[1] Similar to a multicast tree, the PST is a tree and not a graph because we force the events to always travel the same route between the root node and any subscriber node.

[2] The PST is a spanning tree and not a Steiner tree because, as explained next, all nodes participate in the publish/subscribe protocol.

[3] This is a fundamentally different and much less limiting assumption than assuming that all nodes are part of a particular multicast tree in traditional multicasting. While there may be many concurrent active multicast groups at any time and only a small portion of nodes are interested in any particular one, we envision a single publish/subscribe session where different node interests are represented by different subscriptions.

subscription. We believe that the above is a reasonable assumption if publish/subscribe is used as a fundamental underlying communication mechanism for such a system. The assumption also implies that our algorithms will not rely on a separate ad-hoc unicast protocol to work. However, if the assumption does not hold, additional steps may be necessary in the algorithms, which will be considered in extensions of this work. For example, new subscribers may need to perform expanding ring searches to look for the nearest node that is also part of publish/subscribe.

Suppose node i is the parent of another node j in a PST. Then any event that j subscribes to will need to pass through i first. We define node i's *effective subscription* (S_i) as the "combined" subscription of itself and all its children. Specifically, let s_i' denote i's *proxied subscription*, which is the disjunction of the (effective) subscriptions of all i's children. Then S_i is simply the disjunction of i's inherent subscription (s_i) and proxied subscription (s_i'). In other words, $S_i(e) = s_i(e) \vee s_i'(e)$, where $s_i'(e) = S_{i1}(e) \vee S_{i2}(e) \ldots$ for all of i's children $i1, i2, \ldots$

When a new event e is published at the root node, the wireless nodes run a distributed *publish/subscribe protocol* to forward e along the branches of the PST until it reaches all nodes that have subscribed to it. Under the protocol, every node i listens for new events broadcast by its parent node in the tree. Then, depending on an event's content, some or all of a set of actions can be performed on it, as shown in Table 1. In particular, receiving an event means capturing it in the air and transferring it to an internal buffer, logging it, etc. Processing implies that an event has matched i's inherent subscription and will be displayed to the user in a pop-up alert window, for example. Finally, forwarding an event involves calculating the set of children nodes that will be interested in the event and re-broadcasting it in the air for them. For example, the first row of Table 1 shows that events satisfying $s_i(e) \wedge \neg s_i'(e)$ are of interest to node i only, and not to its children. Consequently, these events will be received and processed, but not forwarded.

Satisfies	Of interest to	Recv	Proc	Fwd
$s_i \wedge \neg s_i'$	i only	√	√	X
$\neg s_i \wedge s_i'$	i's children only	√	X	√
$s_i \wedge s_i'$	both i & children	√	√	√
$\neg S_i$	neither	X	X	X

Table 1: Classification of new events seen by node i and whether they will be "received", "processed", and/or "forwarded".

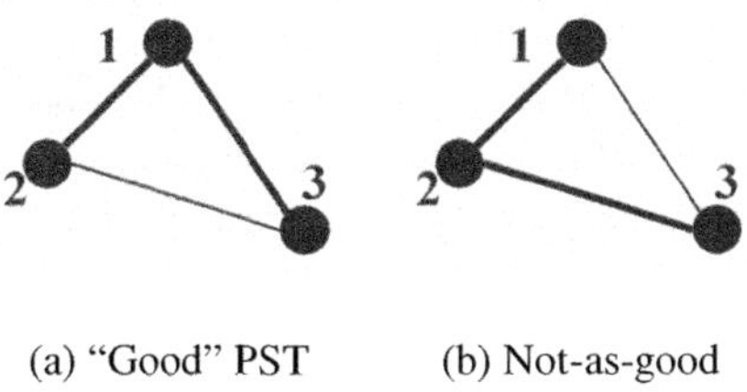

(a) "Good" PST (b) Not-as-good

Fig. 2: Two PSTs of the same connectivity graph. Lines represent the connectivity graph, while thick lines constitute a PST.

We do not focus on reliability in this paper. Instead, we develop best-effort algorithms where nodes may occasionally miss events, especially when they move around geographically. If desired, reliability can be added to our algorithms via logging and retries, etc.[7] However, we suspect that such a guaranteed delivery system may be costly in a dynamic wireless ad-hoc environment, and we leave a detailed study to future work.

2.1 PST Evaluation Metric

There usually can be many possible PSTs in any given system (actually, as many as there are spanning trees in the connectivity graph G). Some trees can be "better" than others, however, as the following example shows.

Example 1. Figure 2 shows a simple system with three nodes. Assume both nodes 2 and 3 can receive events directly from the root, node 1. If they both do, the resulting PST is given by the thick lines in Figure 2(a). If, on the other hand, node 3 chooses to rely on node 2 to forward events, we have the PST in Figure 2(b). Intuitively, the PST in Figure 2(a) is "better" than 2(b) since the extra work by node 2 to forward events is avoided.

In order to meaningfully compare different PSTs, we define a metric called the *cost* of a publish/subscribe tree. A PST's cost with respect to a set E of events, denoted $C(E)$, is the total amount of work performed by all tree nodes in order to publish E. Cost measures the "efficiency" of a PST, as a tree with a lower cost will need less overall work to deliver all the events in E. We believe that this is a useful and realistic criterion due to severely limited resources in a wireless ad-hoc environment. We have $C(E) = \sum_i C_i(E)$, where $C_i(E)$ is the amount of work node i has to perform to publish E, called the cost of i.

Let us assume that receiving an event constitutes r units of work, and processing and forwarding need p and f units, respectively. Let $E(\alpha)$ represent the subset of E satisfying the predicate α, and let $\Phi E(\alpha)$ denote the number of events in that subset. According to Table 1, we get $C_i(E) = (r+p) \cdot \Phi E(s_i \wedge \neg s_i') + (r+f) \cdot \Phi E(\neg s_i \wedge s_i') + (r+p+f) \cdot \Phi E(s_i \wedge s_i')$.

Note that, similar to some previous work on ad-hoc multicast/broadcast [5], we have assumed the cost (f) to forward an event to be constant, regardless of how many of i's children want this event. This is a reasonable assumption because of the broadcast nature of radio. With one broadcast operation, a node can normally send a message to all its neighbors. In our case, specifically, when node i broadcasts an event in the air, all nodes within i's radio range will be able to hear the transmission. However, only the intended recipients of the event will accept and receive it; others will simply ignore it. Furthermore, since we concentrate on best-effort algorithms as noted before, we do not assume the broadcast primitive to be reliable, which may be expensive to enforce.

The cost calculation not only depends on E, the set of events being published, but also on the connectivity graph and user subscriptions. We will propose models to simulation these factors in Section 4. In general, if tree T_1 has a lower cost than T_2, T_1 will incur less overall work given the user subscriptions and event distribution that are used to calculate the cost.

Although the cost metric can be used directly to compare two PSTs, we observe that a component in the cost formula is unchanged in all possible trees. Specifically, regardless of i's children, the work needed to receive and process the sets $E(s_i \wedge \neg s_i')$

and $E(s_i \wedge s'_i)$ is always needed. Therefore, to highlight the differences between two PSTs and to simplify calculation, we next define another derivative metric, the overhead of a PST.

Node i's *overhead* with respect to E, denoted $O_i(E)$, is the additional amount of work that i needs to perform on behalf of its children in a PST (i.e., extra work in addition to what i would always need to do even without any children). In other words, the overhead is the cost of a node minus the amount of work it needs to do for itself. The *overhead* of a publish/subscribe tree, $O(E)$, is simply the sum of overheads of all nodes, namely, $O(E) = \sum_i O_i(E)$.

Based on above analysis, we get $O_i(E) = (r + f) \cdot \Phi E(\neg s_i \wedge s'_i) + f \cdot \Phi E(s_i \wedge s'_i)$. Without loss of generality, we assume that f corresponds to one unit of work, i.e., $f = 1$. To simply calculation, we also assume that $r = 1$. That is, it takes the same amount of work to receive an event as to transmit one. This is reasonable if "work" is measured in terms of processing time, for example. On the other hand, since radio transmission often consumes much more energy than reception, we will also study the impact of letting $r \ll f$ in Section 5. Given $r = f = 1$, we have, $O_i(E) = 2\Phi E(\neg s_i \wedge s'_i) + \Phi E(s_i \wedge s'_i)$.

Example 2. Referring back to Figure 2, we assume $E = \{e_1, e_2\}$, where e_1 matches node 2's subscription, while e_2 matches node 3's. For the PST in Figure 2(a), we have $O_2(E) = 2\Phi E(\neg s_2 \wedge s'_2) + \Phi E(s_2 \wedge s'_2) = 2 \cdot 0 + 0 = 0$ since node 2 has no children. Likewise, $O_3(E) = 0$. Also, $O_1(E) = 2\Phi E(\neg s_1 \wedge s'_1) + \Phi E(s_1 \wedge s'_1) = 2 \cdot 2 + 0 = 4$. Hence, $O(E) = \sum_i O_i(E) = 4 + 0 + 0 = 4$.

In Figure 2(b), however, node 3 is node 2's child. Thus node 2's effective subscription includes that of 3. We get $O_2(E) = 2\Phi E(\neg s_2 \wedge s'_2) + \Phi E(s_2 \wedge s'_2) = 2 \cdot 1 + 0 = 2$. As before, we also have $O_1(E) = 4$ and $O_3(E) = 0$. Overall, $O(E) = \sum_i O_i(E) = 4 + 2 + 0 = 6$. Hence, the left PST is indeed "better" according to our metric.

3 Algorithms

A *tree construction algorithm* is the protocol run by the nodes of a wireless publish/subscribe system in order to determine which PST to use. The algorithm aims to produce an *optimal PST*, namely, to minimize the overhead of the resulting PST based on the metric presented earlier. In this section, we give several such algorithms.

Our first algorithm, OPT, is a centralized algorithm that is guaranteed to be optimal. Specifically, Algorithm OPT runs on a single node with global knowledge about the system. That is, a controller node gathers all inputs needed, including the entire connectivity graph G and user subscription information, before it calculates the optimal PST, and distributes the tree structure back to other nodes.

When finding the optimal PST, OPT exhaustively searches through all spanning trees of the connectivity graph G, and selects the one with the smallest overhead. Since a PST must be a spanning tree of G, OPT guarantees optimality of the resulting PST. Note that an optimal PST differs from a minimum spanning tree of G in the graph theoretical sense, because the "weight" of an edge is not fixed, but is dependent on what events flow through it, which in turn depends on the shape of other parts of the tree. Hence one cannot use the more efficient minimum spanning tree algorithms in this step. Due to space limitations, we will omit further details of this algorithm.

Because of its exhaustive and centralized nature, OPT is very inefficient, which makes it particularly unsuitable to a wireless ad-hoc operating environment. Nevertheless, we have included it here to establish a basis for comparison, because OPT gives the best case of what any tree construction algorithm can achieve. Next, we follow with a distributed algorithm, which, although not optimal, is efficient and practical.

3.1 Algorithm SHOPPARENT

Algorithm SHOPPARENT is a greedy distributed algorithm that avoids the disadvantages of the centralized OPT. The PST is constructed by each node running one instance of the algorithm and making its own decision about which other node to select as its parent. The parent is chosen from all the node's neighbors according to criteria aimed to minimize overhead. No node needs global knowledge about the system, only information (e.g., subscriptions) about itself and its parent or children.

When a new node i comes into an existing system, it broadcasts a PARENT-PROBE to all its immediate neighbors, looking for the "best" potential parent. If a recipient of the probe, say node k, currently has a route to the root publisher, k will reply with a PARENT-ADVERTISE message. The reply contains information which allows node i to calculate a *routing metric*, which we will discuss promptly. For example, in addition to k's id, PARENT-ADVERTISE may contain k's current route to the root (to avoid a routing loop), and its effective subscription etc.

From all the PARENT-ADVERTISE replies it receives, node i selects the one, say node k's, that gives the smallest routing metric, and connects to that node as its parent in the publish/subscribe tree. Node k, in turn, may need to notify its own parent (or even "shop around" for a new one) if S_k has been expanded by the joining of node i. On the other hand, if node i does not receive any PARENT-ADVERTISE from its PARENT-PROBE, that means it is currently partitioned from the publishing tree, and it will periodically re-attempt to connect later.

To deal with connectivity changes in the network as a result of nodes' moving around or failing, for example, each node i periodically broadcasts its PARENT-ADVERTISE message. At the same time, it constantly listens for these messages from neighboring nodes, and calculates the routing metric and compares against that of the current parent. This periodic "beaconing" serves two purposes. First, if a new potential parent comes into range that offers a "better" route to the root, node i can take advantage of it by initiating a procedure to switch its parent in the PST to the new node. The switching may involve notifying the new and old parents, and possibly i's descendants. Second, if a child does not receive PARENT-ADVERTISE from its current parent during some timeout period, it expires the parent because the parent must have failed or gone out of range. After expiration, the node probes for a new parent, essentially as if it had just joined the network.

We introduce three variations of algorithm SHOPPARENT, each using a different routing metric when "shopping" for the best parent. The first, called SP-NHOP, uses H_k as the routing metric for node k. H_k is k's distance to the root node, measured in number of hops. This algorithm is designed to mimic some existing distributed network routing protocols, and results in a shortest-path spanning tree of the connectivity graph.

Note that for this to work, each node should keep track of its own hop count, which can be propagated and refreshed by PARENT-ADVERTISE messages.

The routing metric used by SP-OVHD is $O_k^{+i}(E) - O_k^{-i}(E)$, where $O_k^{+i}(E)$ is node k's overhead if k takes on i as one of its children, while $O_k^{-i}(E)$ is if it does not. The metric measures the increase in $O_k(E)$ if k becomes i's parent. Since $O_k(E)$ is part of the PST overhead, this metric indirectly estimates the impact of k's parenting i.

Note that in order to calculate the overhead metric above, the algorithm needs the "event distribution" of the set E of events to be published. Essentially, an event distribution allows one to calculate $\Phi E(\alpha)$, the expected number of events in E matching the subscription predicate α. This distribution can be computed fairly accurately in certain applications, e.g., if the events of interest to a user are generated by a sensor with a regular pattern. In situations where the precise distribution is not known a priori, however, we often find that the absolute numbers are not needed to run the algorithm correctly. Because the node only uses this information to choose among several alternative parents, it is usually sufficient to know the relative distribution of events in several categories. For example, even though the exact rate of new postings in each newsgroup varies over time, the relative frequencies of different newsgroups may well stay comparatively fixed, especially over longer periods of time. Thus, for example, the groups in `comp.os.linux` will always have more traffic than `comp.os.mach`. Finally, nodes can also project future event distributions using recorded history of past events. Section 5 studies the resilience of the algorithm to skew in the approximation.

The last algorithm, SP-COMBO, simply uses the product of the two metrics used above $((O_k^{+i}(E) - O_k^{-i}(E)) \cdot H_k)$. Intuitively, SP-OVHD underestimates the impact on the overall PST overhead if k parents i, since not only k, but some of its ancestors as well, are likely going to be affected. SP-COMBO is intended as a better estimate by assuming that all H_k nodes on the path from k to the root are equally affected. Naturally, SP-COMBO may overestimate in some cases. Section 5 studies the actual performance of the three SHOPPARENT algorithms in detail.

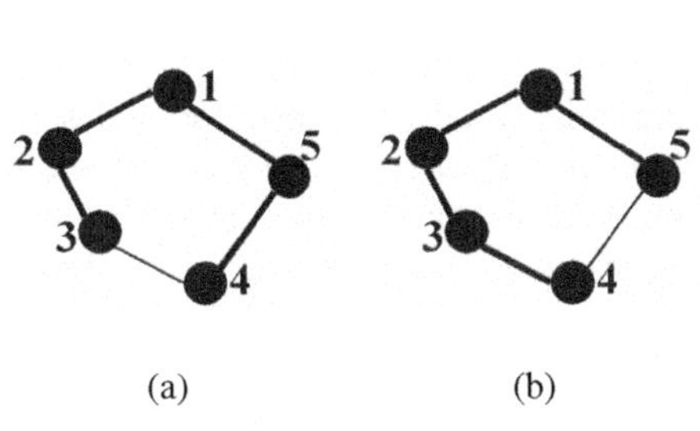

Fig. 3: Two PSTs of the same connectivity graph.

Example 3. Figure 3 shows a simple system with 5 nodes. Node 4 is joining the system, and is trying to select either node 5 (Figure 3(a)) or node 3 (Figure 3(b)) as its parent. Let E contain 10 events, all of which match s_2 and s_4, while half of which (5 events) match s_5. Moreover, assume no events match s_1 or s_3.

When node 4 sends out its PARENT-PROBE, it will receive two replies, from nodes 3 and 5, respectively. Assume SP-COMBO is being used. The routing metric calculated from 3's PARENT-ADVERTISE will be $(O_3^{+4}(E) - O_3^{-4}(E)) \cdot H_3 = (20 - 0) \cdot 2 = 40$ because, for example, $O_3^{+4}(E) = 2\Phi E(\neg s_3 \wedge s_4) + \Phi E(s_3 \wedge s_4) = 2 \cdot 10 + 0 = 20$. On the other hand, $(O_5^{+4}(E) - O_5^{-4}(E)) \cdot H_5 = (15 - 0) \cdot 1 = 15$. Therefore, node

4 will eventually pick 5 as its parent, resulting in the PST of Figure 3(a). (Note that, in this particular instance, all three SHOPPARENT variants give the same outcome.) The reader can verify that this is indeed the best PST in this scenario.

SHOPPARENT is a greedy local algorithm because each node makes its own decision without global information. Omitted here due to space, examples can indeed be constructed to show that the resulting publish/subscribe tree may not always be optimal. Section 5 will investigate how close the algorithm comes to the global optimum.

For comparison purposes, we introduce another simple algorithm called RANDOM. RANDOM is a distributed algorithm similar to SHOPPARENT. However, each node randomly selects its parent from the pool of candidates. Effectively, the algorithm generates a random spanning tree of G as the PST. This behavior is similar in nature to techniques used in some existing ad-hoc multicast tree protocols [8] and fixed network publish/subscribe systems [9, 10, 11], where a new node joining the system often connects to a random existing member.

4 Evaluation Model

We use simulation to study the effectiveness of our algorithms. Recall that the algorithms need as inputs the connectivity graph G, each node's subscription and the distribution of the set E of events to be published. In this section, we present models to simulate these factors.

4.1 Connectivity Graph

To obtain the connectivity graph G, we first place a total of N wireless nodes in a two-dimensional space. Instead of randomly generating an x- and a y-coordinate for each node, we use a clustered model, which we believe is more representative of a real wireless ad-hoc deployment. In particular, the nodes are partitioned into N_c clusters, and each cluster consists of N_n nodes, so that $N = N_c \cdot N_n$. Nodes belonging to the same cluster are expected to be geographically closer to each other in general.

Each cluster has a cluster center, which is a point in the x-y plane. The set of all cluster centers are generated with respect to the origin in a polar coordinate system. The angle ρ of all the centers follows a uniform random distribution between 0 and 2π. The distance r to the origin follows a normal distribution with mean D_c.

Once we have the cluster centers, the node coordinates in each cluster are generated in a similar fashion. In particular, the nodes' ρ relative to their cluster center also follows a uniform distribution, while their r follows a normal distribution with mean D_n.

Finally, we obtain the connectivity graph G by picking a value for R, which denotes the transmission *radio range* of the wireless devices. A pair of nodes are neighbors (i.e., they are adjacent in G) if the distance between them in the x-y plane is less than R. We assume that R is the same for all wireless devices to ensure that G is undirected.

Our evaluation will assume that G is connected, since otherwise, some nodes will not be able to receive all necessary events simply because no route exists to the publisher. Consequently, we only use values of R large enough to guarantee the connectedness of G.

By varying R, we can obtain multiple connectivity graphs with differing degrees of "connectedness." A more connected graph implies a scenario where users are more clustered together, while a less connected one means they are farther apart. Figure 4 shows two G's obtained from the same set of nodes, but with different values for R.

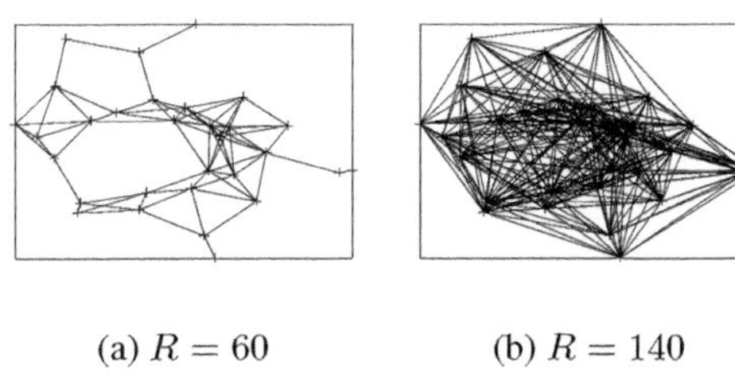

(a) $R = 60$ (b) $R = 140$

Fig. 4: A bigger R results in a more"connected" G.

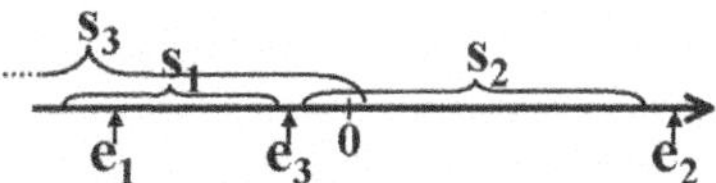

Fig. 5: The Number Intervals subscription model, with 3 sample subscriptions (intervals) and 3 events (points). As shown, e_1 matches s_1 and s_3, e_2 matches nothing, while e_3 matches s_3.

4.2 Subscription Model

We next present models to simulate each node's inherent subscription, s_i. The subscription model should allow for diverse interests for different users. Otherwise, if every user subscribes to the same things, the system essentially performs broadcast flooding. On the other hand, the model should also allow for overlap of interests between different users. Just like in the real world, there is often some sharing of interest between users. Note that we are not trying to model specific applications. We want simple models that, with few parameters, capture the essential features of subscriptions. With such models, we can understand why a particular algorithm works well, or what is the nature of overhead.

Our first subscription model (Figure 5), called Number Intervals (NI), represents a user's subscription by an interval on the number axis, e.g., $[3.0, 4.0]$, or $[-100, \infty)$. An event is represented by a real number. An event e matches node i's subscription if and only if e's number falls within range s_i. For example, if $s_i = [3.0, 4.0]$, and a new event e_1 is published as the number 3.2, then e_1 matches. On the other hand, if $e_2 = 5.2$, then e_2 does not match.

The NI model can be mapped directly to many real world applications. For example, in a nuclear reactor temperature monitoring application, new events report the current temperature of the reactor. Interested users want to receive these temperature readings only when they fall within a certain range. For instance, an overheating warning system is interested in receiving new readings only when they are above 2000 degrees. In such case, the subscription interval of the warning system is $[2000, \infty)$.

Within the NI model, the actual interval of node i is generated by picking a center and a length. The lengths of the intervals are selected randomly according to a normal distribution. The NI model is further divided into several submodels depending on how the centers of s_i's are determined. In the first submodel, called *random-center* (NI-R), we pick a random number (within a certain range) as s_i's center. The *x-center* (NI-X) submodel uses node i's x-coordinate (in the plane used to generate G) as the center. The idea is to introduce some correlation between a user's interest and its geographical location, so nodes that are nearby subscribe to similar events. The impact of such "geographical correlation of subscriptions" will be inspected in detail in Section 5. Finally, the *modified x-center* submodel (NI-Xmod) adds a random offset to i's x-coordinate before using it as the center. The offset is picked randomly from a normal distribution with mean 0 and standard deviation σ. By increasing σ, we can dilute the geographical correlation of subscriptions in NI-Xmod. Therefore, NI-Xmod represents a middle ground between NI-R and NI-X. When comparing the submodels, we use the same set of random numbers as the interval lengths. This way the total overhead of the system under different submodels should be comparable.

We feel that the NI model is fairly general and flexible enough to capture many important tradeoffs. Specifically, different users can have subscriptions of different sizes, corresponding to different lengths of the intervals. Moreover, we can simulate the different amount of overlap between two users' interests by adjusting the overlap of their two corresponding intervals. Nevertheless, our studies also use a second subscription model, the Topic Tree (TT), which simulates a newsgroup type of environment with a hierarchical tree of topics. We omit the details of the TT model in this paper since the conclusions drawn from the TT experiments are similar to those from the NI model.

4.3 Event Distribution Model

For simplification, we assume that an event has an equal probability of being represented by any real number (within a range of possible numbers) in the Number Intervals model. Although we realize that in practice this assumption does not always hold (e.g., a nuclear reactor has a higher probability of being in a certain temperature range), we also believe that a skewed event distribution function will not affect our conclusions.[4] Based on the assumption, the number of events in E matching a subscription α, i.e., $\Phi E(\alpha)$, is proportional to the length of α's interval.

For most of our simulations, we assume that the event distribution used by the nodes when running the algorithm will be the same as that of events actually published during the simulation. In Section 5, however, we will also study what happens if this assumption does not hold.

Table 2 lists all the parameters used in the simulation, and their default values.

5 Results

In this section we study and compare the performance of different algorithms. Due to space limitations, we will not present all of the studies we have done, but only give a few representative results. We have opted to run simulation at a higher level than the detailed MAC-layer simulators used, for example, by some earlier performance

[4] In a sense, the effect of a skewed event distribution can also be modeled by a skew in the subscription interval length, for example.

Sym	Meaning	Base value
N_c	# clusters	3
N_n	# nodes per cluster	10
N	total # nodes	$= N_c \cdot N_n$
D_c	avg. dist. btw. cluster centers & origin	100
D_n	avg. dist. btw. nodes & cluster center	80
R	radio transmission range	80
l	avg. interval length (NI)	25

Table 2: Parameters used in analysis

studies to compare multicast protocols [8]. As a result, we have not studied aspects of the systems such as control overhead or link contention. We believe, however, that our simulator is adequate given the high level issues we are trying to study at this stage, such as the appropriateness of our new metric and the fundamental difference between optimal PST algorithms and traditional multicast protocols.

For every algorithm except OPT, we do several simulation runs and take the average as the final performance number for that algorithm in that particular configuration. During each run, nodes take turn to execute the algorithm, i.e., to send messages and receive replies. The order in which nodes take turn is random, and is different for each run. This simulates, for example, nodes joining the publish/subscribe system in different orders. We run the simulation long enough to reach a steady state, when nodes stop switching to better parents between successive iterations. We then take a snapshot of the PST to calculate our evaluation metric: total tree overhead O. Since we do not simulate node mobility or failures except for our dynamics study in Section 5.6, the tree structure is guaranteed to eventually converge.

5.1 Overhead of RANDOM

Figure 6 compares the performance of RANDOM with the three variants of SHOPPARENT. The y-axis gives the overhead metric (O) of the algorithms, while the x-axis is l, the average interval length in the Number Intervals subscription model. Note that the NI-R submodel is used, where subscriptions are random intervals on the number axis that are of varying lengths. A lower y value, naturally, implies a more efficient PST. A larger l implies that users are interested in more things (larger intervals). Hence the total overhead (overall work) rises with increasing l. Moreover, larger intervals also mean that there is more potential overlap between different user's interests. For example, at $l = 1$, the probability that two random user subscription intervals overlap is about 0.6%. The same probability increases to 36% when $l = 50$.

We see from the figure that RANDOM has a much larger overhead than the others. Because resources are extremely limited in a wireless ad-hoc environment, an efficient PST is especially important, and clearly RANDOM is not a good algorithm for this environment. Instead, a node should be discriminating in choosing its parent, as is the case with the SHOPPARENT algorithms.

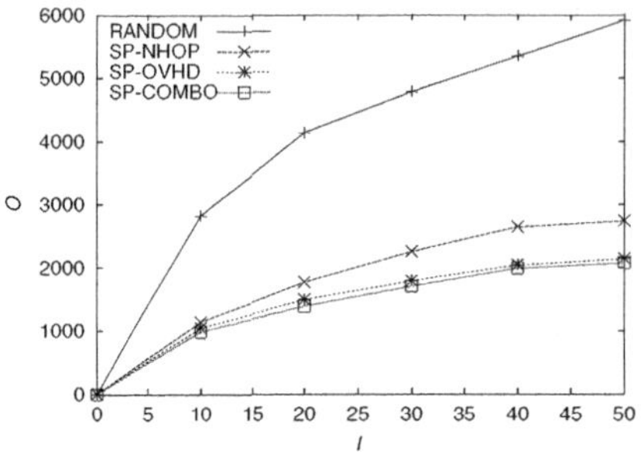

Fig. 6: Total overhead vs. average interval length l of NI-R.

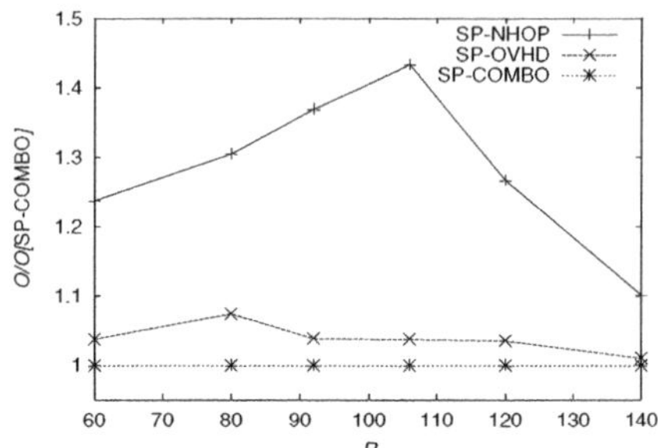

Fig. 7: Relative overhead of three SHOPPARENT algorithms vs. radio range R.

Although all three SP algorithms work much better than RANDOM, SP-NHOP is out-performed by SP-OVHD and SP-COMBO by a significant percentage. (Our next figure highlights these differences.) We use the general term "subscription-aware" to describe algorithms that take into account user subscription information when constructing PSTs. SP-OVHD and SP-COMBO are two such examples, because user subscriptions are used in the routing metric to help a node select a "better" parent. SP-NHOP, on the other hand, is not subscription-aware. Figure 6 suggests that it is advantageous to use a subscription-aware algorithm. Furthermore, the advantage amplifies as l increases. In other words, subscription-aware algorithms are better suited to take advantage of increased overlap of user interests, marked by a bigger l. Intuitively, if a node can find another node which shares much of its own interest as its parent, the overall overhead is often reduced.

SP-COMBO is the best performing algorithm in the pack. We attribute SP-COMBO's superiority to its routing metric, which we believe more accurately estimates the impact of a local decision on the overall global outcome. However, as shown in the figure, the difference between SP-COMBO and SP-OVHD is usually quite small.

We also conducted the same set of experiments using different values for r in the overhead metric formula. At $r = 0$, for example, we ignore the cost of receiving a message, and concentrate only on the cost to transmit (f). We have found that, as expected, the curves all have smaller y values with smaller r. Moreover, the relative difference between them decreases (e.g., the difference between SP-COMBO and SP-NHOP is reduced from about 30% at $r = 1$ to about 23% at $r = 0$). However, the shape of the curves remains almost identical with varying r, so our conclusions are still valid.

5.2 SHOPPARENT Variants

Figure 7 compares the three SHOPPARENT algorithms as network connectivity varies. Each curve represents the ratio of O of the corresponding SP algorithm to that of SP-COMBO, plotted against R, the radio transmission range. We plot the overhead ratio instead of absolute overhead numbers to more clearly show their relative performance. SP-COMBO is chosen as the "base" algorithm because it has the best performance. For example, we see in the graph that SP-NHOP generates about 30% more overhead than SP-COMBO at $R = 80$. We still use the NI-R model. As before, we observe that the curves follow the order SP-NHOP, SP-OVHD, and then SP-COMBO from worst to best performance.

Recall that the radio range R affects the connectivity graph G, with a larger R implying a more "connected" G. For example, at $R = 60$, each node in the 30-node system has 5.2 immediate neighbors on average. The average increases to 21.3 at $R = 140$. A striking feature of Figure 7 is that the performance gap first widens and then narrows as R increases. Initially, as G becomes more "connected", it contains more spanning trees, which are candidate PSTs, thereby giving a superior algorithm a better chance at excelling. However, as R continues to grow, G approaches a fully connected graph. In such case, nodes can simply broadcast events, and all algorithms can perform equally well.

5.3 Near-Optimality of SP-COMBO

Figure 8 compares SP-COMBO to OPT, plotting their overhead against l, again using the NI-R model. Algorithm OPT is globally optimal, always producing the best PST. Because of the inherent intractability of OPT, we use, for this experiment only, a smaller scale system setup. Specifically, we let $R = 180$ to limit the potential number of spanning trees in G, thus to allow OPT to finish in a reasonable time. The figure shows that in many cases SP-COMBO comes fairly close to the optimal in terms of performance, with difference not more than 8%. In fact, after running simulations for several configurations, the most overhead we have observed is less than 15%. Of course, in larger scenarios, the difference could be larger. Yet, we expect SP-COMBO to continue to do very well, at a much lower computational cost.

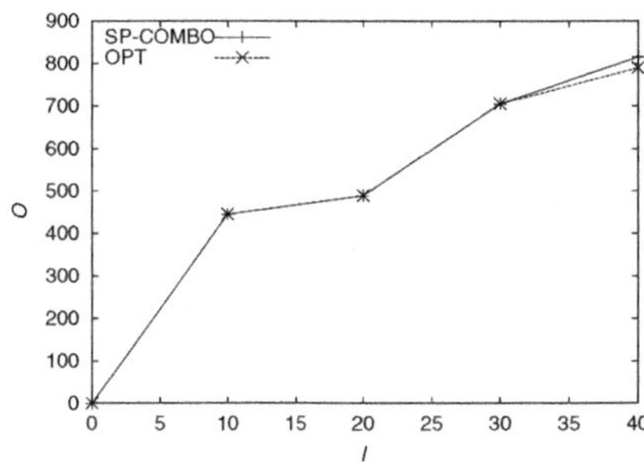

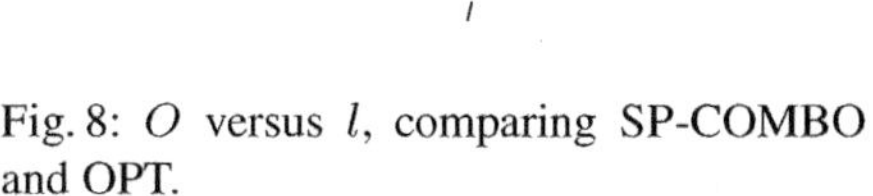

Fig. 8: O versus l, comparing SP-COMBO and OPT.

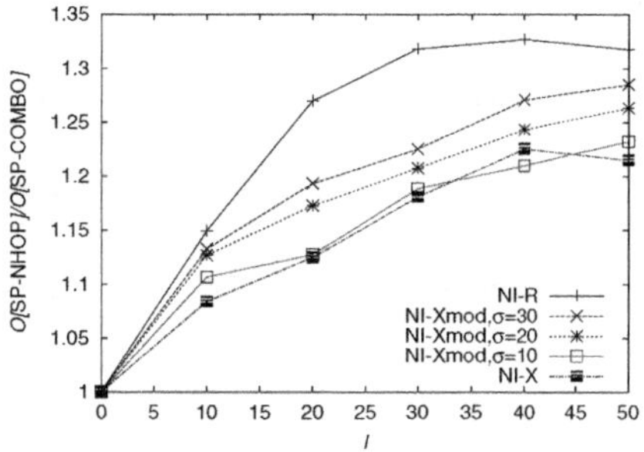

Fig. 9: Relative overhead of SP-NHOP over SP-COMBO, with different NI submodels.

5.4 Subscription Geographical Correlation

Figure 9 plots the overhead ratio of SP-NHOP over SP-COMBO, against average interval length l. The curves use different submodels of the Number Intervals subscription model. The y value indirectly reflects the advantage of subscription-aware algorithms, represented by SP-COMBO, over non subscription-aware ones, represented by SP-NHOP. A higher y value implies a bigger performance gap between them. In general, performance difference magnifies as l increases, confirming the trend we had observed in Figure 6.

However, Figure 9 shows a surprising phenomenon that ran contrary to our intuition. We had originally expected the performance advantage of SP-COMBO to be bigger (larger y values) with the NI-X model than with NI-R. Since subscriptions are geographically correlated in NI-X, nodes closer together would also share more common interests. It seemed to us that the subscription-aware SP-COMBO should better be able to take advantage of the situation, resulting in a bigger edge over SP-NHOP. However, Figure 9 shows exactly the opposite. We explain this phenomenon next.

With geographical correlation of subscriptions, such as that provided by NI-X, a node has a much higher chance of finding that nodes around it share more or less the same interests as itself. This is a valid assumption for certain applications, e.g., one where a user subscribes to events pertaining to its local surroundings. Since the set of neighboring nodes is also where a node "shops" for a parent, geographical correlation increases the chance that a node's subscription overlaps with that of its parent. However, although this correlation results in less total overhead in the PST, surprisingly, it also *reduces* the advantage of a subscription-aware algorithm such as SP-COMBO over a non subscription-aware algorithm such as SP-NHOP. In other words, SP-NHOP is actually able to benefit more from geographical correlation. The reason is that, because there is no longer as much variation between the subscriptions of a node's neighbors, (after all, they are all similar to this node's) the potential benefit that can be reaped by a subscription-aware algorithm diminishes. In particular, picking a candidate with the fewest hop to the root is often just the right thing to do.

The above explanation is further validated by the three middle curves in Figure 9, which represent NI-Xmod submodels with different σ parameters. An NI-Xmod model is the same as NI-X except that the geographical correlation is diluted by subjecting the subscriptions to an arbitrary perturbation with standard deviation σ. The bigger the σ, the less geographical correlation there remains. As predicted by our hypothesis, the advantage of subscription-aware algorithms increases (larger y values) with decreasing amount of geographical correlation (larger σ). Moreover, as correlation gradually diminishes, the curve for NI-Xmod approaches that of a random model like NI-R.

5.5 Event Distribution Skew

An event distribution is used by certain SHOPPARENT algorithms such as SP-COMBO to calculate the routing metric, which in turn determines the best parent to attach to. So far our simulations have assumed that the projected event distribution is the same as the real published one. Next we study the resilience of SP-COMBO to skew in the projected event distribution, which can occur if, e.g., the characteristics of published events suddenly changes.

We introduce a variation of SP-COMBO called SP-COMBOerr, which is the same as SP-COMBO except that we deliberately inject a random error to the event distribution projection. Specifically, if the real value of an $O_k^*(E)$ term is v, we will substitute

$v(1+\epsilon)$ with a random ϵ. The percentage error, ϵ, follows a normal distribution with mean 0, and standard deviation σ_ϵ. For example, if $\sigma_\epsilon = 0.1$, there is only a 68% chance[5] that ϵ will lie between $[-0.1, 0.1]$, and thus the actual value used to calculate the routing metric will be between $[0.9v, 1.1v]$.

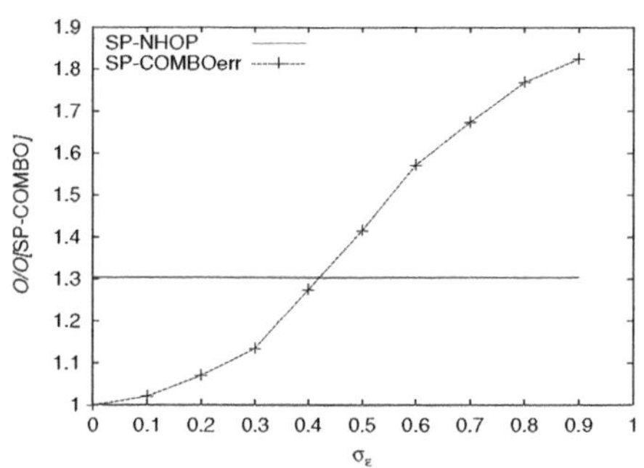

Fig. 10: Relative overhead of SP-COMBOerr and SP-NHOP over SP-COMBO vs. σ_ϵ, the standard deviation of injected error.

Figure 10 plots the relative overhead of SP-COMBOerr over SP-COMBO, against σ_ϵ. We also plot SP-NHOP over SP-COMBO for comparison. When $\sigma_\epsilon = 0$, there is no error, hence no difference between SP-COMBOerr and SP-COMBO (y value is 1). At small errors, the curve rises slowly with σ_ϵ. At $\sigma_\epsilon = 0.2$, e.g., SP-COMBOerr is still within 10% of SP-COMBO. The two curves cross at about $\sigma_\epsilon = 0.4$. Therefore, for a relatively wide range of errors (σ_ϵ between 0 and 0.4), SP-COMBO outperforms SP-NHOP. That is, even when expected traffic predictions can be off quite a bit, the predictions can still be helpful. On the other hand, if $\sigma_\epsilon > 0.4$, one is better off switching to SP-NHOP instead. Note that, however, even if significantly incorrect information is used at first with SP-COMBO, adaptive prediction methods can allow a node to detect shifts in event distribution based on observed events, and the ShopParent algorithm enables a node to easily switch to a better parent when conditions change.

5.6 System Dynamics

We conducted experiments to gauge how the algorithms cope with dynamism in the wireless ad-hoc environment, such as node mobility and failures. Here we report on our experience with mobility, as the conclusion is similar for other types of tests. We envision a mobility pattern where a node can move around occasionally, but it settles down for a period of time between moves. The settled period should be long enough (e.g., on the order of minutes) to allow the publish/subscribe tree to adapt and stabilize, and also to allow the nodes to take advantage of the new efficient tree. We do not consider, for example, scenarios where most of the nodes move constantly at high speeds. In fact, it has been shown in multicast research [12] that flooding usually is the most effective method of communication given such "fast mobility."

For our experiment, after all the nodes have joined the system and the tree has stabilized, we randomly select one node and move it to another random location. Naturally, while the move is taking place, old links time out, and some nodes will likely miss events as a result. Event losses are acceptable since, as stated before, we only deal with best-effort algorithms. For each experiment, we try to answer the following questions:

[5] According to definition of standard deviation.

whether the system eventually stabilizes after the node moves to its new location; how long it takes the system to converge to a new PST; and whether an efficient new PST is always formed.

We inject mobility as described above to a range of systems with different parameters. We find that in all cases the new system eventually converges to a new PST. The stabilization time varies, but is roughly proportional to the timeout period in SHOPPARENT (i.e., the time it takes for a child to realize that the parent has gone out of range), and also to the distance between the node in question and the root (i.e., the more hops the node is from the root, the longer it takes the PST to recover). Additionally, we are able to verify that in all cases, the overhead of the newly formed PST is comparable to that if the mobile node had originally been placed at its new location when it joined the system. In summary, we find that our algorithms perform very well in our target mobile environment.

6 Related Work

Our work combines a publish/subscribe mechanism with wireless ad-hoc networking. Naturally, we draw upon previous research in both areas. Yet, our work also differs from each in significant ways.

Publish/subscribe systems [13, 14, 10, 11, 15, 16] have been researched and developed for many years, but as far as we know in fixed networks. Making them work in a wireless ad-hoc environment introduces new challenges [1, 2, 3]. Our system differs from fixed network systems in two major aspects. First, our algorithms take full advantage of the broadcast nature of wireless radio by building directly on top of lower level radio broadcast primitives. Traditional publish/subscribe systems are often built as overlay server networks over an IP-like networking infrastructure, which allows them to assume universal connectivity between any two nodes in the system, and at a roughly constant cost. Although this assumption is reasonable in fixed networks, it is wasteful in our environment because it hides the fact that a unicast is actually implemented with multi-hop broadcasts. Second, the goal of our system is different from most traditional publish/subscribe systems. Our algorithm tries to find the most "efficient" tree so that the least amount of work is needed to publish events. This is a direct result of the characteristics of a wireless ad-hoc environment. In contrast, publish/subscribe systems in the past have not traditionally been too concerned with this measure because of the relative low cost of communication in a fixed network.

Multicast routing protocols for traditional networks have been well studied. In recent years protocols have also been proposed for multicast in wireless ad-hoc networks [8]. Due to the "many-cast" nature of publish/subscribe, it is not surprising that our algorithm resembles some of these protocols to a certain extent, especially tree based protocols without unicast support [6]. The optimality criterion of [5] resembles to a certain extent the metric defined in this paper, while in many other cases, especially in fixed networks, the efficiency of the resulting multicast tree has not been a primary design consideration. While the combination of the publish/subscribe paradigm and wireless ad-hoc networks has allowed our algorithm to be simpler in certain respects (e.g., every node participates in publish/subscribe), it has also raised new challenges which have not been fully addressed in prior work. For example, our algorithm is able to take advantage of information unique to the publish/subscribe paradigm, such as user subscription information. As our simulation shows, subscription-aware algorithms can have significant benefits in certain situations.

Some ad-hoc multicast protocols use a mesh-based approach instead of tree-based, to increase route reliability. These systems trade efficiency of the routing structure for resilience in a highly dynamic and mobile environment. Our algorithms can deal with mobility as shown in Section 5.6, but we are not focusing on "unstable" systems. That is, our target operating environment is one with occasional reconfigurations, followed by periods of stability. In such an environment, our experiments have shown that a good routing tree can yield significant advantages.

Distributed sensor networks [4, 17] coordinate a large array of small wireless sensors that monitor and interact with the physical world. Because such systems face many of the same challenges as our target environment, such as energy constraints and lack of existing infrastructure, their solutions have similarities with our algorithms.

7 Conclusion

In this paper we have studied the tree construction problem in wireless ad-hoc publish/subscribe systems. We have proposed a distributed algorithm ShopParent, and developed an evaluation metric, total overhead of a publish/subscribe tree, to study it. We have found via simulations that our algorithm, despite being a non-optimal greedy algorithm, works quite well in normal situations. We have also found that subscription-aware variants of the algorithm give further performance improvements by considering user subscription information during tree construction. Furthermore, the effectiveness of this subscription-awareness depends on multiple factors such as the overall amount of overlap between user subscriptions, and the existence of geographical correlation.

References

[1] Huang, Y., Garcia-Molina, H.: Publish/subscribe in a mobile enviroment. In: Proceedings of the Second ACM International Workshop on Data Engineering for Wireless and Mobile Access. (2001) 27–34

[2] Cugola, G., Nitto, E.D., Picco, G.P.: Content-based dispatching in a mobile environment. In: Workshop su Sistemi Distribuiti: Algoritmi, Architetture e Linguaggi. (2000)

[3] Meier, R., Cahill, V.: STEAM: Event-based middleware for wireless ad hoc networks. In: Proceedings of the International Workshop on Distributed Event-Based Sytems. (2002) 639–644

[4] Estrin, D., Govindan, R., Heidemann, J., Kumar, S.: Next century challenges: scalable coordination in sensor networks. In: Proceedings of the Fifth Annual ACM/IEEE International Conference on Mobile Computing and Networking. (1999) 263–270

[5] Lim, H., Kim, C.: Multicast tree construction and flooding in wireless ad hoc networks. In: Proceedings of the 3rd ACM International Workshop on Modeling, Analysis and Simulation of Wireless and Mobile Systems. (2000) 61–68

[6] Wu, C., Tay, Y., C.-K.Toh: Ad hoc Multicast Routing protocol utilizing Increasing id-numberS (AMRIS) functional specification. Internet Draft (1998)

[7] Garcia-Molina, H., Kogan, B.: An implementation of reliable broadcast using an unreliable multicast facility. In: Proceedings of the 7th IEEE Symposium on Reliable Distributed Systems. (1988) 101–111

[8] Lee, S.J., Su, W., Hsu, J., Gerla, M., Bagrodia, R.: A performance comparison study of ad hoc wireless multicast protocols. In: Proceedings of the IEEE Conference on Computer Communications (INFOCOM). (2000) 565–574

[9] Kantor, B., Lapsley, P.: Network News Transfer Protocol: A proposed standard for the stream-based transmission of news. Request for Comments: 977 (1986)

[10] Banavar, G., Kaplan, M., Shaw, K., Strom, R.E., Sturman, D.C., Tao, W.: Information flow based event distribution middleware. In: Proceedings of the 1999 ICDCS Workshop on Electronic Commerce and Web-Based Applications. (1999)

[11] Carzaniga, A., Rosenblum, D.S., Wolf, A.L.: Achieving scalability and expressiveness in an Internet-scale event notification service. In: Proceedings of the 19th Annual ACM Symposium on Principles of Distributed Computing. (2000) 219–227

[12] Ho, C., Obraczka, K., Tsudik, G., Viswanath, K.: Flooding for reliable multicast in multi-hop ad hoc netowrks. In: Proceedings of the 3rd International Workshop on Discrete Algorithms and Methods for Mobile Computing and Communications (DIAL-M'99). (1999) 64–71

[13] Oki, B., Pfluegl, M., Siegel, A., Skeen, D.: The Information Bus - an architecture for extensible distributed systems. Operating Systems Review **27.5** (1993) 58–68

[14] Segall, B., Arnold, D.: Elvin has left the building: A publish/subscribe notification service with quenching. In: Proceedings of the 1997 Australian UNIX Users Group Technical Conference. (1997) 243–255

[15] TIBCO Software Inc.: TIBCO Rendezvous. (`http://www.tibco.com/solutions/products/active_enterprise/rv/`)

[16] Cabrera, L.F., Jones, M.B., Theimer, M.: Herald: Achieving a global event notification service. In: Proceedings of the Eighth Workshop on Hot Topics in Operating Systems (HotOS-VIII). (2001)

[17] Bonnet, P., Gehrke, J., Seshadri, P.: Towards sensor database systems. In: Proceedings of the 2nd International Conference on Mobile Data Management. (2001) 3–14

syhwang@mis.nsysu.edu.tw

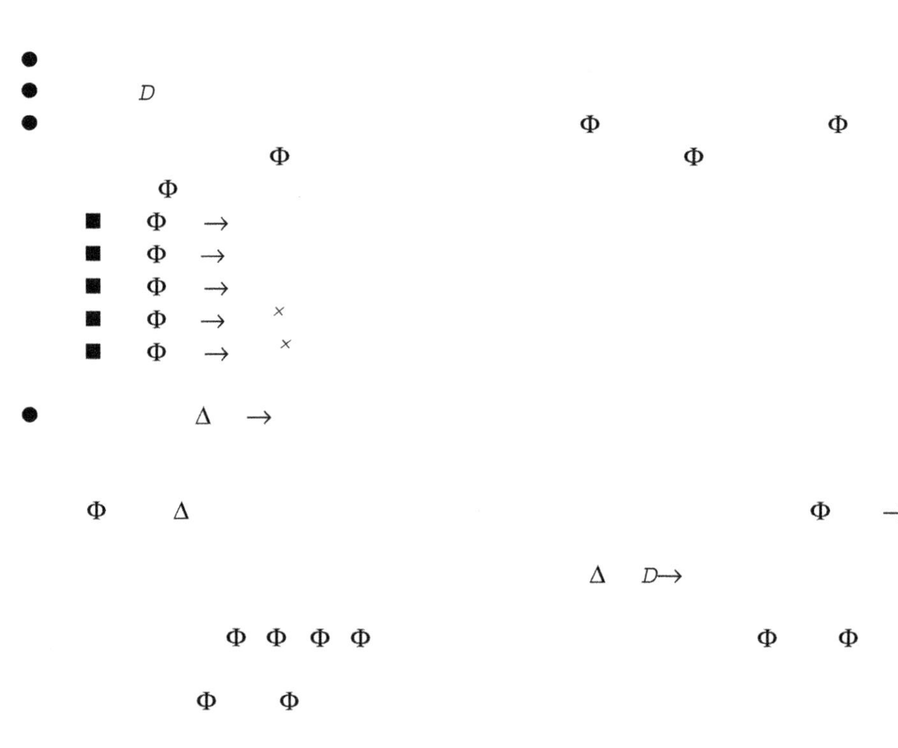

∈ ⊆

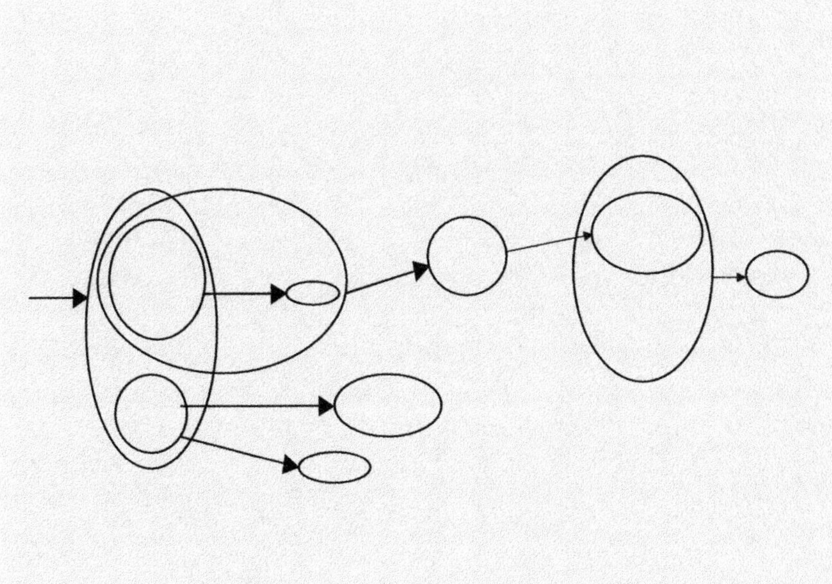

$\cap$ $\cup$ $-$

σ

$\times$

$\times$ $\times$

- $\times$ $\rightarrow$

- $\times$ $\rightarrow$

- $\times$ $\rightarrow$

- $\rightarrow$

$\circ$

$\circ$

$\circ$ $=$

$\in$ $\exists$ $\in$ $\exists$ $\in$ $\exists$ $\in$

$\circ$

$\circ$

$\circ$ $=$

$\in$ $\exists$ $\in$ $\exists$ $\in$ $\exists$ $\in$

$\rightarrow$

$\rightarrow$

$\rightarrow \quad = \quad \in \quad \subseteq$

$\rightarrow$

$\rightarrow$

$\subseteq$

$=$

$\Sigma_{\in}$

$\uparrow$

$\uparrow$

$\uparrow \quad \equiv \cup \quad \in \quad - \cup \quad \in$

$\uparrow$

$\uparrow$

$\uparrow \quad \equiv \cup \quad \in$

$\cup \cap \quad - \quad \sigma$

$\sigma \quad \sigma$

$\sigma_{=} \quad \cup \uparrow \quad \sigma_{=} \quad \cup \sigma_{=}$

$\rightarrow \sigma_{=} \quad - \sigma_{=} \quad - \sigma_{=}$

$\circ \ \sigma_{=}$

$\circ \ \sigma_{=}$

$\uparrow \ \sigma \quad \cup\sigma \quad \cup\sigma$

σ

σ

σ → σ ° °

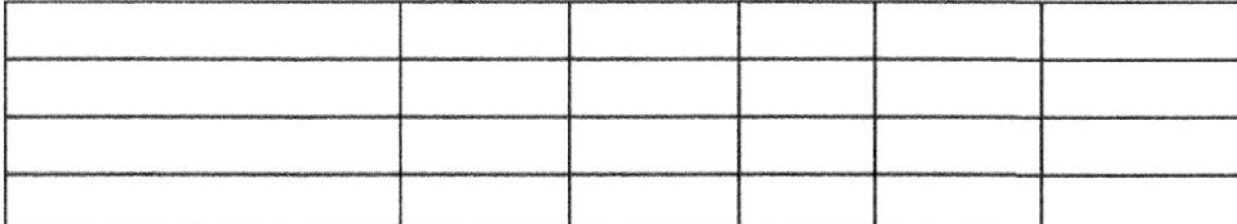

$\leq \leq$

$\leq \leq$

$\sum_{=} \cdot$

$= \quad \sum$

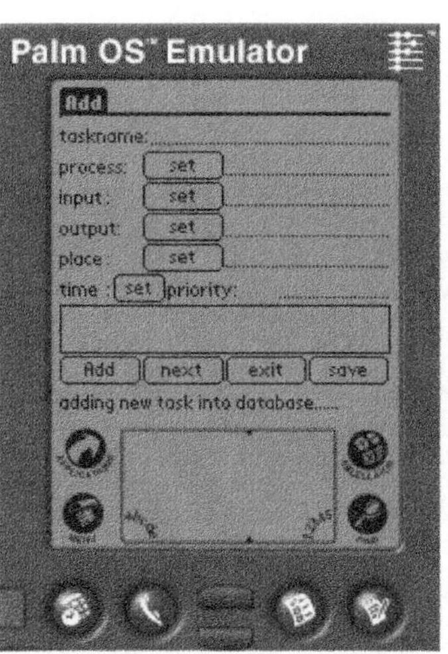
Palm OS™ Emulator
Add
taskname:
process:
set
input:
set
output:
set
place:
set
time:
set
priority:
Add
next
exit
save
adding new task into database......

Palm OS™ Emulator
Set INPUT
INPUT
Set INPUT List
INPUT List in DB
CreditCard
DecorationStuff
Food
InvitationCards
InvitationLetter
ok
next
exit
adding INPUT data set into database......

-Template Recommendation System-

Process Name

Registration

User Status/Demand Query

Is freshman or not? No Submit

Back to Template Recommendation System

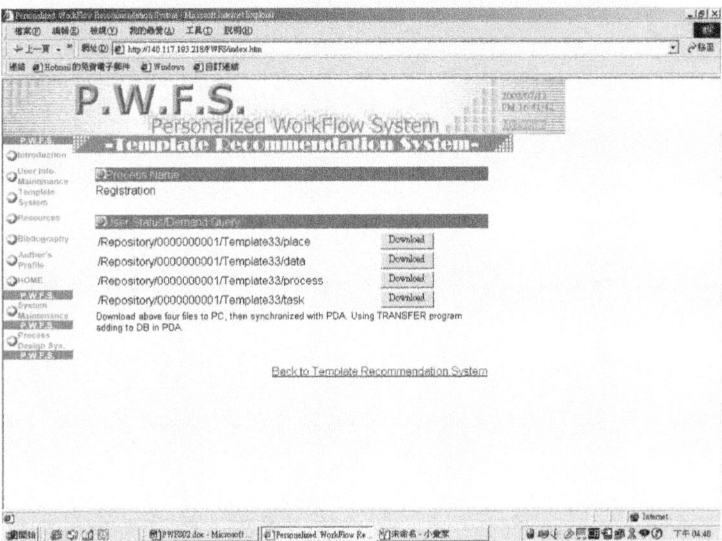

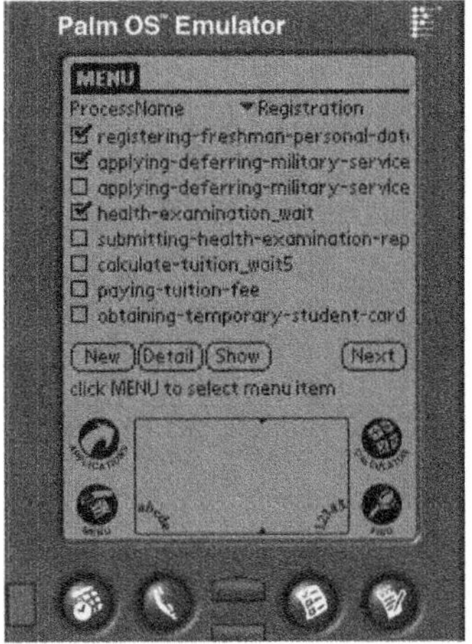

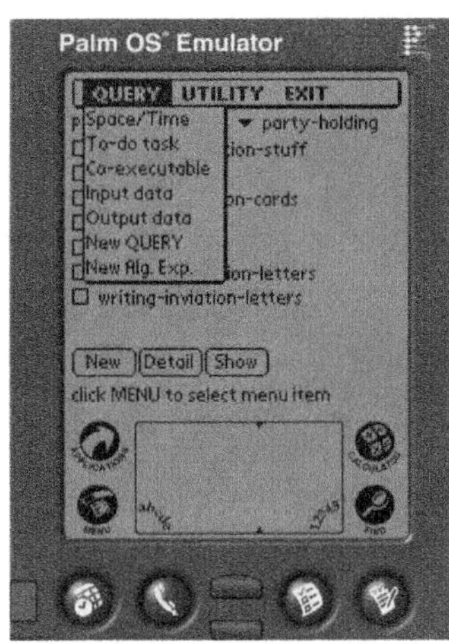

{al, graham, ron, ajm, kevin, yyyang}@dcs.st-and.ac.uk

{firstname.lastname}@cis.strath.ac.uk

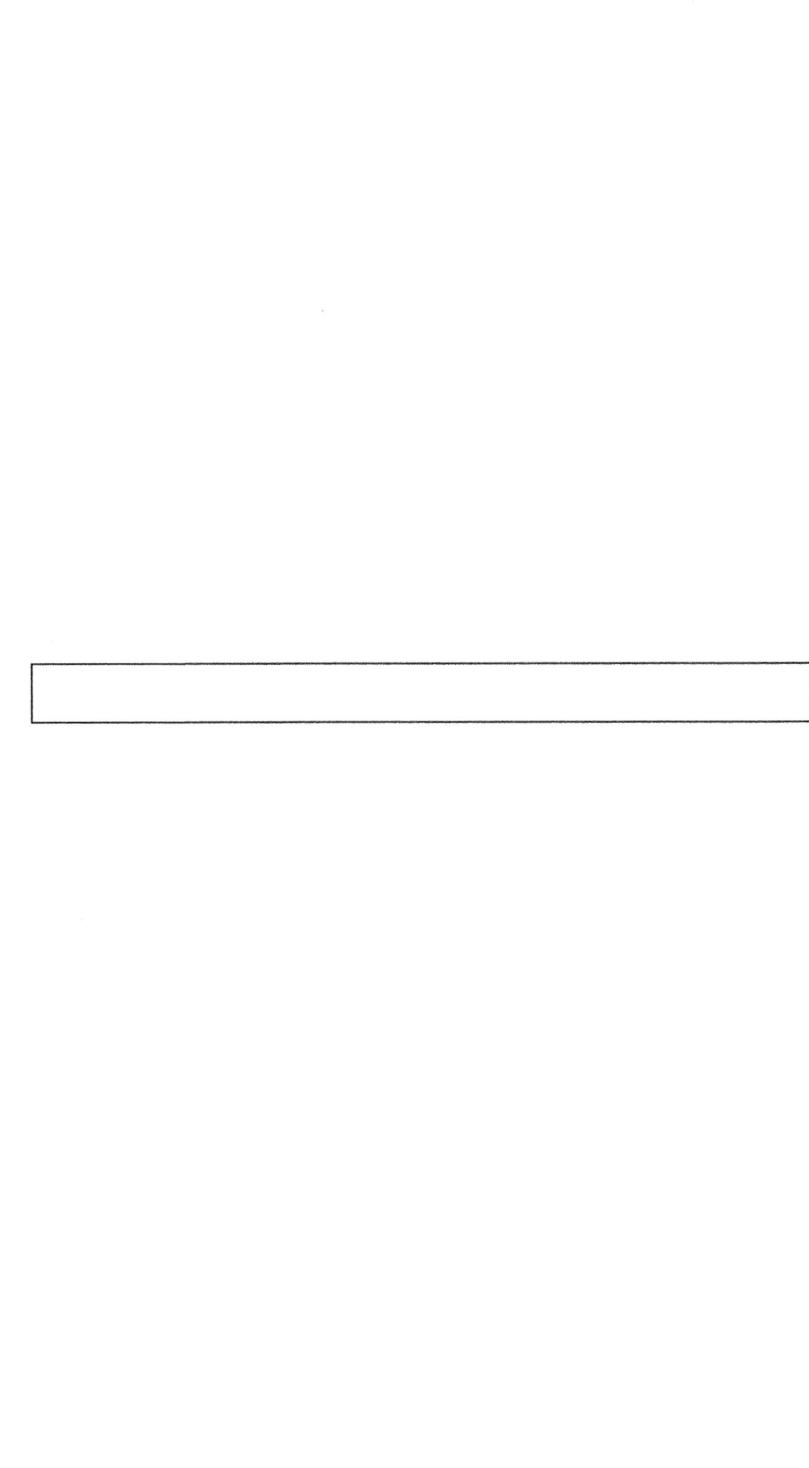

Where X Profile -> Information

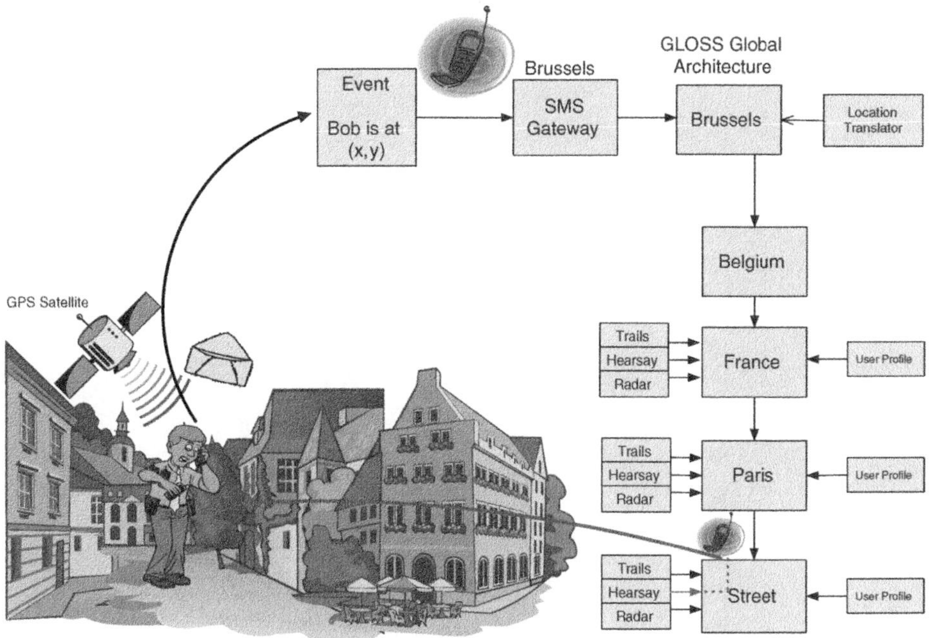
Event
Bob is at (x,y)
Brussels
SMS Gateway
GLOSS Global Architecture
Brussels
Location Translator
Belgium
GPS Satellite
Trails
Hearsay
Radar
France
User Profile
Trails
Hearsay
Radar
Paris
User Profile
Trails
Hearsay
Radar
Street
User Profile

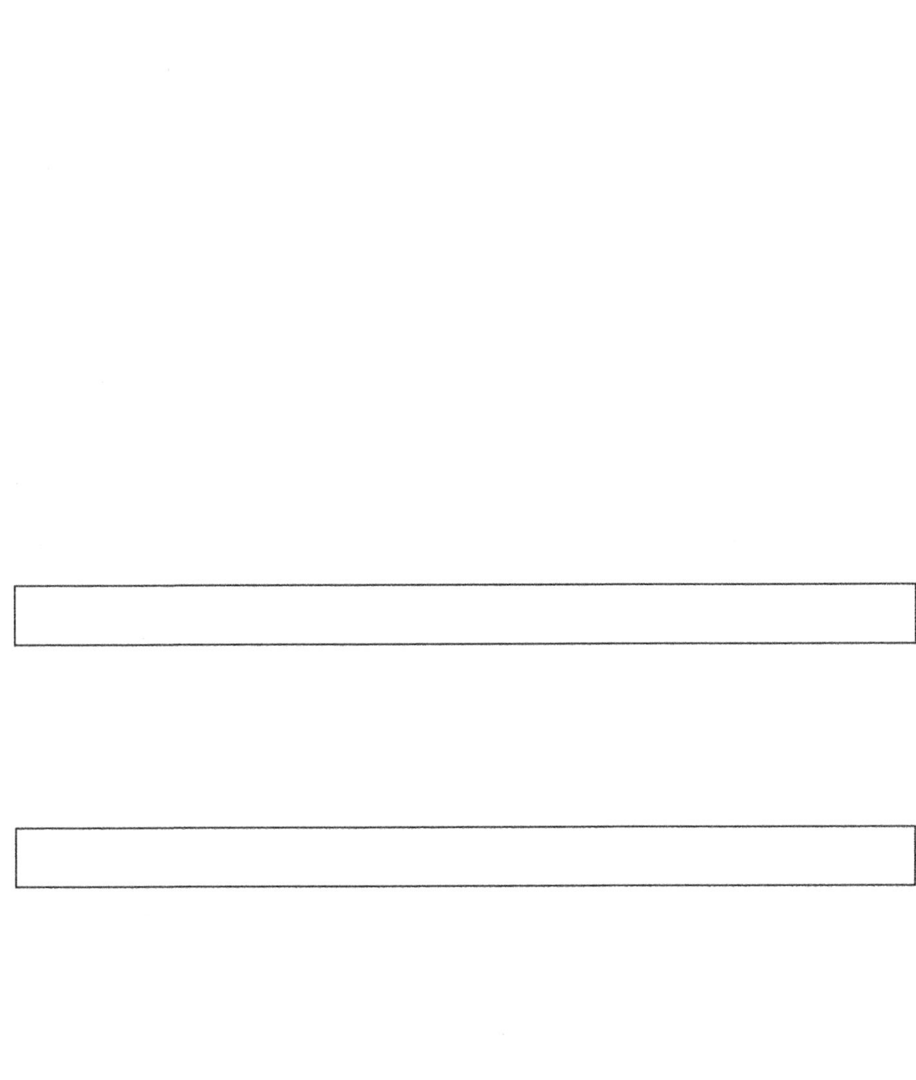

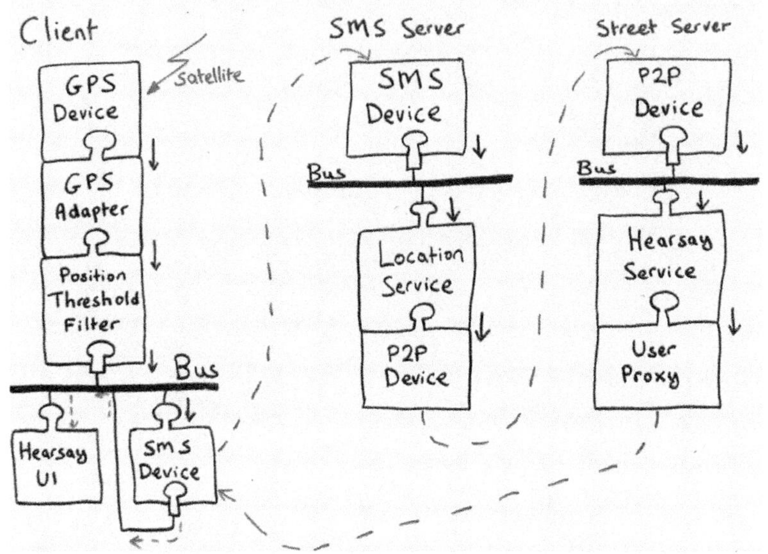
Client
satellite
GPS Device
GPS Adapter
Position Threshold Filter
Bus
Hearsay UI
SMS Device
SMS Server
SMS Device
Bus
Location Service
P2P Device
Street Server
P2P Device
Bus
Hearsay Service
User Proxy

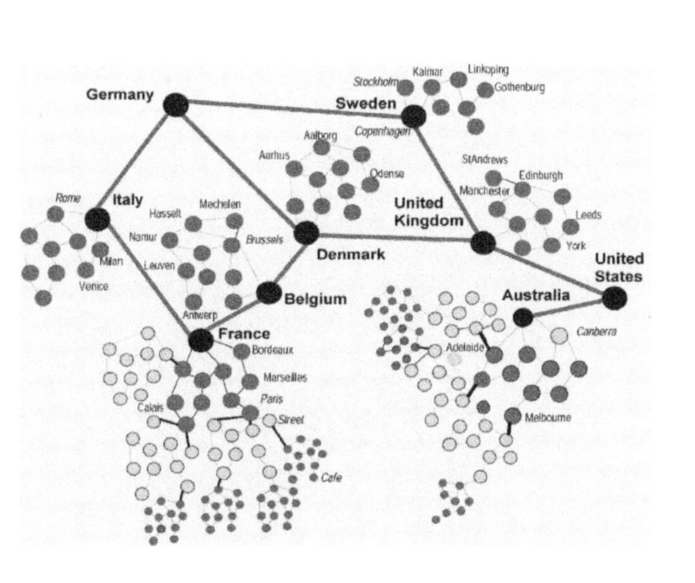

Germany
Sweden
Stockholm
Kalmar
Linkoping
Gothenburg
Copenhagen
Aalborg
Aarhus
Odense
StAndrews
Edinburgh
Manchester
Leeds
York
United Kingdom
Rome
Italy
Hasselt
Mechelen
Namur
Brussels
Leuven
Milan
Venice
Denmark
United States
Belgium
Australia
Antwerp
France
Canberra
Bordeaux
Adelaide
Marseilles
Paris
Calais
Street
Melbourne
Cafe

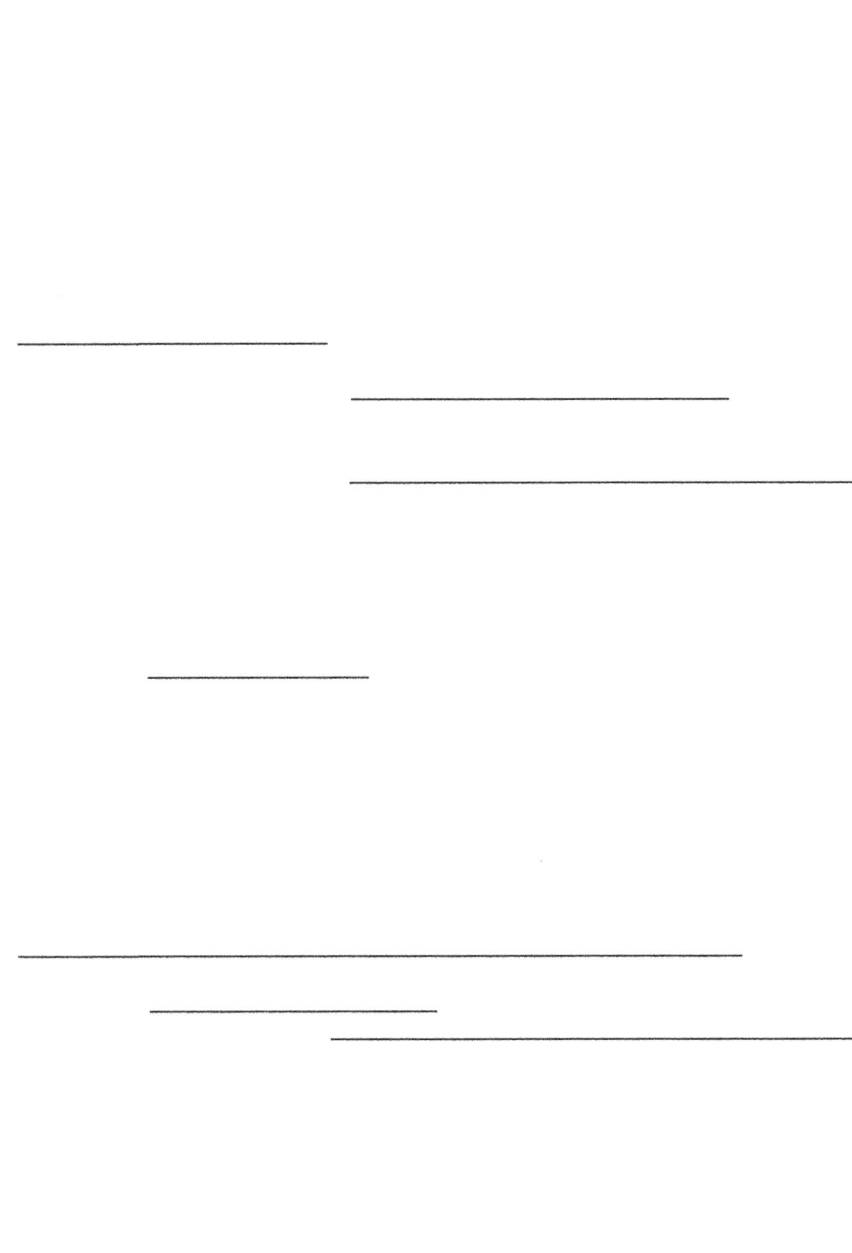

FATES: Finding A Time dEpendent Shortest path*

Hae Don Chon[1], Divyakant Agrawal[2], and Amr El Abbadi[2]

[1] Samsung Electronics
hd.chon@samsung.com
[2] Department of Computer Science, University of California, Santa Barbara
{agrawal,amr}@cs.ucsb.edu

Abstract. We model a moving object as a sizable physical entity equipped with GPS, wireless communication capability, and a computer. Based on a grid model, we develop a distributed system, FATES, to manage data for moving objects in a two-dimensional space. The system is used to provide time-dependent shortest paths for moving objects. The performance study shows that FATES yields shorter average trip time when there is a more congested route than any other routes in the domain space.

1 Introduction

Global Positioning Systems (GPS) have been widely accepted as the technology for locating objects such as vehicles on a highway. One of the features GPS provides, other than locating, is speed tracking so that we can get speed information without integrating GPS to the speedometer in the vehicle. Wireless communications technology such as Cellular Digital Packet Data (CDPD) [3] and Mobile IP [12] has also gained popularity. With these technologies in place, finding and managing the current locations of moving objects have become possible. Since keeping track of the location of many, continuously moving objects is hard to achieve, there are significant data management problems that need to be addressed. The focus of much of the research on this issue has been on how to manage location information efficiently without updating the location information every time the moving object moves. One basic and fundamental observation on moving objects is that if a moving object maintains a constant speed for a certain period, future locations of that moving object can be modeled as a linear function of time during that period. Then the location information can be managed by maintaining the parameters of the linear functions instead of constant updating [17, 7, 4]. There is no known alternative to this approach so far. We also base our research on this observation and make further observations as follows: (1) Moving objects with a few exceptions follow certain paths such as a route on a highway system. (2) Such paths on which moving objects move have

* This research was partially supported by the NSF under grant numbers EIA-0080134, EIA98-18320, IIS98-17432, and IIS99-70700.

M.-S. Chen et al. (Eds.): MDM 2003, LNCS 2574, pp. 165–180, 2003.

physical limitations such as the number of lanes and/or route conditions. Based on our everyday experience, we can expect that as the number of vehicles in a section of a route increases, the average speed of the vehicles decreases. This is supported by a report from the Texas Transportation Institute [16]. The report observed that as the daily traffic volume (the average number of vehicles observed in a day at a location) increases, the average speed of vehicles decreases. (3) Due to these limitations, a real trajectory of a moving object is a polyline instead of a line segment. Note that before the insertion, the trajectory of a moving object is a line segment, which then could become a polyline depending on the number of moving objects on the way the moving object follows. For this reason, if we are to manage future location information about moving objects on a network of routes, the routes need to be broken into smaller sections, which then are associated with information such as the number (or list) of objects and the maximum velocity allowed in the section. There are many multi-dimensional index structures such as R^*-Tree [2] that could be used to store and retrieve the line segments. However, if those index structures are used, sectional information should be either maintained in a separate structure or deduced from the index structure while the line segment is being inserted. One way of deducing sectional information directly from an index structure is to issue a range query for a section and for a complete insertion a series of range queries are needed. Furthermore, to compute a time-dependent shortest path, cost information of sections (which can be deduced from the number of moving objects in the sections) is needed, for which a series of range queries is needed. This series of range queries, however, is unnecessary overheads if a separate structure is used to maintain sectional information. By using a grid model as proposed in [5], we can process range and k-nearest neighbor queries as well as insertion and deletion of moving objects.

The best example for moving objects is vehicles on a road although it is not the only example. The moving objects considered are not particles, meaning that they do not move irregularly such as in a Brownian motion. Instead, they follow a certain path such as a route on a highway system. In addition, they are sizable physical entities with the following characteristics: ability to locate their position, wireless communication capability, and some computer such as a PDA. Although there are several other technologies to locate an object, GPS is by far the dominant technology and it is getting better and cheaper. As a result, we expect that in the future many (if not all) moving objects, especially vehicles, will be equipped with a GPS, wireless communication capability, and a computer.

Among many possible applications that could benefit from this research is *Time-Dependent Shortest Path Problems*. In transportation applications, *Time-Dependent Shortest Path Problems*, where the cost (travel time) of edges between nodes varies with *time*, are of significant interest [11]. Traffic congestion is a huge problem, especially in metropolitan areas. In recent years, in order to avoid traffic congestion a growing number of vehicles are equipped with a navigation system which provides a shortest path to their destination as well as other features like roadside assistance [9]. The navigation system uses a digital map, which includes static information such as the distance of a path and the speed limits along the way. However, such information does not reflect the dynamically changing traffic information, hence could lead vehicles to incorrect shortest paths. In contrast to this static approach, if there is a way to obtain the cost in real time and then apply a time-dependent shortest path algorithm, it would result in a better solution for the shortest path queries. In this paper we develop FATES, a system that could be used to gather this dynamically changing information in order to run a time-dependent shortest path algorithm.

The paper is organized as follows: Related work is discussed in Section 2 and we provide a summary of the grid model developed in [5] in Section 3. In Section 5, we propose FATES to handle moving objects in a two-dimensional space using multiple one-dimensional components. A performance study is given in Section 6 followed by the conclusion in Section 7.

2 Related Work

Most of the early work [17, 18, 20, 13, 7, 15, 4] did not take the following fact into account: the trajectory of a moving object is the result of the interactions among other moving objects yielding a polyline as a trajectory. Recently, however, Pfoser et al. [14], Chon et al. [5], and Vazirgiannis et al. [19] worked on the assumption that the trajectory of a moving object is a polyline. One difference between [14] and [5, 19] is that in [14] the polyline is the result of recording a moving object's past locations and only past information can be queried instead of the future. We believe that this assumption will be the basis for future research that will address issues of managing and querying future locations of moving objects.

The shortest path problem is one of the problems that have been researched extensively and thoroughly [1] and the time-dependent shortest problem is a variation of the problem [10, 8, 11]. In this paper, we do not attempt to solve or propose yet another variation of the shortest path problems. Instead, we adapt a time-dependent shortest path algorithm developed in [11] to the context of moving objects. There is one more interesting related work by Handley et al. [6]. The notion of a cost function in [6] is somewhat different from those in [10, 8, 11]. More precisely, Handley et al. do not address the type of time-dependent shortest path problems solved by FATES. Instead, they address the following problem: Data from sensors embedded in a freeway in the vicinity of San Diego, California are collected. Such data includes the average velocity of vehicles in sections that the sensors cover and traffic density in those sections over some period of time. Then, some number of trips with different routes and time information are randomly generated, and for each of the trips, a duration of the trip is computed and stored in a database using the collected data. When a query for a duration of a trip is issued, a nearest neighbor query is executed against the computed trips, and the duration of the trip is that of a trip in the answer set. Handley et al. show that their approach gives better prediction than digital maps which use static information. These digital maps are available on car navigation systems. In contrast to this approach, FATES computes the time-dependent cost function from the information about the future locations of moving objects. Then a time-dependent shortest path algorithm is run to provide a shortest path to a moving object.

3 Background

In this section, we provide a summary of the model developed in [5], which maintains dynamically changing information about moving objects assuming that a moving object moves on a one-dimensional path. It is assumed that each section of a route is associated with a list of moving objects in the section and a predefined function of the maximum velocity at which moving objects can move. This maximum velocity function returns the maximum velocity a moving object can move in the section depending on the number of moving objects currently in the section. This function can be derived from statistical data.The domain space $T \times Y$ is then partitioned, where T corresponds to the time domain and Y corresponds to the space domain, into an array of cells such that $Y = [y_0, y_1) \cup ... \cup [y_{p-2}, y_{p-1}) \cup [y_{p-1}, y_{max}]$ and $T = [t_0, t_1) \cup [t_1, t_2) \cup ... \cup [t_{q-1}, t_q)$, where $t_q = t_0 + \Delta T$. A cell $g[i,j]$ is defined as a rectangle $[t_i, t_{i+1}) \times [y_j, y_{j+1})$, for some $t_i \in T, y_j \in Y$. Note that although time is increasing indefinitely, only the information about some predefined ΔT period of time is maintained. As time evolves, the time domain is shifted forward. It is also assumed that on its entry to the system, a moving object m is responsible to provide initial information such as the starting location (s), the starting time (t_s), the destination (e), and the intended velocity (v), which is the maximum velocity at which it intends to move. When a moving object m appears in the system providing its initial information, it is inserted into the system as a polyline. Before we explain the insertion algorithm, we introduce a notation $\triangleright$ and a term *entry point* as follows: If the corresponding polyline of m intersects with a cell g, m is said to intersect with g and denoted by $m \triangleright g$. An entry point of m to a cell g is the point where m's corresponding polyline and g intersect for the first time.

With the initial information $m = (s, t_s, e, v)$, the first cell g is found such that $m \triangleright g$. The following steps are then repeated: the maximum velocity allowed in g is computed depending on the number of moving objects in the cell, and m's velocity is adjusted accordingly. When there is a velocity change, the corresponding entry point is added to the linked list associated with m which represents a polyline. The next cell to intersect is then found by computing the entry point to the cell, and m's identifier along with the adjusted velocity is inserted into g. These steps are repeated until the cell to which the destination e of m belongs is processed. In that case, the insertion process is completed by inserting m into the Object Store, and the polyline information is returned to m.

There are, however, cases where one insertion may cause a series of updates. Consider the situation shown in Figure 1(a) where three moving objects, $\{m_1, m_2, m_3\}$, are in the system and a new object, m_4, is about to be inserted.

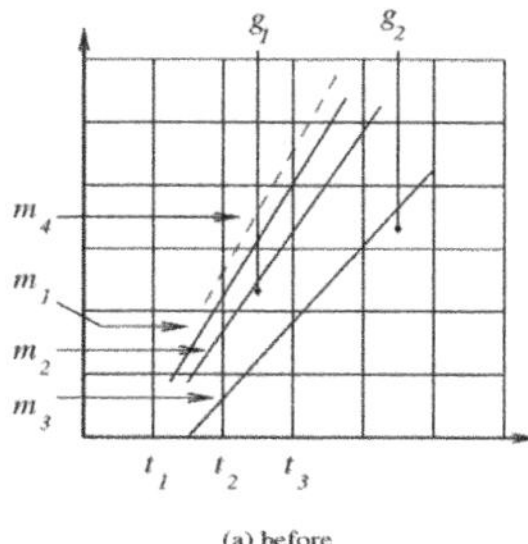
(a) before

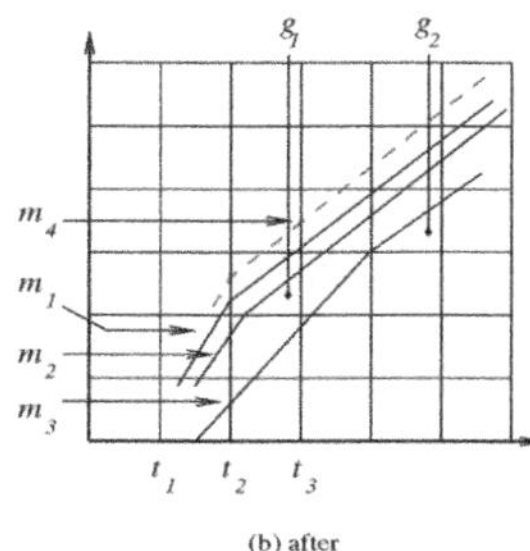
(b) after

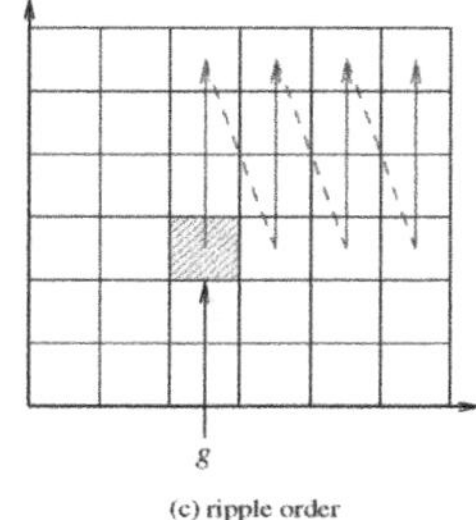
(c) ripple order

Fig. 1. Ripple Effect

The moving object m_4 in this case is said to cause a *ripple effect*. Assuming that three moving objects is a threshold for moving objects to reduce their velocities to a certain degree, cell g_1 is about to have three objects so that existing objects as well as the new one will need to reduce their velocities based on the maximum velocity function associated with g_1. Furthermore, the updates of m_1 and m_2 cause the cell g_2 to have three moving objects. As a result, another update on m_3 has to be executed. Figure 1(b) shows the result of the insertion of m_4. For this reason, we need to check if the number of objects in g reaches a threshold, before inserting a moving object m into g, by checking $f_g(|g|) > f_g(|g|+1)$, where $f_g(n)$ is the maximum velocity function associated with g. If a threshold is reached, an algorithm called *RippleInsert* is invoked with a (row, column) pair of g. Otherwise continue the insertion of m until it is finished.

The RippleInsert algorithm works as follows: suppose g is the first cell whose maximum velocity needs to be changed because of the insertion of a moving object. The information about all moving objects in g is then adjusted according to the new velocity allowed in g and continue this process with the cells whose time and location intervals are greater than or equal to those of g. Other cells need not to be processed since an insertion would not cause any update to the past or to previous locations. Figure 1 (c) shows the order in which cells are updated. The details of this algorithm can be found in [5].

4 Time Dependent Shortest Paths

Shortest path problems are among the most studied network flow problems and one of the most important application areas is transportation. In transportation applications, *Time-Dependent Shortest Path Problems*, where the cost (travel time) of edges between nodes varies with *time*, are of significant interest [11]. As in transportation applications, where the cost changes dynamically, the correctness of the shortest path very much depends upon the correctness of the cost model. For example, typical drivers choose a route based on static information that, in general, is heavily weighted by the length of the route. Unfortunately, as many drivers in metropolitan areas are aware, superiority of a route cannot entirely be based on the length of the route. In particular, current and future conditions of routes such as the severity of congestion should play an important role in determining which route would incur minimum delay. To see why the time-dependent shortest path problems are of interest, we illustrate a simple scenario in Figure 2. The bottom of the figure is a snapshot of a network of

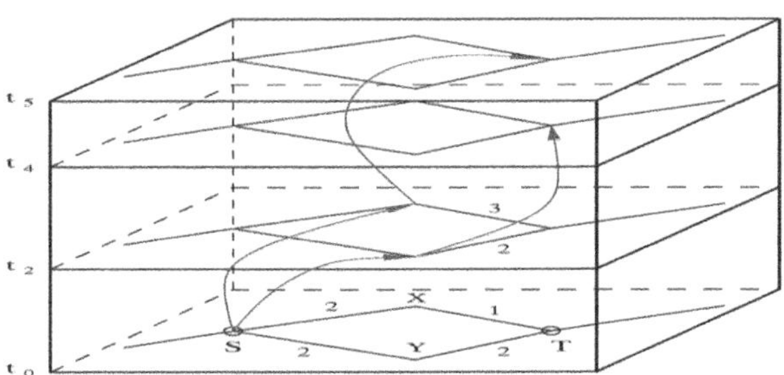

Fig. 2. Time-Dependent Shortest Path

routes with cost information at time t_0. The shortest path from S to T based on the current information (available at t_0), is $S \rightarrow X \rightarrow T$. However as shown in the figure, by the time (t_2) when X is reached, the cost information changes, so reaching T from X takes 3 time units instead of 1 unit as it was expected at t_0. As a result, T is reached at t_5. In contrast, if the path going through Y is followed, T would have been reached at t_4. This example illustrates that when the cost varies with time, using only static information to find a shortest path is not enough. Cost variations should be taken into consideration.

In general, the time-dependent cost (normally, travel time) is given arbitrarily [10, 8, 11]. We noticed that this time-dependent shortest path problem is an interesting application for the data management for moving objects. Extending the system developed in [5] to two-dimensional space will provide dynamically changing cost estimation to the time-dependent shortest path algorithm [11] leading many vehicles to real shortest paths. This will result in better utilization of the road network and efficient traffic management. We develop the extended system in the next section.

5 FATES: Finding A Time dEpendent Shortest path

5.1 Overview

FATES is a distributed system that consists of multiple managers - Java RMI servers at possibly different machines. An overview of the system architecture is shown in Figure 3. Note first that any paths in a network of one-dimensional lines

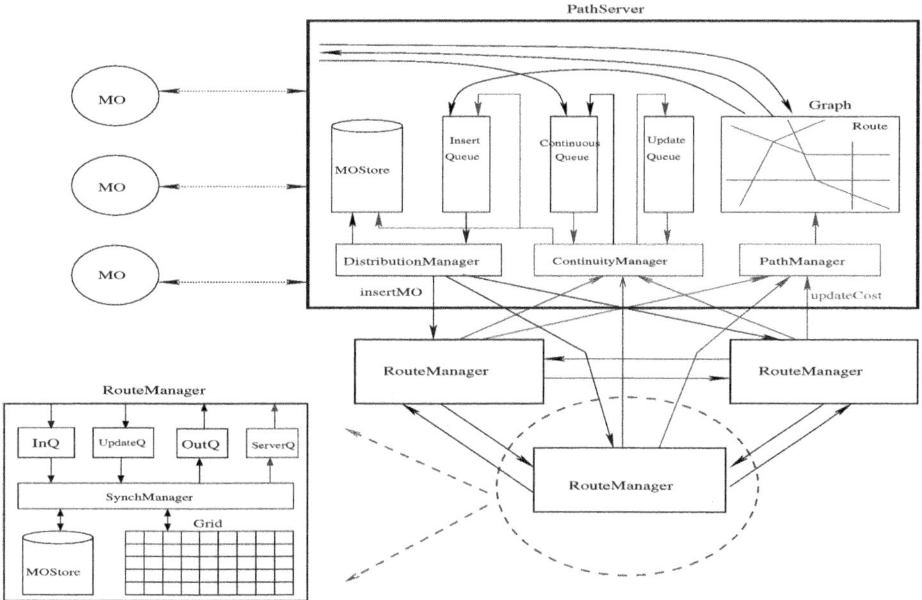

Fig. 3. FATES: System Architecture

in a two-dimensional space can be divided into subpaths, each of which belongs to a one-dimensional line. Based on this, we build a two-dimensional system as a network of one-dimensional sub-systems. The one-dimensional system developed in [5], called Route Manager, is associated with a route. Each Route Manager maintains a local view of the corresponding route. A local view is the dynamically changing cost information for sections on a route. Based on these local views, a frontend interface to moving objects, called Path Server, maintains a global view of the network as a directed graph. In this directed graph, a section on a route corresponds to an edge. When a moving object m joins FATES, the Path Server finds a time-dependent shortest path p for m, enqueues the (m, p) pair into a queue, and returns the path information p to m. The queue is constantly checked by a thread, Distribution Manager. If the queue is not empty, an element (a pair of the id of a moving object and the corresponding 2D path) is dequeued and broken into 1D subpaths. The Distribution Manager then distributes the 1D subpaths to corresponding Route Managers. For the Path Server to keep up-to-date information about the routes, each Route Manager constantly sends an updated local view to the Path Server. Then a thread in the Path Server, called Path Manager, revises the cost information in the graph using the updated local views sent by Route Managers. This way, the two-dimensional system is able to use an already developed one-dimensional system (Route Manager) and take advantage of parallelism. Note that a *ripple effect* can occur in any Route Manager after the Distribution Manager distributes 1D subpaths to the corresponding Route Managers. Once a ripple effect occurs, the moving objects affected will arrive at subsequent routes later than the routes anticipate. Hence, the Route Managers need to communicate with each other exchanging information about what moving objects will be late and by how much. In subsequent subsections, we discuss each manager in more detail.

5.2 Path Manager

To compute time-dependent shortest paths for moving objects, FATES needs to maintain global information about the severity of the congestion on the routes that the system monitors. A route in the system is represented as a series of edges in a directed graph. Each edge corresponds to a section of a route and is associated with an array of maximum speeds allowed in the corresponding section. Note that the size of the array (the number of elements) is the number of partitions of the space domain. An element of the array is the maximum speed allowed in the corresponding cell in the grid, i.e., the maximum speed allowed in a section at a given time period. The Path Manager is responsible to update such information in the graph. To do so efficiently, the Path Manager maintains an array of local views of the routes. A local view is a two-dimensional array corresponding to the maximum speeds allowed in a route, i.e., an element of the two-dimensional array corresponds to the maximum speed in a cell in the grid maintained by a Route Manager. Each Route Manager periodically sends the update information of the corresponding local view. Once a local view is updated, the Path Manager uses the information to update the cost information in the directed graph.

5.3 Distribution Manager

Once the Path Server computes and returns a time-dependent shortest path to a moving object, the path is inserted into a queue, InsertQ, along with the ID of the moving object. The Distribution Manager then dequeues the (ID, path) pairs from the InsertQ one by one, breaks them into one-dimensional subpaths, and sends insertion requests to the corresponding Route Managers. Note that it is possible for a moving object to appear in a route twice or more. For instance, suppose that a moving object m is going from a location A to a location B and there is a route R passing through A and B. However, there may be congestion somewhere between A and B and there may be a detour D around the route. Therefore m may start at A on R, take the detour D, and finish its trip at B on R. Consequently, m appears on the route R twice in separate occasions. When this happens, the Distribution Manager appends a character '\$' to the ID of the moving object to uniquely identify moving objects in a Route Manager. Besides the Path Server, the Continuity Manager could also insert a path into the InsertQ as will be explained in the next subsection.

5.4 Continuity Manager

The Continuity Manager maintains two queues: UpdateQ and ContinuousQ. The UpdateQ is for the messages from Route Managers. There are three types of messages that a Route Manager can send to the Continuity Manager. When a moving object m reaches its destination on a route, there are two possibilities: m can either go to the next route or finish its trip on this route. These events are distinguished by a flag *nextRoute*. If the flag is set to false, m finishes the trip on this route and the corresponding Route Manager sends a $FULL$ message to the Continuity Manager. Otherwise, m goes to the next route sending a $PART$ message to the Continuity Manager. Due to the ripple effects that could occur in any route, the arrival time of a moving object could be changed from what was expected in the beginning when the Path Server computed the time-dependent shortest path. Whenever there is a change in the arrival time, each Route Manager sends a message $ARRIVAL$ along with the new arrival time to the Continuity Manager. Upon receiving these messages, the Continuity Manager inserts them into the UpdateQ and processes these messages appropriately.

The other queue that the Continuity Manager manages is the ContinuousQ. When a path of a moving object is inserted into the InsertQ, the ID of the moving object is also inserted into the ContinuousQ. The Continuity Manager then periodically checks the ContinuousQ, to determine whether a moving object m has a better alternative in terms of time. If that is the case, m is removed from the Route Managers that currently contain m. Then the new path is inserted into the InsertQ, which will then be processed by the Distribution Manager in the same way as it would have been in the first place.

5.5 Route Manager

The Route Manager maintains four queues: InQ, UpdateQ, OutQ, and ServerQ as well as a two-dimensional array of cells called the Grid. When the Distribution Manager sends an insertion request to a Route Manager, the request is inserted into the InQ. The Route Manager, then, inserts moving objects in the InQ into its Grid one by one. Due to the ripple effects that could occur when the number of moving objects in a cell reaches some predefined thresholds, some moving objects' arrival time could be different from what the subsequent Route Managers or the Path Server anticipates. For this reason, whenever a ripple effect occurs, the affected moving objects' information needs to be sent to either a subsequent Route Manager or the Path Server. The OutQ holds this information for a Route Manager. The UpdateQ receives this information from other Route Managers. As pointed out in [5], since a ripple effect is a performance bottle neck, we process the UpdateQ in a way that can avoid invoking the RippleInsert method too frequently. To this end, the moving objects in the UpdateQ are removed from the Grid without invoking the RippleInsert method. Then, at the end of the process the RippleInsert method is called once. Lastly, the ServerQ is the queue that holds information bound to the Path Server. As mentioned in the previous subsection, there are three types of messages: $FULL$, $PART$, and *ARRIVAL*. A Route Manager periodically processes these queues in the order of (InQ, OutQ, UpdateQ, ServerQ) and sends its local view (the severity of congestion) to the Path Server. The local view is in the form of a two-dimensional array of maximum speeds allowed in sections of the route.

In addition to managing queues, a Route Manager is also responsible for managing time information and the grid accordingly. When a Route Manager sends its local view to the Path Server, it receives a timestamp. The Route Manager then compares it with its own timestamp and if there is an increment, it increases its timestamp. Since the number of columns (cells) in the system is finite, the amount of time covered by the grid is also limited. Therefore, we need to reuse the columns that covered the past. When the number of columns covering the present and the future becomes half the size of the entire columns, we invoke a recycling algorithm. The recycling algorithm first empties the cells in the columns to be reused and adjusts cell information for those cells to cover the future.

6 Performance Study

In this section, we first present the experimental setting and report a performance study of FATES. For the experiments, we create a network of one-dimensional lines as shown in Figure 4. The network is a simplified version of a highway map

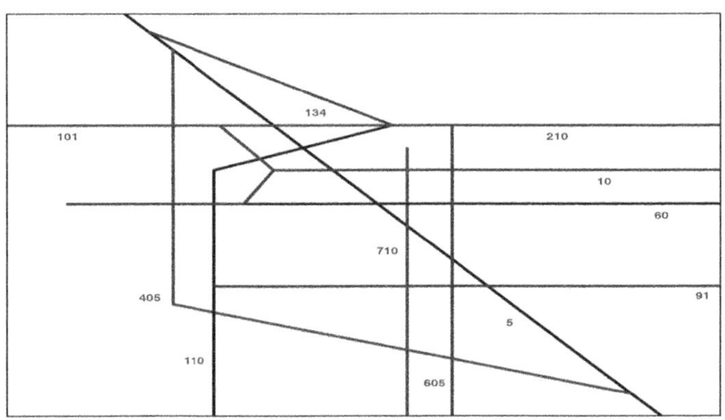

Fig. 4. A network of one-dimensional lines

of Los Angeles area and consists of 22 routes along with 102 vertices and 242 edges (the numbers in the figure refer to major route numbers in the Los Angeles area). Note that (north, south) and (east, west) pairs of routes are looked as a line, but treated separately. These vertices and edges comprise the directed graph maintained by FATES. Upon entry to the system, a moving object is responsible to provide its ID, start location, destination, start time, and initial speed to the system. These values are generated as follows: A unique ID is assigned to each moving object. For maximum speeds, 1.2 is assigned to two thirds of the moving objects and 0.9, 1.0, 1.1, and 1.3 are evenly distributed to the rest of the moving objects. Speed 1.0 can be thought of as 60 miles per hour. Start locations and destinations are restricted to be one of the vertices in the directed graph, thus we generate a random number (identifier of one of vertices) to represent a vertex.

To control the load of the system, a notion of an interarrival time is introduced. The time between the arrival of a moving object and the next moving object at a source vertex is called the interarrival time and is drawn from an exponential distribution. The average, λ, of the exponential distribution represents the severity of the load in the system. The greater the average, the more the load, hence the severer congestion. For the experiments, we vary λ from 12 to 28 increased by 4. In addition to the interarrival time, the notion of a skew rate (r) is introduced to create more congestion on a specific route than any other routes. Route 5 (south and north) is chosen to be the most congested route. By making one route more congested than others, we aim to see how the system distributes the load on a congested route to other routes. If r equals to 1, the load is distributed almost uniformly throughout the system, and if r is 2, then the load on the Route 5 is twice as much as the one on other routes. Given λ and r, five sets of data are generated, tested, and averaged. To generate the data in a skewed way, we divide the set of vertices into 4 categories: Boundary is the set of vertices that are on the boundary of routes, and this is further divided into two groups: Route 5 Boundary (R5BD) and Non Route 5 Boundary (NR5BD). Inside is the set of vertices that are not on the boundary of routes, and this is also further divided into two groups: Route 5 Inside (R5IN) and Non Route 5 Inside (NR5IN). With this categorization, moving objects are generated in the following manner: For vertices in NR5BD, moving objects are generated according to a given λ. For vertices in NR5IN, 2λ is used to generate moving objects at those vertices, therefore, approximately one half the number of the moving objects are generated in comparison to the vertices in NR5BD. Given a skew rate r, λ/r and $2\lambda/r$ are used to generate moving objects at vertices in R5BD or R5IN, respectively. Therefore, on Route 5, r times more moving objects are generated than on any other route.

For the destination distribution, we use the following policy: one half of the moving objects generated at the vertices in NR5IN, NR5BD, and R5IN are randomly assigned to inside vertices (either NR5IN or R5IN) and the other half are randomly assigned to boundary vertices (either NR5IN or R5IN). For the moving objects generated at the boundary of the Route 5 (R5BD), inside vertices are assigned as a destination to one quarter of the objects, boundary vertices other than the Route 5 boundaries are assigned to another one quarter of the objects, and finally the Route 5 boundaries (R5BD) are assigned to the rest of the generated moving objects. This way we make sure that Route 5 is more congested than any other routes.

Based on a report from Texas Transportation Institute [16], 5 maximum speed levels (2, 0.82, 0.69, 0.63, 0.58) with corresponding thresholds (11, 17,

22, 30, 34) can be deduced. However, those maximum speeds in the report are the average speeds of the vehicles observed over a long period of time. They do not necessarily reflect the dynamic situations that could happen at any given time of the day. For instance, as is often experienced, when a route becomes really congested the average speed could be down to less than 10 MPH (0.17) as opposed to 34.8 MPH (0.58). As a result and in the absence of real statistical data, we employ the following maximum velocity function in a uniform manner: 11 speed levels with maximum speeds (2, 1, 0.9, ..., 0.1) and the corresponding thresholds (10, 20, ..., 110).

We run two separate experiments for every data sets: one with the Path Manager and Continuity Manager (PMCM) and the other without the both managers (NPNC). In other words, one with the technique developed in this paper and the other without it. For fair comparisons, the generated data sets are stored in files, each of which is in the order of starting time and is read later by a thread to create a Java object and be inserted into the system. In every data set, the starting time ranges from 1 to 1200 time unit. One time unit can be thought of as a second or two. The system starts at time 1 and runs until all the inserted moving objects complete their trip. Since in the initial period of the running time the system does not have enough objects to reflect the real world situations, we only gather information of moving objects whose starting time is between 300 and 900 units, roughly speaking, we monitor the system for an hour or two. When a moving object completes its trip, its id, trip time, and trip distance are logged on a file so that they can be compared with its counter parts of the other experiment (either with or without the Path Manager and Continuity Manager).

Figure 5 shows the average trip time for all the moving objects that completed their trip between 300 and 900 time units for various skew rates. Figure 5 (a) shows the average when the program is running without the PM and CM (NPNC) and Figure 5 (b) shows the average when the program is running with the PM and CM (PMCM). As the figures show, given the same interarrival rate λ and skew rate r, PMCM yields shorter trip time than NPNC does. Furthermore, as λ increases, the trip time decreases for all cases due to less congestion. Also, as r increases, the difference between the two trip time averages widens because Route 5 becomes more and more congested, therefore the system reroutes more and more moving objects to alternatives. The percentage of reduction in trip time is shown in Figure 6 (a). The shorter average trip time does not come for free. The alternative routes that the system suggests to moving objects have longer distance than the original routes. Consequently, the average trip distance for all objects is increased. Figure 6 (b) shows the percentage of the increments.

To understand the system behavior better, we divide the moving objects into two groups: those who had benefit from using the Path Manager and Continous Manager (Winners) and those who had to pay extra time on routes (Losers). Figure 7 shows the average trip time for Winners under NPNC and PMCM. The Winners group benefits because they are rerouted from a very congested route to a less congested route paying extra trip distance. However, due to these

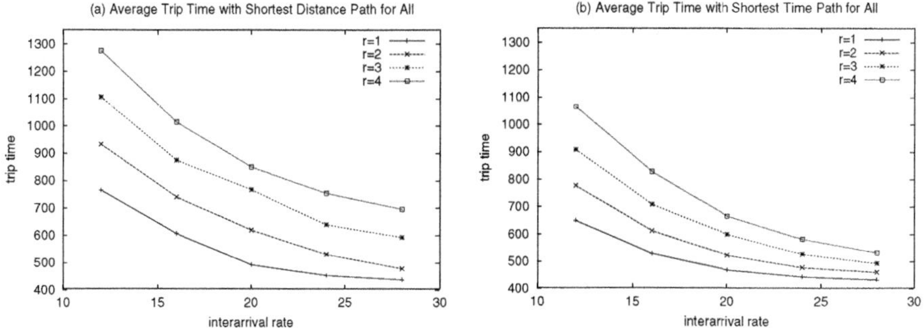

Fig. 5. Average Trip Time for All Objects

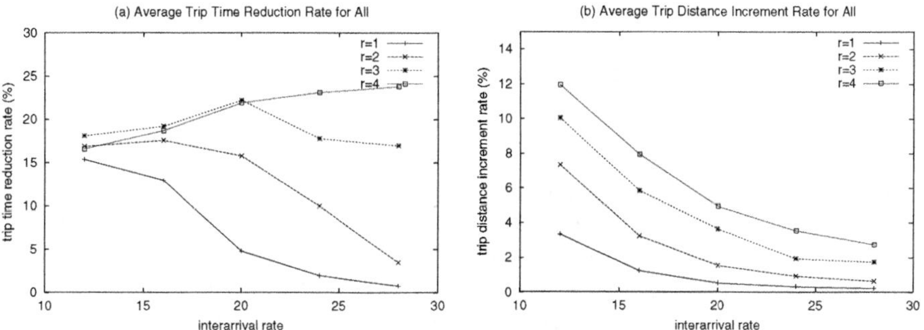

Fig. 6. Ratio for All Objects

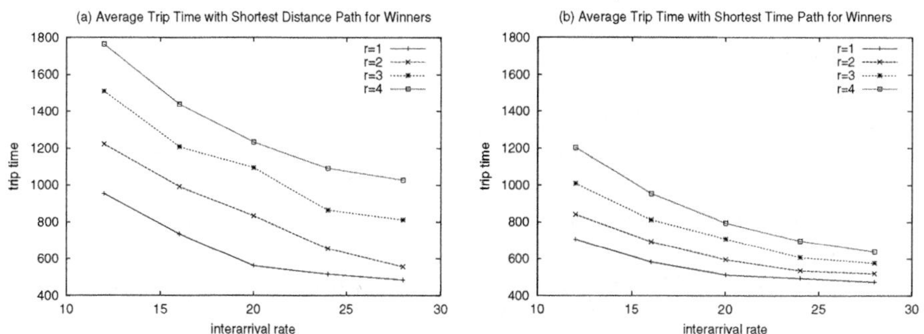

Fig. 7. Average Trip Time for Winners

rerouted moving objects, the previously uncongested or less congested routes have become more congested to some extent. As a result, the moving objects that were taking these routes were forced to slow down because more moving objects come into their routes causing more trip time. Figure 8 shows the average trip time for Losers. In Figure 9 (a), we show the number of winners and losers.

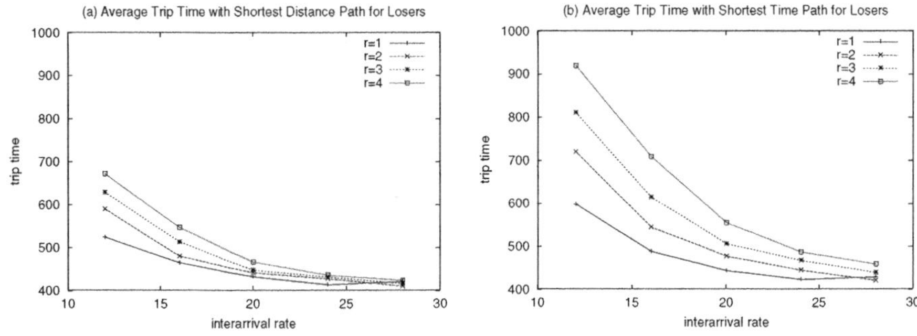

Fig. 8. Average Trip Time for Losers

Note that the sum of the winners and losers does not necessarily equal the total number of moving objects that completed their trip due to the existence of neutral objects, whose trip time did not change. In all cases, there are always more winners than losers. However, when $r = 2$ and $\lambda = 20$, losers and winners are about the same size. Nevertheless, the system yields less overall trip time (See Figure 6 (a)). The reason is that as Figure 7 and Figure 8 show, the winners are getting much more benefit than the penalties that the losers pay. Due to this reason, the standard deviation of the trip time also decreases as shown in Figure 9 (b). This deviation decrement means that everybody gets their fair share of the routes, which is another benefit of using FATES.

7 Conclusion

In this paper, we considered moving objects that are sizable physical entities following certain paths which have certain physical limitations. We have observed that the real trajectory of a moving object is the result of interactions among moving objects in the system yielding a polyline instead of a line segment. Based on a space-time grid model, we have implemented a two-dimensional extension of the grid model, FATES. FATES is a distributed system and manages the information about moving objects. We show how FATES can be used to provide time-dependent shortest paths to moving objects. The time-dependent shortest problem is of particular interest in transportation applications.

The performance study shows that given the same interarrival rate λ and skew rate r, FATES reduces the average trip time when there exists a more

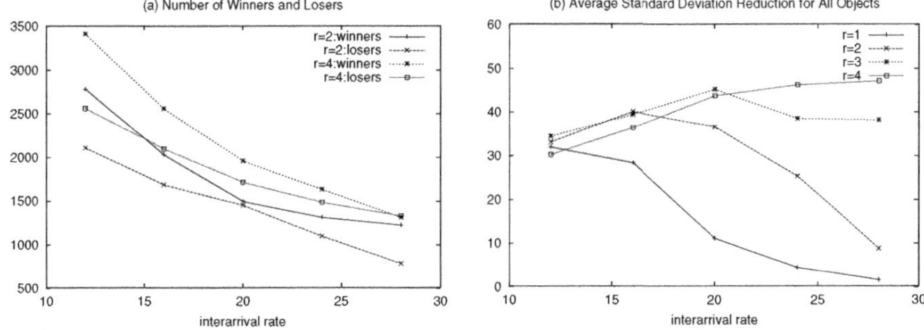

Fig. 9. Winners vs. Losers and Deviations

congested route than any other routes. Furthermore, as λ increases, the trip time decreases for all cases due to less congestion. There are, however, tradeoffs for shorter average trip time. The alternative routes that the system suggests to moving objects have longer distance than the original routes. Further analysis of the experiments shows that the winners group benefits because they are rerouted from a very congested route to a less congested route paying extra trip distance. However, due to these rerouted moving objects, the previously uncongested or less congested routes have become more congested to some extent yielding the losers group. Nevertheless, the winners benefit more than the penalties that the losers pay. The overall decrement in the standard deviation of the average trip time results in all moving objects getting their fair share of the routes.

One of the assumptions made in this paper is that the moving objects are responsible for providing their information. Although this may become true in the not-too-distant future, this assumption does not reflect the real world at the present time. Therefore, an interim solution is needed based on the fact that not 100 % of the moving objects provide their information to the server. A part of our future work is to modify FATES so that it can still provide time-dependent shortest paths based on partial information.

References

[1] R. K. Ahuja, T. L. Magnanti, and J. B. Orlin. *Network Flows: Theory, Algorithms, and Applications*. Prentice Hall, 1993.

[2] N. Beckmann, H.-P. Kriegel, R. Schneider, and B. Seeger. The R^*-Tree: An Efficient and Robust Access Method for Points and Rectangles. In *Proc. ACM SIGMOD Int. Conf. on Management of Data*, pages 322–331, 1990.

[3] CDPD Forum. Cellular Digital Packet Data System Specification, Release 1.1. Technical report, Jan. 1995.

[4] H. D. Chon, D. Agrawal, and A. El Abbadi. Storage and Retrieval of Moving Objects. In *Proceedings of the Int. Conf. on Mobile Data Management*, pages 173–184, 2001.

[5] H. D. Chon, D. Agrawal, and A. El Abbadi. Query processing for moving objects with space-time grid storage model. In *Proceedings of the Int. Conf. on Mobile Data Management*, 2002.

[6] S. Handley, P. Langley, and F. Rauscher. Learning to predict the duration of an automobile trip. In *Proceedings of the Int. Conf. on Knowledge Discovery and Data Mining*, pages 219–223, 1998.

[7] G. Kollios, D. Gunopulos, and V. J. Tsotras. On Indexing Moving Objects. In *Proceedings of ACM Symp. on Principles of Database Systems*, pages 261–272, 1999.

[8] K. Nachtigall. Time depending shortest-path problems with applications to railway networks. *European Journal of Operational Research*, 83:154–166, 1995.

[9] OnStar. http://www.onstar.com.

[10] A. Orda and R. Rom. Shortest-path and minimum-delay algorithms in networks with time-dependent edge-length. *Journal of the ACM*, 37(3):607–625, 1990.

[11] S. Pallottino and M. G. Scutella. Shortest path algorithms in transportation models: classical and innovative aspects. In *In Equilibrium and Advanced Transportation Modelling, Kluwer*, pages 245–281, 1998.

[12] C. E. Perkins. Mobile IP. *IEEE Communications Magazine*, pages 84–99, May 1997.

[13] D. Pfoser and C. S. Jensen. Capturing the Uncertainty of Moving-Object Representations. In *Proc. of the Int. Symposium on Spatial Databases, SSD*, pages 111–132, 1999.

[14] D. Pfoser, C. S. Jensen, and Y. Theodoridis. Novel Approaches to the Indexing of Moving Object Trajectories. In *Proceedings of the Int. Conf. on Very Large Data Bases*, pages 395–406, 2000.

[15] S. Saltenis, C. S. Jensen, S. T. Leutenegger, and M. A. Lopez. Indexing the Positions of Continuously Moving Objects. In *Proc. ACM SIGMOD Int. Conf. on Management of Data*, pages 331–342, 2000.

[16] D. Schrank and T. Lomax. The 2001 Urban Mobility Report. Technical report, Texas Transportation Institute, 2001.

[17] A. P. Sistla, O. Wolfson, S. Chamberlain, and S. Dao. Modeling and Querying Moving Objects. In *Proceedings of the Int. Conf. on Data Engineering*, pages 422–432, 1997.

[18] J. Tayeb, O. Ulusoy, and O. Wolfson. A Quadtree Based Dynamic Attribute Indexing Method. *The Computer Journal*, 41(3):185–200, 1998.

[19] M. Vazirgiannis and O. Wolfson. A Spatiotemporal Model and Language for Moving Objects on Road Networks. In *Int. Symposium on Spatial and Temporal Databases*, pages 20–35, 2001.

[20] O. Wolfson, B. Xu, S. Chamberlain, and L. Jiang. Moving Objects Databases: Issues and Solutions. In *Proceedings of the 10th International Conference on Scientific and Statistical Database Management*, pages 111–122, July 1998.

Search K Nearest Neighbors on Air*

Baihua Zheng[1], Wang-Chien Lee[2], and Dik Lun Lee[1]

[1] Hong Kong University of Science and Technology
Clear Water Bay, Hong Kong
{baihua,dlee}@cs.ust.hk
[2] The Penn State University, University Park, PA 16802
wlee@cse.psu.edu

Abstract. While the K-Nearest-Neighbor (KNN) problem is well studied in the traditional wired, disk-based client-server environment, it has not been tackled in a wireless broadcast environment. In this paper, the problem of organizing location dependent data and answering KNN queries on air are investigated. The linear property of wireless broadcast media and power conserving requirement of mobile devices make this problem particularly interesting and challenging. An efficient data organization, called *sorted list*, and the corresponding search algorithm are proposed and compared with the well-known spatial index, *R-Tree*. In addition, we develop an approximate search scope to guide the search at the very beginning of the search process and a learning algorithm to adapt the search scope during the search to improve energy and access efficiency. Simulation based performance evaluation is conducted to compare sorted list and R-Tree. The results show that the utilization of search scope and learning algorithm improves search efficiency of both index mechanisms significantly. While R-Tree is more power efficient when a large number of nearest neighbors is requested, the sorted list has better access efficiency and less power consumption when the number of nearest neighbors is small.

Keywords: KNN, location-dependent search, wireless broadcast, index structure, mobile computing.

1 Introduction

The advance of wireless technologies and mobile devices allows users to obtain information anywhere and anytime. Location-dependent information services (LDISs) are services that return results based on some location information. Examples of LDISs include returning local traffic reports and the nearest restaurants with respect to the user's current location.

A K-Nearest-Neighbor (KNN) query returns a specific number of data objects, sorted by their distances from a given position. The KNN problem, well

* Research supported by Research Grants Council, Hong Kong SAR, China under grant number HKUST6079/01E and AoE/E-01/99.

M.-S. Chen et al. (Eds.): MDM 2003, LNCS 2574, pp. 181–195, 2003.

studied in the traditional disk-based, client-server computing environment, represents a very important class of queries in LDISs [1, 4, 7]. With the anticipated dramatic increase of the mobile user population, the scalability of LDISs will be a major challenge. A wireless broadcast system capable of answering KNN queries is considered a promising solution because it can serve a virtually unlimited number of users within its coverage.

Although studies on wireless broadcast alone and KNN problems in the disk-based environment are well documented in the literature, this paper, to the best of our knowledge, is the first attempt to address the KNN problem in a wireless broadcast environment. The linear property of wireless broadcast media and power conserving requirement of mobile devices make the problem particularly interesting and challenging. Considering the specific features of wireless broadcast environments, a new data orgranization scheme and a new index structure are proposed, along with several related algorithms. The main contributions of this paper are three-folded: 1) proposed a simple but efficient index structure to support linear transmission of location dependent data and processing of KNN queries; 2) developed a search radius estimation function to provide approximated search scope for KNN query processing; 3) developed a learning algorithm to dynamically adapt the search scope based on objects' distribution and the ratio of requested k.

The rest of this paper is organized as follows. Section 2 provides a brief overview of related work. A search radius estimation function is introduced in Section 3 to provide an initial search scope, along with a learning algorithm devised to dynamically adapt the scope according to the real situation. Section 4 explains the detailed index structure proposed in this paper. Performance evaluation results are shown in Section 5. Finally, Section 6 concludes this paper and discusses the future work.

2 Background and Related Work

Related work on KNN search and wireless data broadcast are reviewed, respectively, in the next two sections. Then, we discuss the problem and revision of R-Tree for broadcasting in a wireless channel.

2.1 K-Nearest Neighbor Search

K-Nearest Neighbor (KNN) search returns a specified number of data objects, sorted by their distances from a given query point. KNN has been addressed mostly in the context of spatial databases, though its applications can also be found in pattern recognition, image processing, CAD, and multimedia indexing [2, 8].

With a large candidates set, answering KNN via scanning through the whole set becomes extremely expensive. Index structures and related search algorithms have been proposed to provide efficient processing of KNN queries. The main idea is to use heuristics to detect and filter unqualified paths, thus reducing the search cost. Existing algorithms for KNN queries can be divided into two categories based on how the candidates set is scanned, namely, single-step search and multi-step search.

Single-Step Search With the support of index structures, algorithms in this category search for KNN by scanning the candidates only once. There are several methods documented in the literature. Branch-and-bound algorithms, e.g., R-Tree, use heuristic distances to choose the next node for visiting and pruning. Various algorithms differ in the search order and the heuristics used to prune branches [4, 10]. Incremental algorithms report the objects one by one to allow the algorithm to operate in a pipelined fashion. They are especially suitable for complex queries involving proximity [3].

Multi-Step Search Methods in this category scan the candidates multiple times until the proper answers are obtained. Korn, et. al. proposed an adapted algorithm [8]. First, a set of k primary candidates was selected based on stored statistics to obtain the upper bound d_{max} which can guarantee that there are at least k sites within the distance d_{max} from the query point q. Next, a range query was executed on the site set to retrieve the final candidates. An extended version of this algorithm was proposed in [11], in which d_{max} was adapted every time a candidate object was checked.

2.2 Wireless Data Broadcast

Generally speaking, there are two approaches for mobile access of location-dependent data. For a **on-demand access** mode, the server locates the appropriate data and then returns the answer according to the query submitted by a client, in which data is transferred in a point-to-point connection. For a **broadcast** mode, server broadcasts information in the wireless channels periodically and the client is responsible for filtering its desirable data.

Compared to on-demand access, a major advantage of broadcast is that it allows simultaneous access by an arbitrary number of mobile users. Disseminating information via broadcast is a very efficient and scalable method for a large client population. LDISs must anticipate demands from a large number of mobile users. It is envisaged that many LDISs such as region-wide tourism information will utilize broadcast for the dissemination of information to the rapidly increasing population of mobile users. The focus of this paper is to investigate the feasibility of answering KNN queries on a wireless data broadcast environment.

Power conservation is a critical issue for mobile devices. In the wireless broadcast environment, power consumption can be reduced by interleaving auxiliary index with data [6]. By looking up the index, mobile devices are able to anticipate the arrivals of the desired data and stay in the doze mode until the data of interest arrives. To interleave the index and the data on the broadcast channel, we employ the $(1, m)$ interleaving technique [6]. That is, the whole index is replicated m times and broadcast before every $\frac{1}{m}$ fraction of the broadcast cycle. *Index search time* is frequently used to evaluate the effectiveness, in terms of power consumption, of an index and data organization method for broadcast channel. Thus, we use it along with *access time* as metrics for performance evaluation.

In wireless communication, the data stream is normally delivered in *packets* (or *frames*), to facilitate error-detecting, error-correction, and synchronization. Thus, both index search time and access latency are measured in terms of *number of packet accesses* [5, 6]. The drawback of air indexing is that broadcast cycles are lengthened due to the additional indexing information. Hence, the index size should be kept as small as possible.

2.3 Spatial Index on Air?

Indexes for KNN search have been well studied, however, sequential access feature of broadcast systems introduces new challenges. Original indexes are designed for random access storage (e.g., disks), therefore they may not work well in a wireless broadcast environment. Figure 1 and Figure 2 illustrate an example. Assume that a query looking for the nearest neighbor of point q_2. The query can be answered by scanning a R-Tree (see Figure 1) in the following order: first visits root node, then R_2 (which contains object o_2 and o_4), and finally R_1 (which contains object o_1 and o_3). This works well for random accessed disk index but renders a problem for air index: when a mobile device tries to retrieve R_1 after visiting R_2, it has to wait until the next broadcast cycle when R_1 is broadcast again (denoted by the second arc in Figure 2).

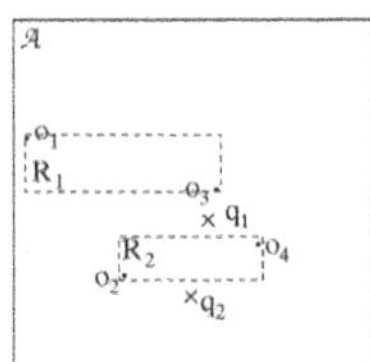

Fig. 1. Example MBR Structures

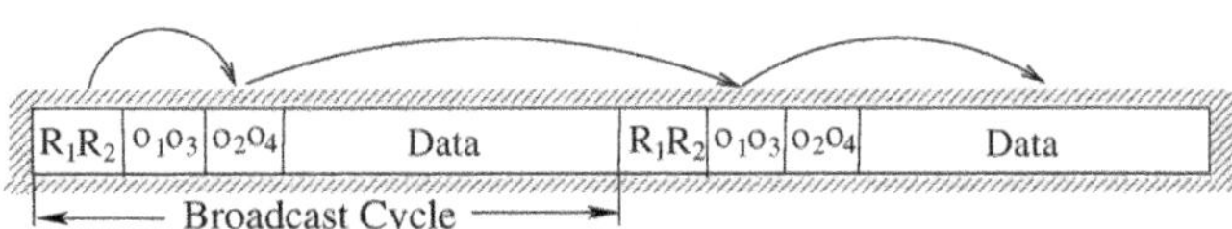

Fig. 2. Sequential Access In Wireless Broadcasts

As observed in the above example, single-step linear scan is a major requirement for any efficient algorithm in wireless broadcast environments. The R-Tree is broadcast in a depth-first order. To perform KNN search on broadcasted R-Tree, a mobile device maintains a buffer for the top-k objects. It sequentially traverses the R-Tree from the root and compares its maximal distance to objects in the current buffer, M, with its minimal distance to the next branch to visit, m. If $m > M$, then the incoming nodes bounded with this unqualified branch can be pruned (by turning the device into doze mode and wake up only when the subsequent branch arrives). Otherwise, the device stays awake and recursively traverses the branches. Since it has to pass through the whole index in order to access the searched objects, the index size has a great impact on access time.

3 Approximate Search Scope

Since requested parameter k, compared to the number of whole candidates set, is really small, some guidence exists to remove the impossible objects. An intuitive guidance is the distance of the kth nearest neighbor to the query point q that defines the radius of the necessary search space, named $radius_{needed}$. Since $radius_{needed}$ is dependent on both the position of the query and the parameter k, it is impossible to provide an exact value. In this work, a search radius estimation function is devised to approximate the search boundary within which the top k objects to a given query point are expected to be obtained. Based on that boundary, some unqualified candidates are detected at the very beginning of the search process. Although it is not guaranteed to be 100% accurate, the accuracy is in the limit approach 100%.

The objective of a search guidance is to facilitate intelligent filtering by trying to fix a search scope as early as possible so that objects outside the search scope will be disqualified. By assuming a uniform object distribution, a search radius approximation function is given in Eq. (1). Due to the space limitation, its detailed derivation is omitted and the complete proof can be found in [12]. Here, n is the number of objects, k is the requested number and c is the control constant factor to increase flexibility considering different paramters k and objects distributions.

$$r = c \times \ln(n) \times \sqrt{\frac{k}{(\pi \times n)}} \tag{1}$$

R-Tree can be modified with the help of our search radius estimation function to further reduce unnecessary traversal. No matter where the query is, only the MBRs within the expected search radius are accessed. Liu et al. proposed a similar approach in which one or more window queries were used to retrieve the KNN objects of a query point [9]. Two estimation methods were provided, using statistical knowledge based on density and bucket. However, their objective is to satisfy the KNN problem by applying 3 or 4 windows queries. This kind of multiple-scan introduces long tuning time and also a large access latency. Our algorithm tries to provide a relatively more accurate estimation in order to satisfy the request.

3.1 Learning Algorithm

In Equation (1), a constant c is introduced to increase the flexibility of the radius approximation. For a large value of k and n, the constant should be smaller than the small of k and n by intuition. Besides, the search radius is approximated based on the assumption that the objects are uniform distributed. Consequently, it is not surprising that the expected search radius is not accurate when the distribution is skewed. Parameter c in Equation (1) services for this purpose, and a learning algorithm is introduced to adapt it according to real situations dynamically.

Originally, the server assigns a static value for constant c and all the clients use the same default configuration. In broadcast environments, it is difficult for the server to obtain the feedback information about the accuracy of the approximated search radius. Consequently, the clients should adapt the setting of c based on their situations. The detailed description of the algorithm is provided in Algorithm 1.

Algorithm 1 Learning Algorithm of Constant Assignment

Input: requested k, SAT_k, objects number n, $CONS_k$;
Procedure:

1: **if** n >= k **then**
2: $SAT_k.num++$; $SAT_k.value+=\frac{n}{k}$;
3: **if** $SAT_k.num ==$ Num **then**
4: $CONS_k=(1-\alpha)\times CONS_k+\alpha\times CONS_k\times SAT_k.num/SAT_k.value$;
5: $SAT_k.num=0$; $SAT_k.value=0$;
6: **end if**
7: **else**
8: $cur\_accu=\frac{n}{k}$;
9: **if** $n==0$ **then**
10: $cur\_accu=0.01$;
11: **end if**
12: $CONS_k=(1-\alpha)\times CONS_k+\alpha\times(CONS_k/cur\_accu)$;
13: $SAT_k.num=0$; $SAT_k.value=0$;
14: **end if**

The basic idea is to adapt the search scope in accordance with the degree of contentment. For any specific k, if the number of objects currently returned is smaller than the requested one, the search scope should be increased accordingly. Here, we use an idea similar to the analogous aging function to obtain the access probability which is shown in line 12, with *cur_accu* denoting the current accuracy. If the search scope always contains adequate number of objects for some specific k, the search scope could be reduced. In our algorithm, a client can use an accumulator to keep the frequency that its requests are satisfied, which is reset to zero once its accuracy is not 100%. Also, the *satisfied degree*, which is defined as the ratio of the number of objects in the candidate set to the requested number k, is maintained. Once the client obtains Num times of accurate answers continuously, it means the current setting is stable enough and can be reduced correspondingly. Here, Num is predefined. Lines $2-5$ show the detailed action employed to reduce the sufficient search radius. Parameter α is the constant factor to weight the importance of the current setting of c.

4 A New Air Index for KNN Search

Based on the guidance of search radius, a simple index structure is devised. In the following, we explain the details.

4.1 Sorted List

In wireless environments, the dimension is usually low, i.e., two or three in accordance with the real world situation. Hence, sorted in each dimension, the objects can be represented by two or three sorted lists, each corresponding to one dimension. The following description is based on a two-dimensional space which is easy to be extended to a three dimensional space.

Represented by two lists, the objects in the first one are sorted in x-dimension. Each object is represented by its x-coordinate and the pointer pointing to its position in the second sorted list. In the second list, each object's y-coordinate and the related pointer pointing to the real data are recorded. In other words, the coordinates and pointer information together provide sufficient information to access the objects.

Given a query point q and the corresponding radius r, the objects to be examined should satisfy the condition: $x \in [q.x-r, q.x+r]$ and $y \in [q.y-r, q.y+r]$, where $(q.x,\ q.y)$ denotes the coordinates of the query point. By listening to the channel, a client can detect two sets of possible candidates that satisfy the above conditions and their intersection provides the candidate answer set. Then the top k objects are returned. In case there are only k' ($k' < k$) objects in the candidate set, this query is not satisfied and the ratio of k' to k is defined as the accuracy of the corresponding search radius. Although the accuracy of this algorithm is not guaranteed to be always 100%, no false answer is returned and the returned k' objects are guaranteed to be the top-most k' nearest neighbors.

4.2 Packing the Sorted List

In wireless environments, information are transmitted to the clients in the unit of *packet*.[1] Therefore, all the data has to be packed into packets.

Considering the broadcast of a sorted list, there are two kinds of information. In the lower level, the packets contain the objects position information and the related pointers to the data packets containing the real data of the objects. In the upper level, the packets contain the index information of the lower level packets for detecting the packets needed for query processing. Given a query

[1] Similar to the concept of *page* in traditional databases.

point q and a search radius r, $[q.x - r, q.x + r]$ defines the x-dimensional scope within which the objects should be checked. With the help of the upper level index, the packets containing the objects whose x-coordinates are within this scope can be obtained. Similar action is done for the y-dimensional sorted list. After downloading the upper level packets, the top k objects can be detected by examining the objects coordinates. Then the lower level packets provide the pointers to the packets containing the real data information of those objects.

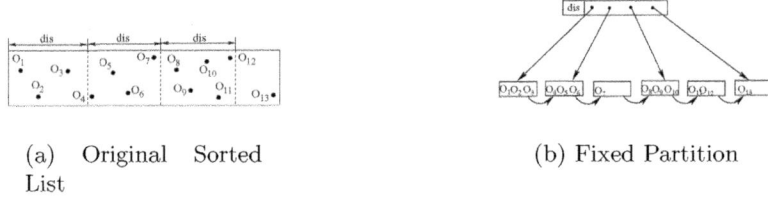

(a) Original Sorted List (b) Fixed Partition

Fig. 3. Fixed-Partition Sorted List for the Running Example

To pack the sorted list for broadcast, two different schemes, namely *fixed-partition sorted list* and *full-occupancy sorted list* are proposed, respectively. The first one partitions the sorted list by a fixed distance, while the other one tries to maximize usage of the packet capacity. In [12], a detailed theoretical analysis is provided to estimate the index search time of these paging schemes. It is omitted here due to space limitation.

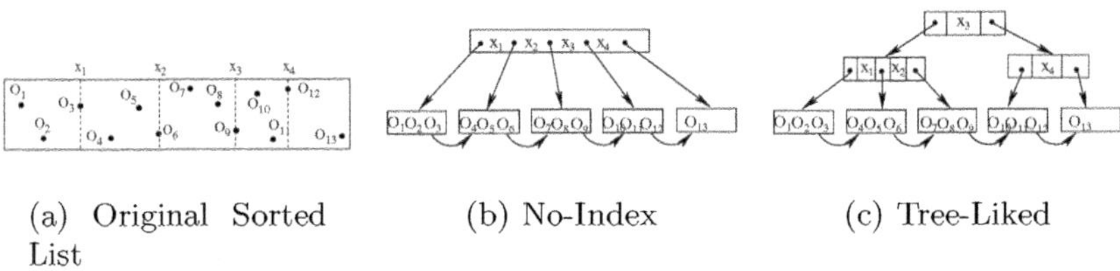

(a) Original Sorted List (b) No-Index (c) Tree-Liked

Fig. 4. Full-Occupancy Sorted List for the Running Example

Fixed-Partition. As illustrated in Figure 3, the whole sorted list is divided by a fixed distance, *dis*. The fan-out of a packet is three for the example. There are several distance chosen metrics, such as the capacity of the packet. In this paper, a simple algorithm is used to choose the suitable fixed partition distance. Based on the capacity of a packet, the number of packets, denoted as num, required for packing the objects can be computed. The basic distance, denoted as b_dis, can be obtained by partitioning the whole list into num parts, i.e., $b\_dis = width(height)/num$. Based on b_dis, a possible range of dis (e.g., $[b\_dis/4, 4b\_dis]$ for our simulation) can be decided. With the help of the theoretical analysis of the index search time, greedy search can be employed to check the possible distance within the range. Thus, the distance that produces the index having the best search performance is obtained.

The advantage of this partition is to simplify the upper-level index. A client can locate the packets by hashing the search scope, i.e., $\lfloor (q.x - r)/dis \rfloor$ and $\lceil (q.x + r)/dis \rceil$, into the lower-level packets. However, due to the fixed partition distance and the packet's capacity, it is not unusual to have many packets with a low utilization rate. As shown in Figure 3, some packets only contain one object.

Full-Occupancy. This partition tries to maximize the utilization of the lower-level packets, enabling the packet to contain the maximal objects information, i.e., the packets in lower level are full-occupied. For the upper-level index, there are two alternatives. One is to store the whole information about the lower-level packets and no index is provided. The clients have to read all the upper-level index to locate the lower-level packets needed for filtering. The other one is to build a tree-liked index. Given the packet capacity, the number of lower-level packets can be obtained. Therefore, the search time in the upper-level index can also be approximated. The one providing more efficient search time is employed. Figure 4 shows an example, assuming each packet contains three objects. The advantage of full-occupancy sorted list is the high utilization of lower-level packets.

4.3 Discussion

For every deletion and insertion of objects, the corresponding index has to be updated. Since sorted list is used to represent the objects, the updated objects should be inserted or deleted from the lists. Considering *full-occupancy* sorted list, the lower-level packing is processed according to the packet capacity. For the *fixed-partition* sorted list, the fixed distance used to partition the objects in the lists needs not be changed when the number of updated objects is small compared to the total number of objects. However, its performance is affected and the re-construction of the index should be done later. The time for carrying out the re-construction of index information should be determined based on update cost and requirements of the applications.

5 Performance Evaluation

Performance evaluation has been conducted to compare sorted list with the air version of R-Tree. Two datasets are used in the evaluation. In the first dataset (UNIFORM), we uniformly generate 5,000 points in a square Euclidean space. In the second dataset (SKEW), a skewed object distribution is generated as follows. First, the square space is equally divided into 5×5 subspaces. Then, we generate 10,000 objects, with the probabilities of falling in the subspaces follow a Zipf distribution and the skewness parameter is set to 1.2.

Based on our discussion in Section 2, R-Tree is revised for air indexing. Due to the linear property of wireless broadcast, the nodes of R-Tree are sequentially accessed while unqualified branches are pruned using the distance heuristics. In the following, *R-Tree (AirIndex)* denotes the revised scheme to cater for air indexing, which also combines the guidance provided by the search radius. While

R-Tree denotes the original algorithm devised for disk index, without any modification.

The system parameters are set as follows. In each packet, two bytes are allocated for the packet id. Two bytes are used for a pointer and four bytes are for a co-ordinate. The packet capacity is varied from 64 bytes to 2048 bytes. Queries are produced randomly in the search space, and the results are obtained by the final statistics of about 1,000,000 queries. For the parameter of k, three different settings are provided, $k = 1$, $k \in [2,4]$ and $k \in [21,30]$. Due to the space limitation, only partial results are depicted. Observation from the rest of results will be summarized in text.

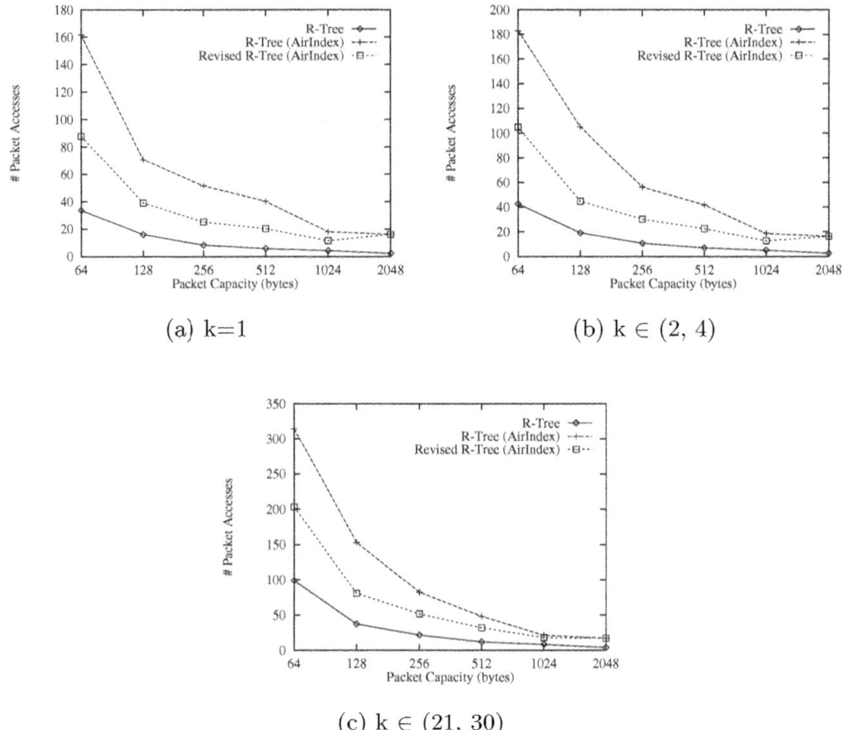

(a) k=1

(b) k ∈ (2, 4)

(c) k ∈ (21, 30)

Fig. 5. Improvement Brought by Search Boundary of UNIFORM Dataset

5.1 Improvement Brought by the Search Scopes

Search scope is shown to be useful in pruning the search space and thus improving performance. Some experiments are done to study the accuracy of the search scope estimation and the accuracy is close to 100% as the value of c increases. Parameter c is set to 0.5 by default in the later simulations. In the following, we show that it is also useful for branch-and-bound algorithms. A revised R-Tree which adopts the search scope is compared to the original R-Tree.

As shown in Figure 5, the number of packets accessed in air-index is extended distinctively, compared to the performance in disk-index environments. The reason is the sequential access property of air index. When applying the approximated search scope, the performance is improved significantly. Considering the UNIFORM data, the improvement achieved by the revised algorithm over the original algorithm is 37.9%, 37.3% and 28.0% for the three different situations, respectively. For the SKEW dataset, the improvement is about 19.3%, 19.8% and 19.2%, respectively. Of course, the revised algorithm sometimes cannot satisfy a KNN request, while the original algorithm always returns the exact answers. However, for the simulations shown, the accuracy is 100% in all cases since the requested k is relatively small compared to the total number of objects.

In summary, the performance of R-Tree is decreased by the specific properties of air indexing. Combining the search scope guidance, the tuning time is significantly improved.

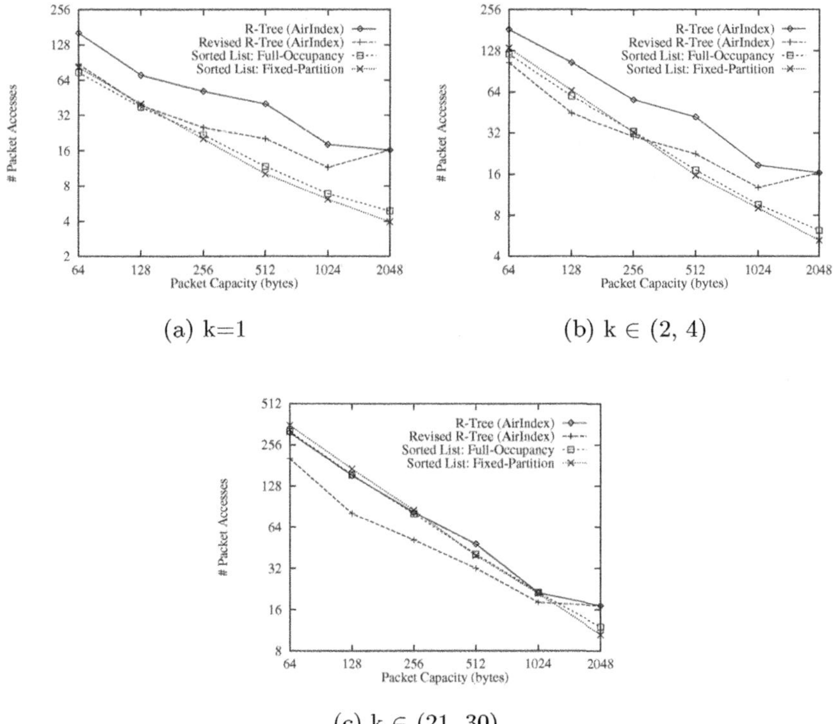

(a) k=1

(b) k ∈ (2, 4)

(c) k ∈ (21, 30)

Fig. 6. Index Search Time for UNIFORM Dataset

5.2 Performance of Sorted List Index

In this section, we compare the performance of sorted list and R-Tree air indexes, in terms of average number of packet accesses, and access latency. Since we are mainly interested in supporting KNN search in broadcast, the performance of the original branch-and-bound algorithm in traditional disk index environments is omitted. In the following simulations, the accuracy of the indexing schemes is 100% accurate, according to the simulation results.

As discussed before, in wireless broadcast, improving the index search time saves power and connection cost. Figure 6 shows the index search time for the various indexes of the UNIFORM dataset. From the result, we found that the sorted list indexes improve performance when the value of k is relatively small compared to the population of objects. When k is set to one in the UNIFORM dataset, the *Fixed Sorted List* outperforms *Revised R-Tree* for about 33.7% in average. When k is increased to between two and four, the performance improvement only occurs for the large packet capacity. When the packet capacity is smaller than 256, the performance of the *Fixed Sorted List* is worse than the revised R-Tree for about 27.5%. When the packet capacity becomes larger, the *Fixed Sorted List* can provide a better performance, about 42.5% better than the one based on R-Tree. Considering the SKEW dataset, the improvement is more significant. It is 43.0% when k is one and 22.1% when k is between two and four. However, when k becomes larger, the revised algorithm based on R-Tree works best, about 29.0% better than the *Fixed Sorted List* one.

Considering two different packing schemes of sorted lists, their performance is decided by the object distribution. For UNIFORM distribution, *fixed-partition* can provide better service in most situations. While the distribution is clustered, its performance is reduced. For the SKEW dataset, *full-occupancy* scheme outperforms *fixed-partition* scheme by about 4.5%, 9.1% and 11.0% for the three different settings of k.

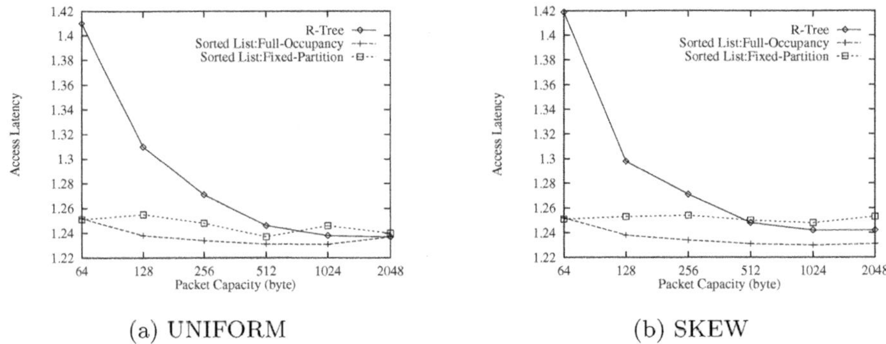

(a) UNIFORM (b) SKEW

Fig. 7. Access Latency for Two Datasets

By observing the simulation results, the performance of sorted list indexes is sensitive to the ratio of k over the total number of objects. When the ratio is really small, *Sorted List* can provide better performance than the existing ones. When the ratio becomes larger, the approximated search radius is also larger. Therefore, more objects have to be checked and the search radius somehow loses its power of providing accurate guidance. Since it only considers the distance in one dimension, the larger the search scope is, the more the false objects checked. Consequently, the performance becomes worse. However, considering real applications in mobile computing environments, the clients usually have interest in a small k. Therefore, the sorted list index structure is expected to be the first choice in many situations, in terms of efficient tuning time, along with a much smaller variance.

The access latency is affected by the index size. The larger the index size, the longer the access latency. Figure 7 shows the access latency for the index methods. The optimal value of m depends on the index size and is calculated for each index structure separately based on the technique presented in [6]. In the figures, the latency is normalized by the expected access latency without any index (i.e., half of the time needed to broadcast the database). It is obvious that R-Tree always performs the worst due to its large index size overhead. Thus, it is safe to conclude that the index overhead for the sorted list indexes is maintained at an acceptable level, and only introduces a limited variation to the access latency.

Comparing the two packing schemes for the sorted list index, *fixed-partition* incurs more size overhead for nearly all the settings due to the poor utilization ratio of packets. Therefore its expected access latency is also larger.

5.3 Performance of Learning Algorithm

The impact of *learning algorithm* on the performance is examined in this section. In our simulation, the parameter α is set to 0.10, 0.15 and 0.25 for $k = 1$, $k \in [2, 4]$ and $k \in [21, 30]$, respectively. The reason for different settings of α is motivated by the intuition from different values of k. We do not consider the relationship between the value of α, k, and n, which deserves further studies. Figure 8 shows the simulation result. The related accuracy result of the simulation results has not been shown, while it is almost 100% for the first two settings of k and about 99% for the last one.

From the simulation results, fixing the value of c is not suitable for different values of k. Thus, dynamic assignment of c is more reasonable and improved the performance significantly. There are several observations obtained. First, the improvement is reduced for the SKEW dataset, compared to the UNIFORM one, since the statistical knowledge, derived based on the uniform distribution assumption, can not provide accurate guidance. Second, the improvement becomes more distinct as k becomes larger. The reason is that originally we assign the same value to c, while in real case the larger k is, the smaller c value is needed.

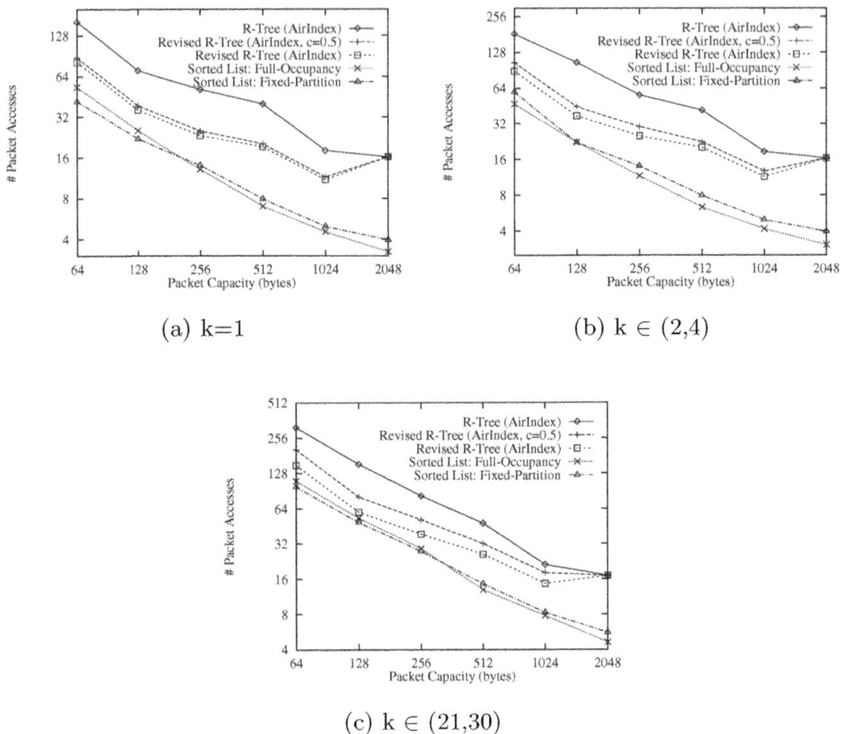

Fig. 8. Performance Comparison of Learning Algorithm for UNIFORM Dataset

6 Conclusion

KNN search is a very important and practical application in mobile computing. Although it has been well studied in the traditional disk-based, client-server computing environments, more research is needed for the wireless broadcast platform. In this paper, we address the issues involved with organizing location dependent data and answering KNN queries on air.

To reduce the search space, a guidance is provided based on the approximation of the necessary search scope for any given k. With the guidance, existing algorithms can be revised to satisfy the specific properties of air indexing. A simple index structure, *sorted list*, is proposed to take advantage of the search guidance. Two packing algorithms for the index are also proposed. Besides, we use a learning algorithm to provide dynamic adaption of the search radius, according to the real situation.

Performance evaluation shows that the search scope can provide valuable guidance for processing KNN queries. Applying the search scope to R-Tree also reduces its search time and variance. For sorted list, it improves the performance in terms of access latency, page accesses and also the variance significantly, especially when the ratio of k to the total number of objects is relatively small. When the ratio becomes larger, the strength of sorted lists is reduced.

As for future work, we plan to continue the study of KNN on air problem since currently the search scope can only provide suitable guidance but not guarantee a return of exact k objects. We also plan to develop new solutions that take into account various object distributions to provide efficient answers for KNN queries.

References

[1] S. Chaudhuri and L. Gravano. Evaluating top-k selection queries. In *Proceedings of the 25th International Conference on Very Large Data Bases (VLDB'99)*, pages 397–410, 1999.

[2] H. Ferhatosmanoglu, E. Tuncel, D. Agrawal, and A. E. Abbadi. Approximate nearest neighbor searching in multimedia databases. In *Proceedings of the 17th IEEE International Conference on Data Engineering (ICDE'01)*, April 2001.

[3] G. R. Hjaltason and H. Samet. Ranking in spatial databases. In *Proceedings of the 4th International Symposium on Advances in Spatial Databases (SSD'95)*, pages 83–95, 1995.

[4] Gí. R. Hjaltason and H. Samet. Distance browsing in spatial databases. *ACM Transactions on Database Systems*, 24(2):265–318, 1999.

[5] Q. L. Hu, W.-C. Lee, and D. L. Lee. Power conservative multi-attribute queries on data broadcast. In *Proceedings of the 16th International Conference on Data Engineering (ICDE'2000)*, pages 157–166, San Diego, CA, USA, February 2000.

[6] T. Imielinski, S. Viswanathan, and B. R. Badrinath. Data on air - organization and access. *IEEE Transactions on Knowledge and Data Engineering (TKDE)*, 9(3), May-June 1997.

[7] N. Katayama and S. Satoh. The SR-tree: An index structure for high-dimensional nearest neighbor queries. In *Proceedings of ACM SIGMOD International Conference on Management of Data*, Tucson, AZ, May 1997.

[8] F. Korn, N. Sidiropoulos, C. Faloutsos, E. Siegel, and Z. Protopapas. Fast nearest neighbor search in medical image databases. In *Proceedings of the 22th International Conference on Very Large Data Bases (VLDB'96)*, pages 215–226, 1996.

[9] D-Z. Liu, E. Lim, and W. Ng. Efficient k nearest neighbor queries on remote spatial databases using range estimation. In *Proceedings of the 14th International Conference on Scientific and Statistical Database Management(SSDBM'02)*, Edinburgh, Scotland, 2002.

[10] N. Roussopoulos, S. Kelley, and F. Vincent. Nearest neighbor queries. In *Proceedings of the 1995 ACM SIGMOD International Conference on Management of Data (Sigmod'95)*, pages 71–79, May 1995.

[11] T. Seidl and H. Kriegel. Optimal multi-step k-nearest neighbor search. In *Proceedings of the 1998 ACM SIGMOD International Conference on Management of Data (Sigmod'98)*, pages 154–165, July 1998.

[12] B. Zheng, W. C. Lee, and D. L. Lee. K-nearest neighbor queries in wireless broadcasting environments. Technical report, Dept. of Computer Science, Hong Kong Univ. of Science and Technology, July. 2002.

Adaptive Location Management in Mobile Environments

Ratul kr. Majumdar[1], Krithi Ramamritham[1], and Ming Xiong[2]

[1] Laboratory for Intelligent Internet Research
Department of Computer Science and Engineering.
Indian Institute of Technology Bombay
Mumbai, India 400076
{ratul,krithi}@cse.iitb.ac.in
[2] Lucent Bell Laboratories
Murray Hill, NJ 07974
xiong@research.bell-labs.com

Abstract. Location management in mobile environments consists of two major operations: location update and paging. The more up-to-date the location information, the less paging becomes necessary and vice versa. The conventional approach is the location area based approach (LA-based approach), where a location area (LA) consists of multiple cells. When the mobile station (MS) enters a new location area, the MS immediately updates its location information at the new location's visitor location register (VLR) and this update is propagated to the MS's home location register (HLR). The major drawback of the LA-based approach is that it does not consider any mobility patterns, or call arrival patterns. Moreover, MS updates frequently when it roams only within the boundary cells of different location areas, resulting in unnecessary location updates. So, there is a need for an efficient algorithm which can eliminate the drawbacks of the LA-based approach. To this end, an adaptive location management algorithm is described in this paper: an MS dynamically determines whether or not to update when it moves to a new LA, so that each location update becomes a necessary location update, i.e., in each updated location area at least one call is made. We have conducted experiments to capture the effect of mobility and call arrival patterns on the new location update strategy. We have also tested our algorithm with SUMATRA (Stanford University Mobile Activity TRAces), which has been validated against real data on call and mobility traces. Experimental results show that our adaptive location management algorithm considerably reduces the location management cost, by avoiding unnecessary location updates.

1 Introduction

Location management in a personal communication services (PCS) network environment deals with the problem of tracking mobile users. A cellular wireless service area is composed of a collection of cells, each serviced by a base station (BS). A number of base stations are wired to a *mobile switching center* (MSC). The backbone of a PCS network consists of all these BSs, MSCs and the existing wired network (such as PSTN, ISDN etc.) between them. This MSC acts as a gateway for the base stations.

M.-S. Chen et al. (Eds.): MDM 2003, LNCS 2574, pp. 196–211, 2003.

In order to route an incoming call to a PCS user, his (or her) whereabouts in the cellular network needs to be determined within a fixed time, which may be specified as a QoS (quality of service). Location management as a whole involves 2 kinds of activities. An Mobile Station (MS) can report his presence as soon as it enters a new location area. This activity, initiated by the MS to limit the search space is called a *location update*. The system can initiate the search for an MS by paging, i.e., simultaneously polling all the cells where the MS can be possibly found, when a new call is made for the MS.

Both paging and location update consume scarce resources like wireless bandwidth, and each has a significant cost associated with it. The mechanism of paging involves the MSC sending a page message over a forward control channel to all the cells where the MS is likely to be present. The MS listens to the page message and sends a response message back over a reverse control channel. The location update mechanism involves the MS sending a message regarding its current location over the reverse control channel, which helps in limiting the search to route a call to the MS.

1.1 Location Update Strategy in the LA-based Approach

In this section we will discuss the location update strategy, focusing on the different entities of Global System for Mobile Communication (GSM) systems.

Elements and Architecture of GSM The area covered by the mobile radio network is divided into zones, which in turn, are broken into cells, each supported by a Base Transceiver Station (BTS). Several BTSs form what is termed a location area (LA). The GSM network consists of three basic subsystems:

- Mobile Station (MS)
- Base Station Subsystem (BSS)
- Network Subsystems (NSS)

The BSS is further split into base transceiver stations (BTSs) and base station controllers (BSC). In the NSS, two main elements can be identified, mobile switching center (MSC) and location registers (HLRs and VLRs).

Figure 1 shows the main entities of GSM systems. A brief description of functions carried out by the different elements in the GSM network is given below.

BTS: This unit comprises the radio transceivers that make the link with the MS possible. A BTS includes a number of transceivers (TRX) which serve a single cell (omni-directional coverage) or several cells (sectorized coverage).

BSC: The element is responsible for most of the control functions related to a set of base stations. The BSC manages the radio sources of the system, assigning channels, deciding on hand-overs, etc.

MSC: It performs functions similar to those of a conventional telephone exchange, such as call routing. It also acts as an interface with the Public Switched Telephone Network (PSTN) or Integrated Services Digital Network (ISDN). MSC also acts as a bridge between BSCs for call hand-overs. In addition, the MSC is the main element involved in the mobility management of MSs. It also gives access to/from fixed telephone networks. To carry out mobility management and routing functions, MSCs are complemented with the two databases (registers) HLR and VLR.

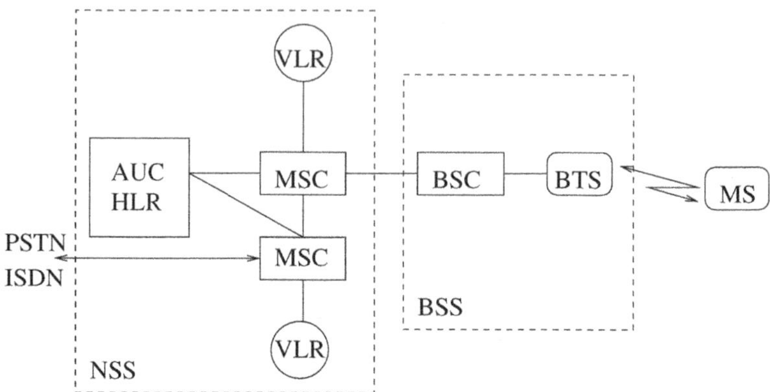

Fig. 1. Main Entities in GSM

HLR: It contains information about the network subscribers and the region in which they are located. This location register is updated when the MS is identified in the coverage of a different MSC. The authentication center (AUC) associated with a HLR contains the required security data for the user authentication process.

VLR: It contains information on all MSs located within its associated LA, so that these MSs may be paged and call may be routed towards their actual position in the network. Information about a visiting MS is deleted from the database when the MS moves to a new LA.

Location Update Location update is initiated whenever one of the following events occurs.

- When an MS moves from an LA to another LA.
- At the time of connection/disconnection (ATTACH/DETACH).

BSS at regular intervals broadcasts the ID of its LA. An MS verifies its Location Area ID (LAI) periodically. In the event of an change, a location update is initiated. If the MS is turned off, it sends a "DETACH" message. When it is switched on again, it sends an "ATTACH" message.

Location update is initiated by the MS when, by monitoring Broadcast Control Channel, it notices that the LAI broadcast is not the same as that previously stored in the MS's memory. An update request and the International Mobile Subscriber Identity (IMSI) or previous Temporary Mobile Subscriber Identity (TMSI) is sent to the new VLR via the new MSC. A Mobile Station Roaming Number (MSRN) is allocated and sent to the MS's HLR (which keeps the most current location) by the new VLR.The MSRN helps to route the call to the new VLR and subsequently translated to the TMSI of the MS. The HLR sends back a cancellation message to the previously visited VLR, so that the previous MSRN can be reallocated. Finally, a new TMSI is allocated by HLR and sent to the MS, to identify it in future paging or call initiation requests.

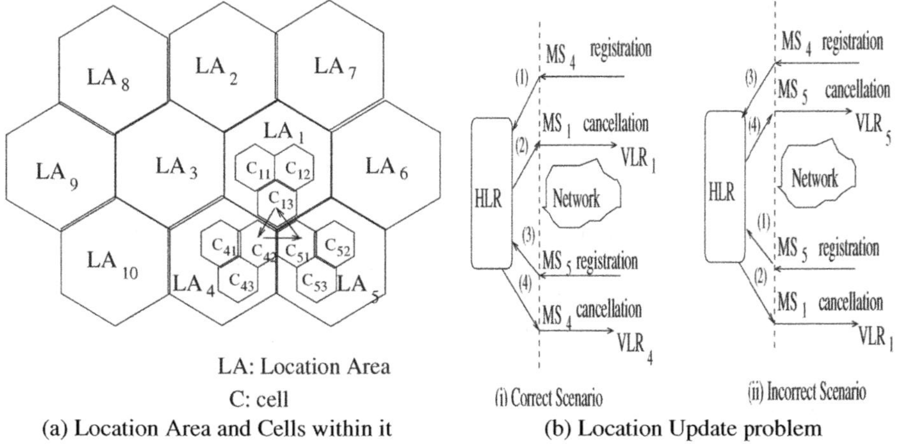

(a) Location Area and Cells within it (b) Location Update problem

Fig. 2. A Sample Mobility Scenario

1.2 Why Do We Need an Alternative to the Conventional Approach?

Consider Figure 2(a). Let us assume an MS starts roaming from location area LA_1 to location area LA_4 and then to location area LA_5. An MS will receive a call if HLR stores the location information about the MS and VLR stores correct location of the MS. In the Figure 2(b)(i) the correct scenario is as follows.

1. While the MS enters location area LA_4 from location area LA_1, MS sends location information to VLR_4. VLR_4 sends location information to HLR.
2. HLR updates this location information and sends cancellation message to VLR_1 via MSC.
3. While the MS enters LA_5 from LA_4, MS sends location information to VLR_5.
4. HLR updates this location information and MSC sends cancellation message to VLR_4 via MSC.

Because of delays in network connecting VLRs and HLR, HLR may receive location update message from MSC of LA_5 before receiving message from MSC of LA_4. As HLR processes the messages in the order of messages received, HLR shows that MS is in location area LA_4 but actually MS is in location area LA_5. Thus, HLR and VLR may contain incorrect location information of an MS (we can refer to these as orphan records), and because of this MS may not receive calls. The order of events of an incorrect scenario is shown in the Figure 2(b)(ii).

Let us once again consider Figure 2(a). It shows several location areas and the cells within them. Location area consists of multiple cells. So, if an MS roams within the boundary cells, we now show that there might be a significant number of unnecessary location updates.

Let us consider the movement of an MS within the cells C_{13} in LA_1, cell C_{51} in LA_5 and the cell C_{42} within LA_4 as in Figure 2(a). We define a location update in an LA to be *necessary* if at-least one call is made in that LA. Since the MS is roaming around the boundary cells of the different LAs frequently, each location update in LA_1, LA_2 and LA_3 may not be necessary, as the MS may not receive a call when it is in an LA, resulting in increase in the number of unnecessary location updates. If an MS is in a cell, which is well inside an LA, this problem does not arise.

So, we address specifically 2 problems (1) avoidance of orphan records from HLR and VLR and (2) optimized location management avoiding unnecessary location updates.

Our contribution lies in the development of 1) the MS-counter approach which ensures the avoidance of orphan records from HLR and VLR and 2) the MS-dynamically-adjustable-counter (MSD) approach, which dynamically determines whether or not to update in each LA so that each location update becomes a necessary location update.

2 Related Work

Most of the practical cellular mobile systems partition the geographical region into location areas (LA) and users update on entering a new LA. In this section we will discuss proposed location update strategies and paging methods.

Work proposed in [9] presents a mobility management scheme that integrates the location area approach with the location prediction idea. It is based on results from traffic flow theory. [5] presents a look-ahead strategy for distance-based location tracking so that the rate of location update can be reduced without incurring extra paging costs. A mobile tracking scheme that exploits the predictability of user mobility patterns in wireless PCS networks is proposed in [1].

A location management algorithm is proposed in [6], which uses the mobility history of individual subscribers to dynamically create individualized location areas, based on previous movement from cell to cell. and the average duration spent in each visited cell.

Paging schemes proposed in [8] provide a way of partitioning the service areas and accordingly minimizing the paging costs. The paging scheme proposed in [2] is based on the semi-real time movement information of individual users.

A mobile user location management mechanism is introduced in [4], that incorporates a distance based location update scheme and a paging mechanism that satisfies predefined delay requirements. Given the respective costs for location update and terminal paging, the average total location update and terminal paging cost is determined. An iterative algorithm is then used to determine the optimal location update threshould distance that results in minimum cost.

A static location Update strategy for PCS networks and its implementation using a genetic algorithm is proposed in [7] . The location update strategy proposed in [7] determines whether or not to update in each LA, and optimizes the average location management cost derived from user-specific mobility model and call generation patterns. But the optimization using genetic algorithms is computationally intensive

Also the ratio of cost of update (C_u) and cost of paging (C_p) is required to be known beforehand as this ratio C_u/C_p is important to determine whether to update or not to update in an LA. As the ratio C_u/C_p may also vary depending on mobile communication network topology it is difficult to get the value of C_u/C_p beforehand.

So, we need an algorithm which can capture user's mobility patterns and call arrival patterns and adjust the location update frequency without the knowledge of the ratio C_u/C_p.

3 Counter Based Algorithms

In the LA-based approach, an MS updates its position when it enters a new location area. The MS-counter approach is a simple implementation of this approach, which handles the first problem, i.e., it deletes orphan records from HLR and VLR. Then MS-counter approach is then extended to develop the MS-Dynamically-Adjustable-Counter (MSD) approach to handle the second problem, reducing location management cost.

3.1 MS-counter

In this approach, the MS updates its position as and when it enters a new Location area. An MS maintains a *c*ounter and increments the *c*ounter on entering each new LA. After the *c*ounter reaches a maximum value, it is reset to 0. In this approach, both location registers HLR and VLR maintain information about the value of the *c*ounter that each MS sends in the process of location updates. On entering a new LA, MS sends an update request with the value of its *c*ounter to the new VLR via the new MSC. A Mobile Station Roaming Number (MSRN) is allocated and sent to the MS's HLR (which always maintains the current location) by the new VLR. If HLR receives a registration message from a VLR with a *c*ounter value less than what it has received earlier, the HLR ignores this registration message and sends a cancellation message to the VLR.

Illustration:

Consider Figure 2(a), which shows that an MS starts roaming from location area LA_1 to location are LA_4 and then to location area LA_5.

Suppose the current *c*ounter value is 3. When the MS enters a new location area LA_4 , it increments the *c*ounter value to 4 and sends it to VLR_4, and VLR_4 sends this to HLR via MSC.

When the MS moves to another location area LA_5, it increments the *c*ounter value to 5 and sends this *c*ounter value to VLR_5, and VLR_5 sends this to HLR via MSC.

Let us consider the following situation, where update information from location area LA_5 reaches HLR before update information comes from location area LA_4 because of delays in the network connecting VLRs and HLR. Normally, the HLR will simply update its database with what it gets from the location area LA_4, but the location update information that HLR stores is wrong.

But in our approach HLR compares the *c*ounter values. Since HLR receives *c*ounter value 5 before *c*ounter value 4, it simply ignores the information regarding the *c*ounter value 4 and sends the cancellation message to VLR_4 via MSC.

3.2 MS-counter(k) Approach

In the MS-counter(k) approach, MS updates its location information after moving through k location areas. The value of k is pre-defined.

Between two successive location updates by MS, it is the HLR's responsibility to find out the location information of MS by paging in the following way.
HLR sends paging message to the last updated LA of the MS (this is the location area where the MS updates the last). If HLR does not receive any response within time-out from MS, HLR finds location of MS following the scheme as described below.

- HLR sends paging message to the most probable LA, if HLR does not receive any response within time-out, it multicasts to all other probable LAs (all the neighbouring LAs of the last updated LA) serially ring by ring till HLR receives response from the MS. We define a ring of distance *d* as a set of LAs, which are separated by (d-1) number of intermediate rings. So, the ring of distance one is the set of LAs which are immediate neighbouring LAs of the last updated LA.

Subsequently, the probability of movement to the boundary cell of different location areas are calculated in the following manner. HLR maintains in the user profiles a table, in which each tuple has the following attribute: ($Previous_{LA}$, $Current_{LA}$, $Future_{LA}$ and probability associated with $Future_{LA}$).

We can consider each LA as a state. While an MS moves from one LA to another we can view this as a state transition. So, we can use a learning automaton to find out the probabilities associated with $Future_{LA}$s.

Generally, the operation of the learning automaton is based on the state transition matrix, which contains the one step transition probabilities P_{ij} from the current state *i* to the next state *j*. When the automaton selects the right response, the positive transition is rewarded while the probabilities of state transition that were not selected are penalized to keep probability 1.

The probabilities associated with the $future_{LA}$s are calculated following linear-reward-penalty scheme:

If HLR does not get any response from MS in its last updated LA, say LA_i and also MS has not sent a "DETACH" message (MS sends this message to HLR when it is switched off), HLR needs to find the location of MS by sending paging message over forward control channel. The HLR pages LA_j for finding MS. If transition from LA_i to LA_j receives a positive feedback (if response comes from the MS) then equations for evaluating the new probabilities of finding MS in the location areas are

P_{ij}= P_{ij}(1-A) + A and P_{ik}= P_{ik}(1-A), if $k \neq j$.

P_{ij} are the probabilities of transition from LA_i to LA_j and P_{ik} are the probabilities of transition from LA_i to LA_k. Here A is a tuning parameter which speeds up this linear-reward-penalty scheme. The value of A between 0 and 1.

If transition from LA_i to LA_j receives a negative feedback (if response does not come from the MS) then equations for evaluating the new probabilities of finding MS in the location areas are

P_{ij}= P_{ij}(1+A) - A and P_{ik}= P_{ik}(1+A), if $k \neq j$.

If P_{ij} becomes negative then it is set to 0. And if P_{ik} becomes greater than 1 then it is set to 1.

If the probability P_{ij} attains the maximum value, MSC pages all the cells in the location area, LA_j, simultaneously as in location area paging method. Otherwise, the MSC starts paging in the probable location areas simultaneously.

Consider Figure 2(a). Let us assume the last updated LA is LA_1. LA_2, LA_3, LA_4, LA_5, LA_6 and LA_7 are the neighbouring LAs to LA_1. The probabilities of transition from LA_1 to LA_2, LA_3, LA_4, LA_5 LA_6 and LA_7 are P_{12}, P_{13}, P_{14}, P_{15}, P_{16} and P_{17} respectively. In order to route a call to an MS, HLR sends a paging message to LA_1. If HLR does not get any response from LA_1, HLR sends a paging message to the most probable LA, say LA_2. If MS is in LA_2, HLR gets a positive feedback and changes P_{12} to P_{12}(1-A) +A. HLR multiplies transition probabilities P_{13}, P_{14}, P_{15}, P_{16} and P_{17} by (1-A) so that P_{12}, P_{13}, P_{14}, P_{15}, P_{16} and P_{17} sum to 1.

HLR gets a negative feedback (If MS is not in LA_2), it changes P_{12} to P_{12}(1+A)-A. HLR multiplies other transition probabilities by (1+A).

4 Performance of Counter-Based Algorithms

In this section we will discuss the simulation environment and metrics used to evaluate the proposed algorithms.

4.1 Simulation Environment

A service area is considered to be made of eight LAs. The model for the network is taken as a graph with eight nodes and eleven edges as depicted in Figure 3(a). This model is also considered in [7]. The diameter of the graph is three. Each node in the graph is considered as an LA and edge as an interconnection between the LAs. We also consider a second network topology, which is considered to be made up of eleven nodes and nineteen edges as depicted in Figure 3(b). The diameter of this graph is four.

Call inter-arrival rate is considered as exponentially distributed with a mean rate λ. varied from .5 to 2.5 with increment of .5. The total number of calls in each experimental set is 1000.

Movement of MS is modeled as discrete time random-walk model with some directional component. This model has been taken as an MS has certain directional component in its movement resulting in higher location probabilities in certain LAs than others.

- At discrete time t, an MS moves to one of the neighbouring LAs with probability q or stays at the current LA with the probability 1-q.

As compared to fluid flow model reported in [3], the random walk model with directional component is appropriate especially when most of the mobile subscribers in a PCN are likely to be pedestrians. Fluid flow model is more suitable for vehicle traffic such that continuous movement with infrequent change in speed and direction is expected.

Location area based paging wherein all the cells in a LA are simultaneously paged is used in the experiments. Paging cost has been assumed to be the same in all the LAs.

As cost of location update (C_u) and cost of paging cost (C_p) are usually not known, we consider different ratios of C_u and C_p from 0.1 to 2 with increment of 0.1 for our experiments.

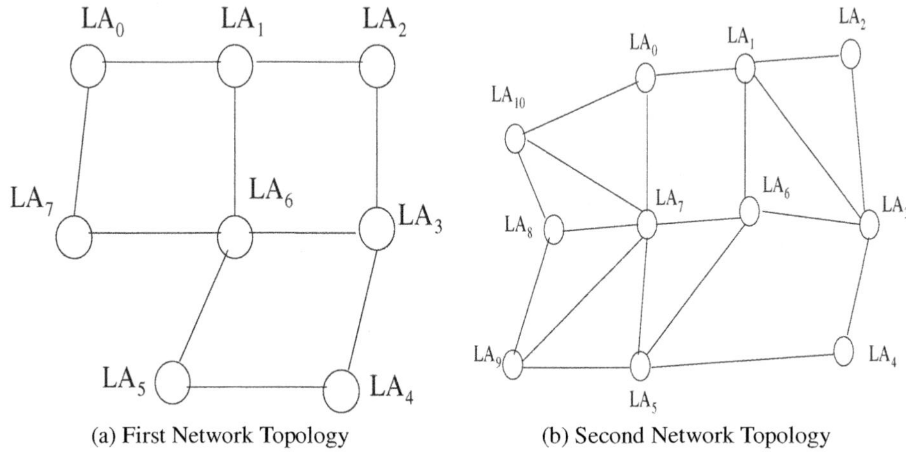

(a) First Network Topology (b) Second Network Topology

Fig. 3. Location Area Graph

4.2 Metrics

1. Location Management Cost (LMC) vs. call-to-mobility-rate (CMR) for constant C_u/C_p.
2. Location Management Cost (LMC) vs. C_u/C_p for constant call-to-mobility-rate (CMR).

The meaning of all symbols used in the metrics are given in Table-1.

Symbol	Meaning
C	Number of calls processed for each set
M	Total number of movements of MS from one LA to a new LA during C number of calls processed
n_u	Number of location updates by MS during C number of calls processed
n_p	Number of LAs paged during C number of calls processed
C_u	Location update cost in a location area
C_p	Paging cost in a location area.
CMR	Call to mobility rate, which is Ratio of C and M
TLUC	Total location update cost, $n_u C_u$
TPC	Total paging cost, $n_p C_p$
LMC	Location Management cost, TULC + TPC

Table 1. Meaning of Symbols used in Metrics

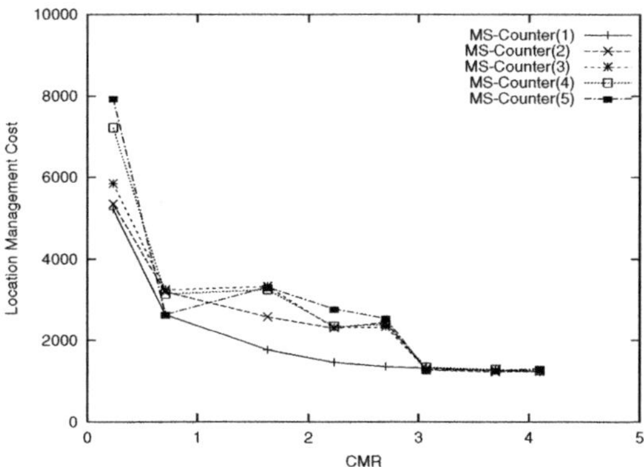

Fig. 4. Location Management Cost of MS-counter(k) for Cu/Cp = 1

4.3 Experimental Results

Experiments have been done with counter(k) with k = 1, 2, 3, 4 and 5. Figure 4 shows that MS-counter(k) for k = 1 produces the least cost for all CMR values. We see the significant variations for k = 2, 3, 4 and 5. Since Figure 4 does not indicate a trend in the location management cost vs. CMR for different values of k, it implies that a static value of k cannot optimize the location management cost. Thus, we need to vary k depending on CMR. MS-dynamically-adjustable-counter approach, which we discuss next, dynamically finds the value of counter, responding to different call arrival patterns and mobility patterns.

5 MSD: MS-Dynamically-Adjustable-Counter Approach

In previous 2 counter based algorithms, the counter whose value is used to decide when to update, takes a static value. But in the MSD approach the counter changes dynamically. The basic principle underlying this approach is to predict the number of useless updates based on history of mobile user's movement pattern.

Let us consider the following example of the combined call and movement pattern of (C, C, M, C, C, C, C, C, M, M, M, M, C) where C denotes a call and M denotes a movement of mobile user from a LA to a new LA. If between two calls there is a single movement then the location update in the new LA is useful. We see that from the combined call and movement pattern that there is single M between Cs, so the number of useless updates is 0. Since for the second sequence of Ms there are four Ms between two Cs, the number of useless updates is three. The protocol is as follows.

MS maintains one variable *c*ounter and one flag *c*all-came.

The value of counter is bounded within ($counter_{min}$, $counter_{max}$). The value of $counter_{min}$ is 1. The value of the $counter_{max}$ is dynamically set to the number of predicted useless updates in future, as calculated by the following formula.

$counter_{max} = \max(U_{min}, \min(U_{max}, U_{est})$

Where U_{min} is the minimum number of the useless updates observed so far, U_{max} is the maximum number of the useless updates observed so far, and U_{est} is the estimated useless updates. The above formula keeps the value of $counter_{max}$ between U_{min} and U_{max}. If the value of U_{est} is greater than U_{min} and less than U_{max} then the value of $counter_{max}$ is set to the value of U_{est}. All the values of U_{max}, U_{min} and U_{est} dynamically change. But if U_{min} is set to 0, there will not be any change in its value. U_{est} is calculated from the following formula.

U_{est} = (pU_{recent} + mU_{avg} + (1-p-m)U_{min})

where U_{recent} is the most recent number of useless updates, U_{avg} is the average of three most recent number of useless updates and *p* and *m* are the tuning parameters.

The prediction of number of useless updates by MS in future needs to track the pattern of changes of number of useless updates. U_{recent} and U_{avg} help tracking these changes in the future.

If a call is not made in an updated LA (where MS updates its location information) counter gets incremented by 1. MS increments counter up to the value of dynamically calculated $counter_{max}$, which actually denotes the predicted number of useless updates in future.

MS starts the first registration with counter being equal to 1. But if a call is not made in this location area, MS increments its counter by 1 on entering a new location area. MS updates its location information after moving through $counter_{current}$ - 1 number of location areas from its last updated location area, or in other words MS skips location updates in the next $counter_{counter} - 1$ number of location areas. $counter_{current}$ is the present value of counter. In this way MS increments its counter linearly, if call is not made in its updated location area till counter reaches $counter_{max}$.

If a call is made in a location area but MS has not updated its location information in this location area then MS sets its counter to 1.

Between two updates by MS, HLR finds the location of MS to route the call as given in MS-counter(k) Approach. (section 3.2).

Illustration:

If $counter_{current}$ is 1 before entering a new location area, MS updates its location information on entering new location area. If call is not made in this updated location area MS increments its *c*ounter value by 1. Thus the value of $counter_{current}$ becomes 2. MS will update its location information after moving through $counter_{current}$ -1 or one location area or in other words MS will skip location update in the next new location area.

If $counter_{current}$ is 3 and call is made in a location area where MS has not updated its location information, then MS sets the *c*ounter to 1. So, MS will update its location information on entering the next new location area.

6 Performance of MSD Approach

For the evaluation of adaptive algorithms the same simulation environment and metrics have been used as mentioned in subsection 4.1 and 4.2.

6.1 Experimental Results with Trace (Random Walk with Directional Component)

Figure 5(a) shows the comparison of location management cost vs. CMR for C_u/C_P =1 for the first network topology as depicted in Figure 3(a). We see from the Figure 5(a) that we can save location management cost in MSD approach for CMR < 2.7. We also see from the Figure 5(a) that we can save as much as 60 % location management cost in MSD approach in the range of CMR from .24 to 1.1. Figure 5(b) shows the comparison of location management cost for the second network topology as depicted in Figure 3(b). We see from Figure 5(b) that we can save location management cost, which is as much as 58%, for CMR < 2.6.

Figure 6(a) shows the comparison of LMC vs. C_u/C_p for CMR =.661. We see from the Figure 6(a) that for $C_u/C_p < 0.3$ both the MSD approach and MS-counter incur almost the same cost. But in the range of C_u/C_p from 1 to 2 we can save significant cost as much as 60 %.

Figure 6(b) shows the comparison of LMC vs. C_u/C_p for CMR =1.1. We see from the Figure 6(b) that for $C_u/C_p < 0.35$ both the MSD approach and MS-counter incur the same cost. But in the range of C_u/C_p from 1 to 2 we can save as much as 60% location management cost.

So, from the Figures 6(a) and 6(b) we can conclude that for $C_u/C_p < 0.35$ MSD approach is as good as MS-counter approach. For larger values of C_u/C_p we can save location management cost as much as 60%.

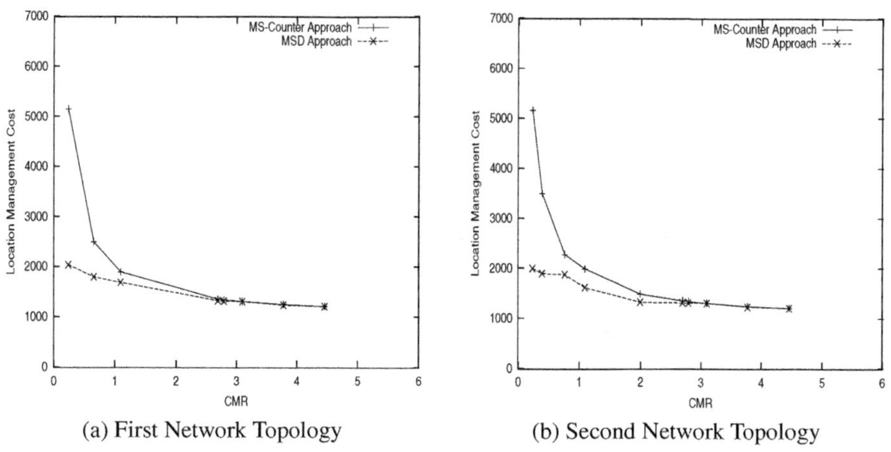

(a) First Network Topology (b) Second Network Topology

Fig. 5. Cost vs. CMR for $C_u/C_p = 1$

6.2 Experimental Results with SUMATRA Trace

SUMATRA (Stanford University Mobile Activity TRAces) is a trace generator that encompasses several calling and mobility models. The unique aspect of SUMATRA is that it is validated against real data on calling and mobility traces.

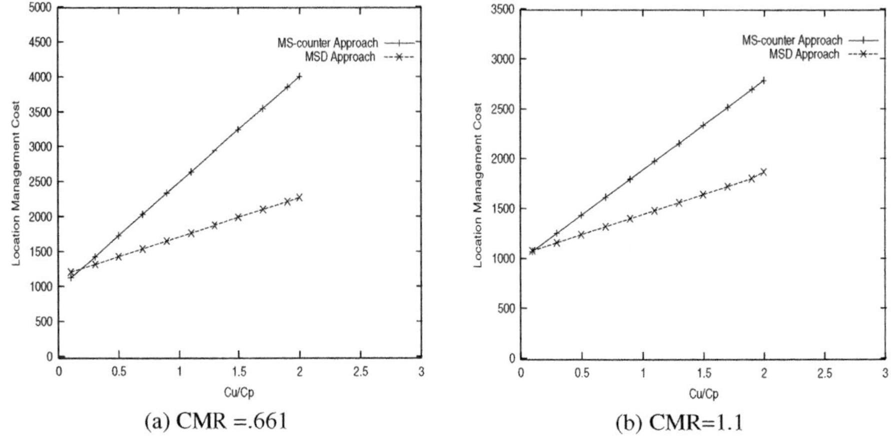

(a) CMR =.661 (b) CMR=1.1

Fig. 6. Cost vs. Cu/Cp for the First Network Topology

We specifically use a trace named BAL1-2 (Bay Area Location Information (real-time) available at [10].The trace duration is 24 hours. Table 2 shows the information about the trace.

#Zones	Users per zone	Calls per hour	Moves per hour	Callee Distribution
90	1, 100	2	0.3	80 % local, 20 % random

Table 2. Value of Trace parameter

Parameter	Definition	Value
C	Number of calls processed for each set	5000
p	weight associated with U_{recent} in MSD approach	0.19 - .2
m	weight associated with U_{avg} in MSD approach	.8 - .85

Table 3. Parameters used in Experiment

Figure 7 shows LMC vs. CMR for C_u/C_p =2. We see from the Figure 7 that we can save LMC using the MSD approach for CMR from 1.59 to 7.34. The saving of LMC can be as much as 16 % in the range of CMR from 1.59 to 4.1. If CMR is less than 1.59, then the saving of cost may be more than 16% .

Figure 8(a) shows LMC vs. C_u/C_p for CMR = 1.59. We see from the Figure 8(a) that for $C_u/C_p < 0.5$ both the MSD approach and MS-counter incur the same cost. But in the range of C_u/C_p from 1 to 2 significant cost saving seen: as much as 18 %.

Figure 8(b) shows LMC vs. C_u/C_p for CMR =4.1. We see from the Figure 8(b) that for $C_u/C_p < 0.5$ both the MSD approach and MS-counter incur the same cost. But in the range of C_u/C_p from 1 to 2 we can save significant cost, which is as much as 13 %.

So, from the Figures 8(a) and 8(b) we can conclude that for $C_u/C_p < 0.5$ MSD approach is as good as MS-counter approach. For larger values of C_u/C_p, we can save as high as 16% of the location management cost.

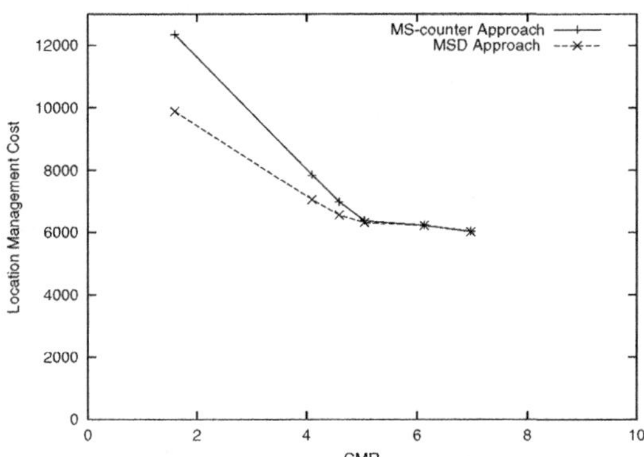

Fig. 7. Cost vs. CMR for C_u/C_p = 2; SUMATRA Trace

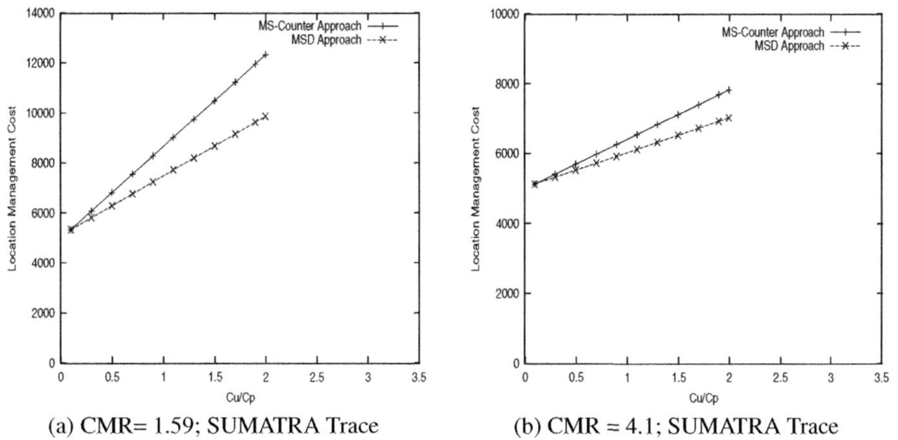

(a) CMR= 1.59; SUMATRA Trace

(b) CMR = 4.1; SUMATRA Trace

Fig. 8. Cost vs. Cu/Cp

6.3 Summary

Typically, location area consists of multiple cells. All MSs update only at the boundary cells of the location area. In the LA-based approach, the update traffic is significant at the boundary cells of the location area. The wireless traffic is generated in the following cases at the boundary cells of location areas.

- Location update traffic.
- call setup traffic, when the call is generated from the boundary cells of the location area.

Due to this considerable amount of wireless traffic at the boundary cells of the location area, the LA-based approach does not distribute the wireless traffic in different cells evenly. This is the one of the bottlenecks of the LA-based approach. Our simulation experimental results show that we can save as much as 60% location management cost for CMR from .24 to 1.1. The expermental results on SUMATRA show that we can save location management cost as much as 16% for CMR from 1.59 to 4.1. So, we can conclude that MSD approach considerably reduces the location management cost by avoiding unnecessary location updates.

7 Effect on Existing Location Management Protocols

The MSD approach is simply a refinement of the LA-based approach, aimed at eliminating unnecessary location update traffic.

The flow of messages in location update strategy approach from an MS required for the implementation of the MSD approach is the same as the existing strategy in GSM as discussed in section 1.1.2.

In MSD approach, HLR sends the paging message in the same manner as in LA-based approach, (i.e., page all the cells in the Location Area). HLR sends the message to the MSC of the future location area to page all the cells of the $future_{LA}$ whose probability of transition from the $current_{LA}$ is maximum. If MS is not found there, HLR sends a message to MSCs to page the neighbouring LAs of the $current_{LA}$ ring by ring serially until the intended MS is found.

MSD approach can be implemented in the existing mobile communication networks by software up-gradation in both HLR and MS. We can implement MSD approach in existing mobile communication networks to save significant amount of wireless traffic, a scarce resource, and transmitting power of MS, by paying the price for software up-gradation.

8 Conclusions and Future Work

Our MS-counter approach ensures the deletion of orphan records in HLR and VLR in LA-based approach. In the LA-based approach unnecessary location updates consume lot of wireless bandwidth and transmission power of MS. Thus, we motivated the need for an efficient algorithm which dynamically evaluates the condition that determines whether or not to update in each LA. MSD approach behaves like MS-counter approach when at-least one call is made in each updated LA. In case of frequent movement of MS between different location areas it saves considerable amount of wireless bandwidth by adjusting the update frequency. From the experiments it can be concluded that MSD approach reduces total location management cost avoiding unnecessary location updates. MSD approach saves location management cost as much as 16% for CMR from 1.59 to 4.1 and 60% for CMR from .24 to 1.1.

We can improve the performance of our approach by predicting 1) number of useless updates in future more accurately using the standard prediction techniques and 2)the most probable LA using hidden markov model (HMM). Exploring these possibilities is left as a future work.

References

1. Ben Liang and Zygmunt J. Haas, Predictive Distance-Based Mobility Management for PCS Networks, *IEEE INFOCOM'99, March 21-25, 1999.*
2. Guang Wan and Eric Lin, Dynamic Paging Scheme for Wireless Communication Systems , *Mobile Computing and Networking, 1997.*
3. H.Xie, S. Tabbane and D.Goodman, Dynamic Location Area Management and Performance analysis , *In Proceedings of IEEE, VTC, May 1993.*
4. Ian F. Akyildiz and Joseph S.M. Ho, A Mobile User Location Update and Paging Mechanism Under Delay constraints, *ACM Wireless Networks, 1995.*
5. I-Fei Tsai and Rong-Hong Jan, The Look-ahead Strategy for Distance-Based Location Tracking in Wireless Cellular Networks , *IEEE Real-Time Systems Symposium, Madrid, Spain, December 2-4, 1998.*
6. John Scourias and Thomas Kunz, A dynamic Individualized Location Management Algorithm , *10th International World Wide Web Conference, Hong Kong, May 2001.*
7. Sanjoy k. Sen , Amiya Bhattacharya and Sajal K. Das, A Selective Location Update Strategy for PCS users, *IEEE Network Magazine, 14(3):6-14, May 2000.*
8. Wenye Wang, Ian F. Akyildiz and Gordon L. Stuber, Effective Paging Schemes with Delay Bounds as QoS Constraints in Wireless Systems , *In proceedings of 1996 USENIX Technical Conference, January 1996.*
9. Yigal Bejerano, Israel Cidon, Location Management Based on Moving Location areas, *IEEE Infocom 2001.*
10. http://www-db.stanford.edu/sumatra/.

{pbellavista, acorradi, rmontanari}@deis.unibo.it

cstefanelli@ing.unife.it

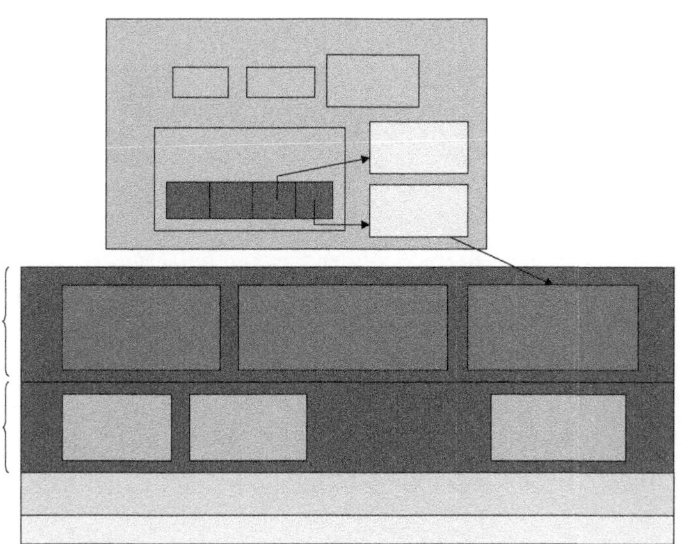

`inst oblig PolicyName "{"`
`on event-specification;`
`subject subject-Expression;`
`target target-Expression;`
`do action-list;`
`when constraint-Expression ;`
`"}"`
	`inst oblig Bind1 {`
	`on Arrival(ProxyID, newNode);`
	`subject s=ProxyID;`
	`target t=Binder;`
	`do t.setAgentBindingType(PrcxyID, "resource movement");`
	`when`
	`MonitoringSystem.getFreeDiskSpace(ProxyID.getLocation())< THRESHOLD;`
	`}`

•

```
class MNSProxy extends ShadowProxy {
...
void init() {
... DataResource resourceID = Scalade.DataResource.get("newspaper");
... }
void run() {
... for (;;) {
        if (isConnected==true) results = resourceID.query(search);
                NITFvisualizer(results);
                sleep(pollingInterval);
        } ... }
... }
```

```
inst oblig Default {
on ProxyInitCompleted(ProxyID);
subject s = ProxyID;
target t = Binder;
do t.setAgentBindingType(ProxyID, "remoteReference");
}
```

Default

```
inst oblig Pol1 {
on disconnectRequest(ProxyID);
subject s = ProxyID;
target t1 = Binder, t2 = ProxyID;
do t2.showWaitingMessage() ||
        t1.setIsConnected(ProxyID, false) ->
        t1.setAgentBindingType(ProxyID, "resourceMove") ->
        t1.updateReferenceObjects(ProxyID) ->
        t2.removeWaitingMessage();
when MonitoringSystem.getFreeDiskSpace(ProxyID.
        getLocation()) > threshold;
}
```

```
inst oblig Pol2 {
on domainArrival(ProxyID, newDomain);
subject s = ProxyID;
target t = Binder;
do t.setAgentBindingType(ProxyID, "copyMove") ||
        t.setIsConnected(ProxyID, true);
}
```

```
inst oblig Pol3 {
on domainArrival(ProxyID, newNode);
subject s = ProxyID;
target t = Binder;
do t.setAgentBindingType(ProxyID, "rebind") ||
        t.setIsConnected(ProxyID, true);
when MonitoringSystem.isDiscoveryAlive() == true;
}
```

kaz@mkg.sfc.keio.ac.jp

funayama@casl.cs.uec.ac.jp

mori@casl.cs.uec.ac.jp

hxt@ht.sfc.keio.ac.jp

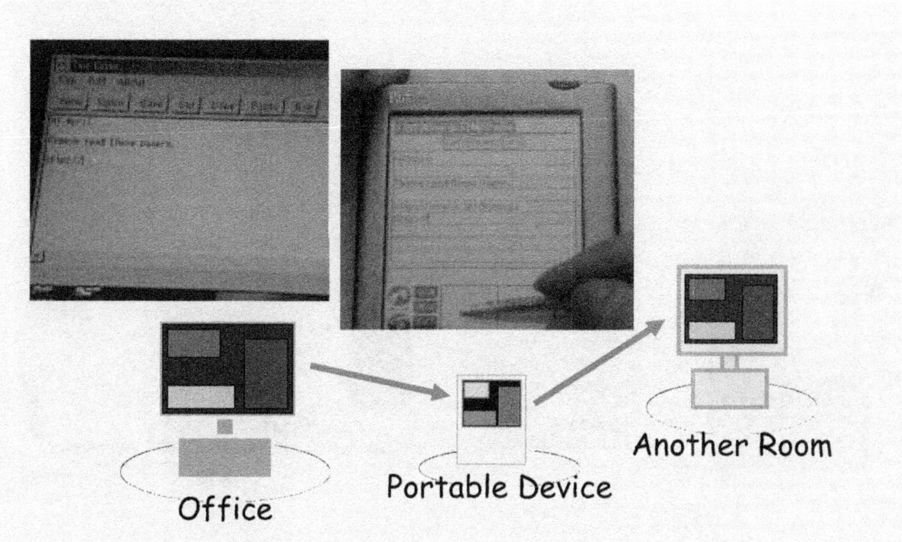
Another Room
Portable Device
Office

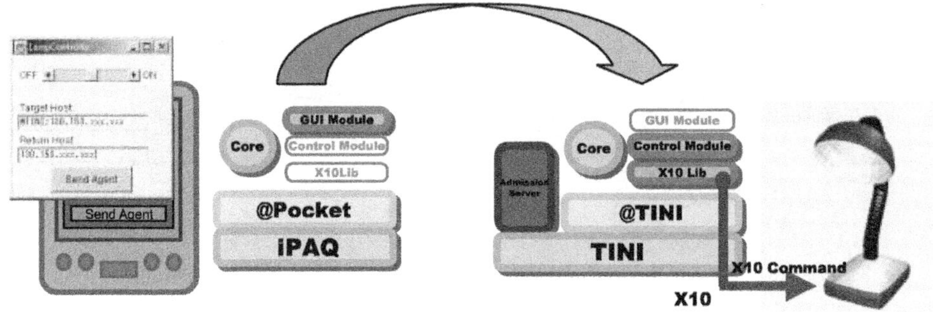

OFF
ON
Target Host
Return Host
Send Agent
Send Agent
Core
GUI Module
Control Module
X10Lib
@Pocket
iPAQ
GUI Module
Core
Control Module
X10 Lib
@TINI
TINI
X10 Command
X10

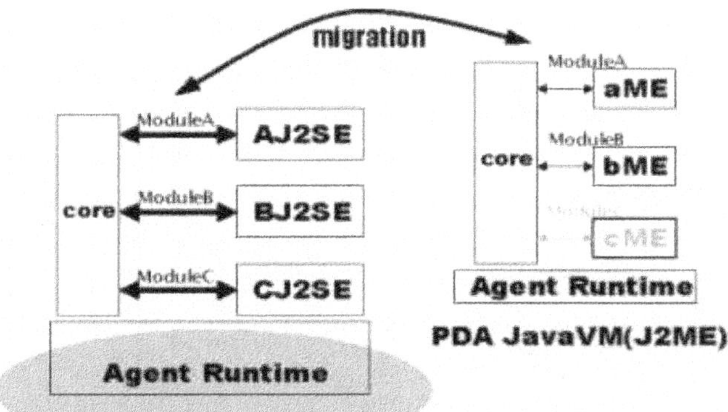
migration
core
ModuleA
AJ2SE
ModuleB
BJ2SE
ModuleC
CJ2SE
Agent Runtime
Desktop PC JavaVM(J2SE)
core
ModuleA
aME
ModuleB
bME
cME
Agent Runtime
PDA JavaVM(J2ME)

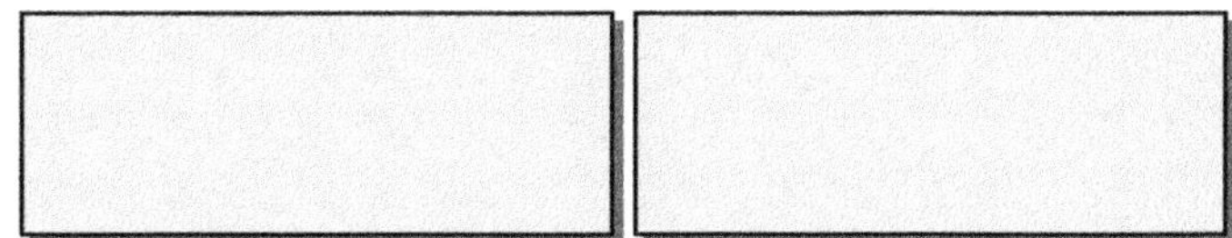

```
<agent_name>@<platform_name>.env
```

```
[module type] = [module class name]
```

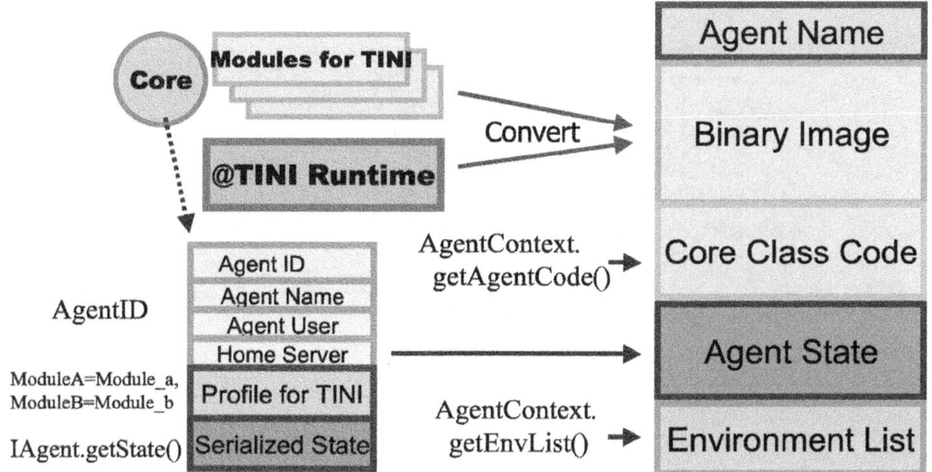

Core
Modules for TINI
@TINI Runtime
Convert
Agent Name
Binary Image
Core Class Code
Agent State
Environment List
AgentContext.
getAgentCode()
AgentContext.
getEnvList()
Agent ID
Agent Name
Agent User
Home Server
Profile for TINI
Serialized State
AgentID
ModuleA=Module_a,
ModuleB=Module_b
IAgent.getState()

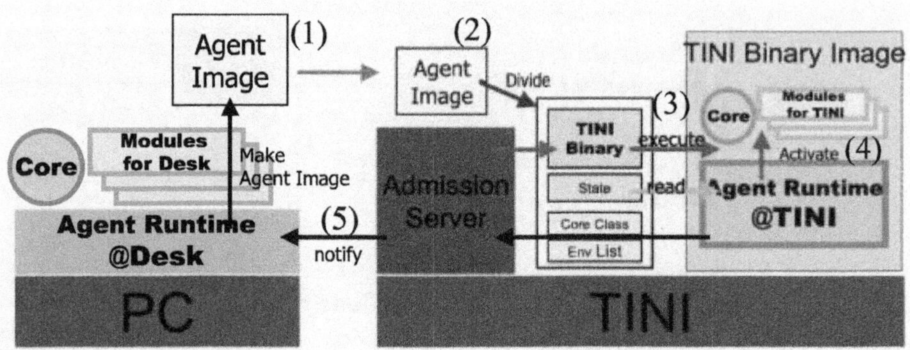

Agent Image
(1)
(2)
Agent Image
Divide
TINI Binary Image
(3)
Core
Modules for TINI
Core
Modules for Desk
Make Agent Image
TINI Binary
execute
Activate
(4)
Admission Server
State
read
Agent Runtime @TINI
Agent Runtime @Desk
(5)
notify
Core Class
Env List
PC
TINI

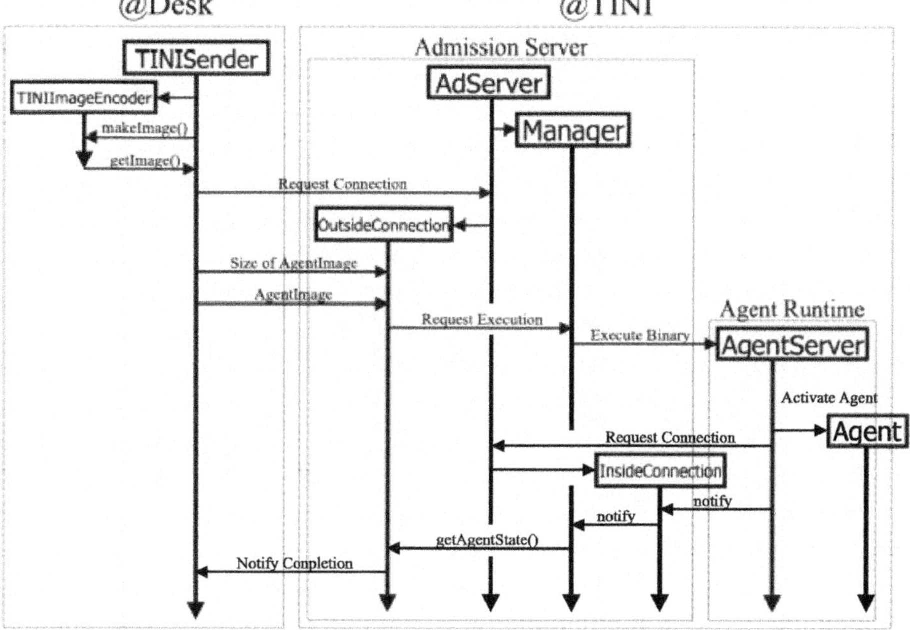

@Desk
@TINI
TINISender
TINIImageEncoder
makeImage()
getImage()
Admission Server
AdServer
Manager
Request Connection
OutsideConnection
Size of AgentImage
AgentImage
Request Execution
Agent Runtime
Execute Binary
AgentServer
Activate Agent
Agent
Request Connection
InsideConnection
notify
notify
getAgentState()
Notify Completion

Experiences in Using CC PP in Context-Aware Systems*

Jadwiga Indulska, Ricky Robinson, Andry Rakotonirainy, and Karen Henricksen

School of Information Technology and Electrical Engineering
The University of Queensland
St Lucia QLD 4072 Australia
{jaga, ricky, andry, karen}@itee.uq.edu.au

Abstract. Future pervasive systems will be based on ubiquitous, often mobile, interconnected devices supporting mobile users in their computing tasks. These systems need to be context-aware in order to cope with highly dynamic environments. In this paper, we present a context model and a context management system able to support a pervasive system infrastructure. This context model is based on the CC/PP standard proposed to support content negotiation between Web browsers and servers. We have defined a set of CC/PP components and attributes that allow to express a variety of context information types and relationships between context descriptions. The paper discusses pros and cons of using CC/PP as a basis for a context model and a context management system.

1 Introduction

The emergence of new types of mobile and embedded computing devices and advances in wireless networking extended the domain of computing from the workplace and home office to other facets of everyday life. This trend is leading towards future computing systems, termed pervasive systems, in which cheap, interconnected computing devices are ubiquitous and capable of supporting users in a wide range of tasks. Pervasive systems need to be able to deal with mobility of users and their devices and also, in the future, with users who change their computing device while running some applications. Pervasive computing environments will differ from current computing environments in many respects, but particularly in scale and complexity. As a result, the success of pervasive computing technologies will rely on a radical design shift. It will require computing applications to become more adaptive, in order to cope with highly dynamic environments and changing user requirements, as well as less demanding on user attention. Accordingly, applications will need to become increasingly sensitive to context. By context, we refer to the circumstances or situation in which a computing task takes place.

* The work reported in this paper has been funded in part by the Co-operative Research Centre Program through the Department of Industry, Science and Tourism of the Commonwealth Government of Australia.

M.-S. Chen et al. (Eds.): MDM 2003, LNCS 2574, pp. 247–261, 2003.

As the programming of context-aware applications is complex and error-prone, there has been a recent emphasis on the creation of infrastructure that assists with tasks such as context acquisition from sensors [1, 2] and context modelling and dissemination [3, 4, 5, 6, 7]. However, the success of context-aware applications also demands appropriate standards for applications to express and exchange context information. Such a standard is currently being developed by the W3C in the form of CC/PP [8, 9]. CC/PP offers an XML-based format for exchanging context descriptions, as well as vocabularies for expressing context information related to device capabilities and user preferences. As CC/PP was designed to support content negotiation between Web browsers and servers, the types of context information it accommodates are limited. In this paper, we present types of context information that are needed in pervasive computing, describe a set of CC/PP components and attributes which we defined to express a variety of context information types and analyse some of the shortcomings of CC/PP. We also describe the architecture of our context management system and evaluate this architecture with respect to scalability.

The structure of the paper is as follows. Section 2 presents a brief overview of CC/PP. Section 3 characterises the types of context information that are important in pervasive computing, while section 4 evaluates the ability of CC/PP to support the types of context identified in section 3, and briefly presents some extensions to the CC/PP vocabulary. Section 5 presents an analysis of the architecture of our context management system and it is followed by section 6 which outlines the lessons learned through our work on CC/PP based context descriptions. Section 7 describes related work on context description and finally, section 8 concludes the paper.

2 Overview of Composite Capabilities Preference Profiles (CC PP)

Composite Capabilities/Preference Profiles (CC/PP) is a W3C proposed standard for describing device capabilities and preferences with a focus on wireless devices such PDAs and mobile phones. The intended use of such a profile is to allow an origin server to deliver content customised to a device's capabilities and preferences. When a device makes a request using a protocol such as HTTP, it sends its CC/PP profile in the request. The origin server can then filter, translate and adapt the content to meet the requirements of the requesting device.

CC/PP is based upon the Resource Description Framework (RDF) [10, 11]. RDF is a way of representing statements, each of which contain a subject, predicate and object. RDF can be serialized in many different ways, but CC/PP uses the XML serialization. A CC/PP profile is essentially a two-level tree containing components and attributes of those components. In this model, statements are made about the components (the subject) that have named attributes (the predicate) and values for those attributes (the object). For instance, a profile may specify that the screen component has a resolution of 640 pixels by 480 pixels. Each profile refers to one or more schemata that govern the types of components that may appear in the profile, and the attributes that belong to each component. It often occurs that the profiles of many devices are the same or similar. For this reason, CC/PP includes the notion of component *defaults*.

These default values may be provided by a hardware or software vendor and stored in a place easily accessible by origin servers. If the profile received from a client device contains its own values for any of the default component attributes then they override the default values. In this way, a client need only specify those components and attributes which vary from the defaults, thereby saving space on the client device and minimizing the use of bandwidth between the device and the origin server.

3 Context Information and Its Management

The goal of pervasive/ubiquitous systems is that computers and computing tasks fade into the background, so that users are less concerned with devices and specific applications. This goal includes "anywhere, anytime" computing which allows users to access the same computing applications and information on their workstation in the office, on their computing devices at home or on a variety of mobile devices when the users are on the move. The pervasive systems need to be able to deal with mobility of users and their devices and also, in the future, with users who change their computing device while running some computing applications. Pervasive computing infrastructure should allow users to move their computational tasks easily from one computing environment to another and should allow them to take advantage of the capabilities and resources of their current environment.

One of the great challenges of pervasive computing is to capture the requirements and current status of computing environments to allow both evaluation of changes in this environment and decisions about neccessary adaptation methods which can be applied to adapt the whole system to these changes. Such adaptations can be carried out at the various layers of the system (e.g. by adapting user applications or applications' communication streams, adapting the distributed computing infrastructure or the behaviour of the underlying operating system or communication protocols).

Therefore the research on context information models is concerned with creating models which describe (i) the features of the whole computing environment (user computing devices, computer network, operating system, distributed computing platform, etc.), (ii) user's computing applications, user requirements and preferences, and (iii) computing application's requirements (e.g, Quality of Service requirements for its communication streams, computational requirements, security) and information about how these requirements are currently met by the computing environment. In addition, some constraints and relationships which exist in the computing environment should also be captured in a context model as they should affect decisions about possible adaptations (these include relationships like device ownership, software licencing [12], but also dependencies between context information, like dependencies between bandwidth and power in a mobile device [13]). The research on context can go beyond supporting "anywhere, anytime" computing in order to provide a better support for user everyday tasks [14]. In the latter case, which is not addressed in this paper, the role of context information is to capture a variety of user needs (tasks) and to provide intelligent computing support for these tasks with minimal user interactions.

The goal of our work was to extend the CC/PP vocabulary which focusses on mobile devices to capture basic classes of context information needed for the infrastructure of pervasive systems. The CC/PP approach was extended to include additional vocabulary for user context information (user profile describing, among others, user applications, user devices, user current location), device capabilities (device profile which includes hardware and software capabilities, device location, networks to which the device is and can be attached, potential and current network QoS on the device's network interfaces), and application requirements and their current status (application profile). The following section shows some examples of these extensions and also shows how relationships and context dependencies were captured in our approach.

Context management in pervasive systems must contain at least the following three separate but complementary functionalities:

- Context sensing (gathering context information) from a variety of hardware and software sensors.
- Interpretation of partial context information gathered from sensors (may involve complex processing and resolution of conflicting context information) to produce context information and represent it in a common format.
- Management of context information represented in common format.

The third functionality needs to provide a persistent storage of context information and allow a variety of clients to access the information or receive notification about context changes. Pervasive systems may be geographically large, have a heterogeneous computing and networking infrastructure, involve a large variety of context sensors and have a large number of clients of context information. Even one type of context information, like location information, may be gathered from a variety of hardware sensors (e.g. GPS coordinates, phone cells) and software sensors (e.g. operating system agent observing user activity on the keyboard, or device's IP address). The location information needs to be processed and conflicts need to be resolved. As users can be mobile such location information needs to be gathered in various domains. Due to these requirements the architecture chosen for a context management system has to be scalable to cope with a large variety of context types, a large variety and number of objects for which context has to be gathered, and a potential large number of clients of context information. In section 5 we discuss the architecture of our context management system and the design decisions which made the the proposed architecture scalable.

4 CC PP Based Context Information

CC/PP provides a model and a core vocabulary for describing device capabilities and user preferences. The goal of our project was to evaluate whether CC/PP can be used to model more general types of context information. Specifically, whether CC/PP is capable of modelling those types of context information described in Section 3.

The core CC/PP schema defines necessary items such as Components, Attributes and the enclosing Profile element. A small example attribute vocabulary is also provided. Third parties, such as device manufacturers and application developers, may introduce their own attribute vocabularies. Furthermore, they may introduce new schemas that contain specialized subtypes of the CC/PP Component class, and define Attributes that apply to these specialized components. For example, the UAProf [15] specification defines HardwarePlaform, SoftwarePlatform and BrowserUA to be subclasses of the generic CC/PP Component class. Then the UAProf specification defines attributes for each of these specialized classes.

Although CC/PP is designed to describe information about device hardware and software, it can describe a much wider variety of context information, as long as that context information can be described in terms of CC/PP Components and Attributes (or subtypes of them). For instance, location information can be modelled in CC/PP by defining a Location component, and perhaps some subclasses such as PhysicalLocation and LogicalLocation. Relevant attributes can then be defined for each of these Location subclasses.

Our extended vocabulary defines many new CC/PP components in order to describe a wide range of context information. Such components include Network Interfaces of devices, Quality of Service, Location, DisconnectionStatus, and Application Requirements. Such context information is used in our pervasive infrastructure to support a wide range of adptation techniques applied to applications when their computational context changes. For example, location information is used to allow usage of nearby devices, provide services which are relevant in a given location, recognise that a user is moving out of network coverage, etc. Quality of Service information describes QoS requirements of applications and is used to make decisions about adapting the application's communication stream or seamlessly moving the communication stream to a different overlaying network. The latter also requires information about network interfaces available on a given device - such information is needed when a decision is made to move the communication stream (seamless handover) to a different network. In this kind of adaptation the QoS requirements are used to map the stream to an appropriate QoS class in a new network, e.g. a QoS class in a GPRS network.

There is also a large number of complex relationships and constraints that need to be captured by the context model, and this is where it becomes difficult and unintuitive to use CC/PP as the underlying context model. The next subsection provides some examples of the Component and Attribute vocabulary we have created to describe more varied kinds of context than those described by UAProf and the basic CC/PP vocabulary. The profile component-attribute trees are shown in a compact notation.

4.1 Examples of CC/PP Extensions

Location There are several components that relate to Location information. These are PhysicalLocation, LogicalLocation, GeodeticLocation, Orientation and Modifications. The last component relates to measurement errors. The following diagram shows the structure of a Location profile.

```
[LocationProfile
    [PhysicalLocation [Country, State, City, Suburb]]
    [LogicalLocation [IPAddress]]
    [GeodeticLocation [Longitude, Latitude, Altitude]]
    [Orientation [Heading, Pitch]]
    [Modifications [VerticalError, HorizontalError, HeadingError,
                    PitchError]]
]
```

NetworkCharacteristics NetworkInterfaces are modelled as a complex attribute of the NetworkCharacteristics component. NetworkInterface also has a complex attribute called QoS, which describes the quality of service offered by the network interface. DisconnectionStatus is also a complex attribute of the NetworkCharacteristics component.

```
[DeviceProfile
    [NetworkCharacteristics
        [DisconnectionStatus
            [ConnectionStatus, DisconnectionType,
             DisconnectionPeriod, ExpectedDisconnectionPeriod]]
        [NetworkInterface
            [QoS [Bandwidth, Jitter, Latency, PacketLoss,
             Throughput], InterfaceType, IPAddress]]
    ]
]
```

There may be many NetworkInterfaces contained in an RDF bag or sequence.

ApplicationRequirements The ApplicationRequirements models the requirements of a particular application installed on a device. The requirements may be related to hardware, software and communication Quality of Service. Each component is of a type defined in the UAProf specification, or of a type already defined above. Therefore, instead of listing the attributes of these types, the type of the component is given. The reader is referred to the UAProf specification [15] for further information.

```
[ApplicationRequirements
    [HardwareRequirements [type=HardwarePlatform]]
    [SoftwareRequirements [type=SoftwarePlatform]]
    [BrowserRequirments [type=BrowserUA]]
    [WapRequirements [type=WapCharacteristics]]
    [PushRequirements [type=PushCharacteristics]]
    [QoSRequirements [type=QoS]]
]
```

CurrentSession The CurrentSession profile captures the description of a user's current computing activities. For instance, the CurrentSession profile describes which device the user is currently using, the application the user is currently using, which network interface is currently being used and so on.

```
[CurrentSessionProfile
    [CurrentUser [CurrentSessionID, URI]]
    [CurrentDevice [URI]]
    [CurrentApplication [URI]]
    [CurrentNetworkCharacteristics [URI type=
     NetworkCharacteristics]]
    [CurrentPhysicalLocation [URI type=PhysicalLocation]]
]
```

The URI attributes point to some resource. Where the type is not constrained to an explicit CC/PP component type, the URI may point to any resource that uniquely describes the component. For example, the URI for the CurrentUser may link to the user's homepage, or it may reference an RDF resource. Where the type is constrained, the URI must point to a CC/PP resource of that type.

Relationships and Dependencies We found it convenient to model some relationships by grouping related components together in their own profile. For example, to state that a particular user is permitted to use only certain devices we can create a profile as follows.

```
[UserDeviceProfile
    [User [URI]]
    [PermittedDevices [Bag of URI]]
]
```

This profile states that the user identified by the resource at the given URI is permitted to use any of the devices in the RDF bag, each of which is identified by a URI. A different approach is to define another attribute for the User component called permittedDevices whose value is an RDF bag of device URIs.

Finally, there are certain dependencies we wish to model. For example, using more bandwidth requires more power. Therefore if the pervasive system infrastructure decides to increase bandwidth of a current application, the battery power has to be checked first. This is modelled at the schema level by introducing a "depends" RDF property between two attributes. First of all, a schema is needed to define the "depends" property. Then, the schema that defines the related attributes (such as bandwidth and batteryPower) asserts the dependency between the two attributes. In our example, the "depends" relation will be defined as a property of the bandwidth CC/PP attribute, whose value will be the batteryPower attribute. This is shown in the XML fragment below.

```
<ccpp:Attribute rdf:ID="bandwidth">
  <rdf:range rdf:resource='#QoS'/>
  <constraints:depends rdf:resource='#batteryPower'/>
</ccpp:Attribute>
```

The above XML assumes that the QoS component type and the bandwidth attribute are defined in the same schema as the batteryPower attribute. It also assumes the existence of a constraints schema that defines the "depends" property.

4.2 A Note on CC/PP Structure

It should be noted that some of the described components have attributes which themselves are components. For example, the NetworkInterface component is modelled as an attribute of the standard UAProf NetworkCharacteristics component. The NetworkInterface component has an attribute called QoS, which itself is a structured attribute. To overcome the two-level tree constraint in CC/PP (as outlined in Section 3 of [8]), the value of each structured attribute is a URI that points to the complete description of the complex attribute. Usually the URI will be a fragment identifier that points to another CC/PP component within the same profile. This is achieved by using the RDF resource property as follows:

```
<prfx:networkInterface rdf:resource='#myNetworkInterface'/>
```

where the myNetworkInterface fragment describes the attribute in its entirity.

5 Context Management Architecture

The architecture of our context management systems consists of a set of interacting components (or modules) organised into three layers as shown in Figure 1. Each layer uses a message based notification service called Elvin [16] to interact with other layers. Elvin uses a client-server architecture for delivering notifications. Clients establish sessions with an Elvin server process and are then able to send notifications for delivery, or to register to receive notifications sent by other components. The Elvin subscription language allows to form complex regular expression. The architecture can be distributed over several locations due to the distribution transparency provided by Elvin.

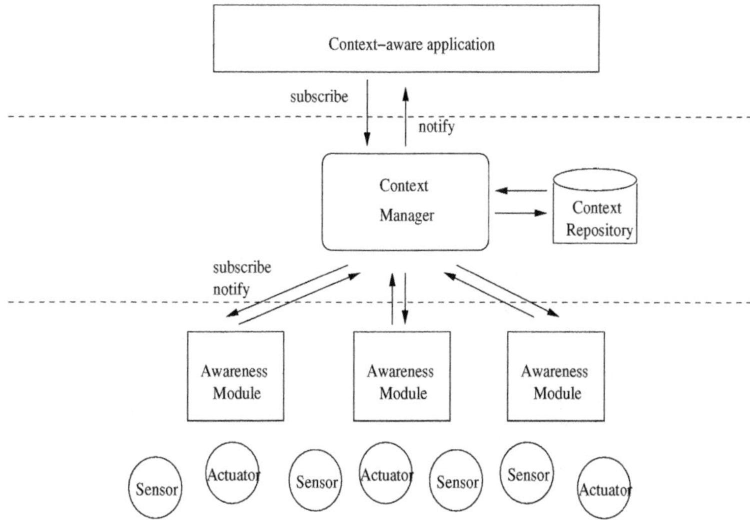

Fig. 1. Context management architecture

5.1 Three Layer Architecture

The ability to dynamically reconfigure the architecture in order to adapt it to new requirements or a changing environment was the main design principle. The architecture is structured into three autonomous layers:

- Context-aware Application layer: In order to be aware of the current context, the adaptation engine of the pervasive system and/or the context-aware application can subscribe to a set of events of interest (e.g. context changes) using Elvin's subscription language. These clients are then notified by the Context Manager upon occurrence of the events of interest.
- Context Manager layer: The Context Manager manages the persistent repository (database) of context information, receives context updates from the Awareness Module layer, detects context changes and provides notifications about the changes to all interested clients. The Context Manager manages events to/from upper and lower layers. The context manager can be seen as a layer that (*i*) co-relates events from different sensors (*ii*) coordinates events to be sent to actuators and (*iii*) cre ates relevant events to be sent to the applications.
- Awareness Module layer: These modules process context information from sensors. They monitor sensors and actuators. An awareness module is a specialised module that provides a particular functionality. For example, it can serve a GPS sensor that provides location coordinates or an actuator that locks a door. Awareness modules transform information from sensors into a format that can be understood by the context manager. They also transform events from the Context Manager to a format understandable by

actuators. The awareness modules use Elvin to send/receive events to/from the Context Manager. The communication between Sensors/Actuators and the awareness modules are often proprietary (e.g a specific protocol that uses RS-232).

5.2 Context Repository

A context management systems needs to provide a persistent storage for context information which was gathered from sensors, processed and converted to a common format. There are efforts underway to define a mapping from RDF to a relational database [17]. The approach taken is to map the RDF model directly to one or more relational database tables. Since CC/PP is an application of RDF, it follows that CC/PP profiles can be stored to a database using one of these mappings. We used this approach to map our CC/PP context descriptions into a relational database. The advantage of such an approach is that even if new vocabularies are added to our CC/PP context model in the future, the mapping does not have to change. Profiles can be reserialised into XML from the database. Although the resulting XML may look different from the original profile (for instance, the mapping application may choose different identifiers for namespaces), the resulting profile will be equivalent to the original.

While there is ongoing work on querying XML files, including the work of the W3C XML Query working group, we used mappings of context information into a database. There are two main reasons for this choice (i) the potentialy large volume of context information stored in a repository, (ii) the work on XML queries is still a work in progress. Performance should be gained by storing profiles in a relational database rather than storing them as XML documents in a repository. The schemas can also be placed into the database. This increases the speed with which a profile can be validated against the schemas it references.

5.3 Awareness Modules

Several awareness modules have been developed for our context management system. The most complex of them is a location manager which receives location information from a variety of physical (GPS, badge system, movement sensors) and logical sensors (IP address, user actions observed by an operating system agent). The module provides mapping of a variety of sensed location information into a physical location. The complexity of the location manager stems from two requirements: (i) the manager needs to resolve conflicting location information (e.g. a reading from user's mobile phone left at home and information about user's current activities, e.g. keyboard typing, in the office) and (ii) privacy of location information has to be protected. We address the first issue by assigning weights to various location readings. To address the privacy issue the system provides authentication of clients requesting location information and we are currently analysing whether the emerging P3P standard can be modified to allow specification of privacy rules for location information.

5.4 Scalability

A context management system can potentially have a large volume of context information gathered from numerous distributed sensors and may need to support many clients requesting context information. To achieve a scalable context management system it is necessary to provide scalable solutions for the system's components. The solutions in the following areas will have high impact on the whole system's scalability:

- reception of context information from the Awareness Module layer and interaction with clients;
- updates of context information in persistent repositories;
- distribution of context information.

While the clients of the Context Manager may query the context database, it was assumed for scalability reasons that the main type of communication will be based on notifications. The same notification system is used to provide interactions between clients and the Context Manager and between the Context Manager and the awareness modules. Elvin has a sophisticated mechanism for the registration of interest in events for which notifications are to be delivered, and therefore it allows clients to specify the granularity of context information notifications, reducing the amount of communication required.

As context information is stored in a persistent repository, a scalable method for updating this repository needs to be developed for sensors which sense context information very frequently (e.g., a sensor providing GPS coordinates of a fast moving car). This is because the organisation of the repository (e.g., a database) is not usually designed for very frequent updates. There are many existing approaches (e.g. spatial indexing techniques) to this problem which need to be employed in the Awareness Module layer in order to limit the number of notifications sent to the Context Manager. As Elvin allows registration for a specified granularity of notifications, we achieved a scalable approach in the following way: the Context Manager registers with the appropriate awareness modules for notifications about changes in particular context parameters (e.g., current location, speed, direction) and the required granularity of this information (e.g., 'notify if the difference between the current location and the previous notification is greater than 100m). The clients of the Context Manager may need different granularities of context information. Therefore, to meet the needs of all the clients, the value of the granularity with which a given Awareness Module sends context information to the Context Manager could be a highest common divisor of the granularities of context notifications requested by the clients.

The proposed architecture, which is very modular and notification based, allows distribution of context managers. Wide-area coverage can be achieved by deploying a number of context managers which gather context information from geographically close awareness modules. On the other hand, if an entity for which context is gathered and evaluated moves a large distance from its home context management system, provision can be made for the creation of a visitor context profile in the context manager at the new destination. In addition, for resource poor and disconnection prone mobile devices, it is possible to retrieve, from the context manager, and locally cache a needed subset of context information. This context information will be available during disconnections and will be updated immediately after the mobile device reconnects. The granularity of context updates may be selected to suit the level of resources in a mobile device.

6 Lessons Learned

We have built a scalable, CC/PP based, context management system which includes basic context types needed to support infrastructures of pervasive systems, provides persistent repository of context information and provides notification about context changes to clients interested in context information.

There are however some limitations in CC/PP which make this model not very suitable as a context model for future pervasive systems. We introduced a variety of context types but future, full fledged pervasive systems will require much more sophisticated context models in order to support seemless adaptation to changes in the computational environment. The context models will need to specify a range of characteristics of context information, including temporal characteristics (freshness and histories), accuracy, resolution (granularity), confidence in correctness of context information, as well as a variety of information types (including various types of dependencies) [7]. However, we have already shown in Section 4 that it was difficult and unintuitive to capture complex relationships and constraints in CC/PP and that the Component-Attribute model becomes unwieldy if there are many layers of attributes.

In addition, there are other limitations which make specification of context information and automatic interpretation of a CC/PP specification difficult:

- CC/PP defines no constraints pertaining to the method of updating an attribute-value in a profile (Should the new value be appended to the old value? Should the new value override the old value? Or no update is allowed?) UAProf embeds these resolution constraints in a comments section of the schema.
- CC/PP defines no relational constraints pertaining to the existence or absence of several attributes within a profile component. For example, we may wish to enforce the constraint that if there is an attribute called "screensize" in a device's hardware profile, then there must also be an attribute called "colour" or "renderingCapability".
- RDF still has some missing functionality like being able to constrain the elements in a container (Bag, Sequence etc) to a specific type. Cardinality constraints on containers are also missing.

We see our project on a CC/PP based context management system as a valuable experience which produced a good prototype context management system which is used in our prototype pervasive system infrastructure to support a variety of adaptations to context changes. We are concurrently working on richer context models required for pervasive systems as described in [7].

7 Related Work

Current research on context-awareness is mostly concerned with providing frameworks that support the processing of context information from sensors and high-level models of context information which can support context-aware applications.

The context toolkit [1] and the sensor architecture of Schmidt et al. [2] address the acquisition of context data from sensors, and the processing of this raw data to obtain high-level context information. The first one is a programming toolkit that provides the developers of context-aware applications with abstract components (e.g. interpreters and aggregators) that can be connected together to gather and process context information from sensors. The second provides a layered model of context processing in which sensor output is transformed to form an abstract context description.

The research carried out by Schilit et al. [18] focuses on modelling context information and delivering it to applications. It proposes the use of dynamic environment servers to manage and disseminate context information for an environment (where an environment might represent a person, place or community). The model of context used is very simple, with context information being modelled as a set of environment variables. The Cooltown project [3] proposes a Web-based model of context in which each entity (person, place or thing) has a corresponding description that can be retrieved via a URL. Entity descriptions may be unstructured and are intended for human (rather than application) interpretation. The context modelling approach proposed by the Sentient Computing project [4] is based on an object-modelling paradigm. A conceptual model of context is constructed using a language based on the Entity-Relationship model, and context information is managed by a relational database.

The Owl context service which is in the early stages of development aims to gather, maintain and supply context information to clients [6]. It aims to address access rights, historical context, quality, extensibility and scalability.

The model under development by Henricksen at el. [7] aims to define a variety of types of context descriptions needed in pervasive systems, relationships and dependencies between context descriptions, and quality of context description (for example, freshness, accuracy, confidence, and so on). In this paper, the first two issues (context types and relationships) are used to examine the applicability of the CC/PP standard to context descriptions in pervasive systems. The quality of context description was not addressed as even the two addressed issues posed some difficulty in the CC/PP vocabulary extension.

8 Conclusions

In this paper, we described our experience in using CC/PP to define several types of context information needed in pervasive systems and building a scalable context management system. The extended CC/PP vocabulary includes, among others, network interfaces of devices and Quality of Service provided by these interfaces, location, disconnection status of devices, application Quality of Service requirements, user preferences, and a variety of relationships and dependencies between context attibutes which are needed to provide various adaptations for applications running in a changing context. Context information is managed by the described context management system which provides persistent storage of context information, context information retrieval and notification about context changes delivered to interested users of the context management system. The users may be context-aware applications and other parts of the pervasive computing infrastucture such as resource discovery protocols or adaptation decision engines.

Our experience with CC/PP revealed that although it is possible to use CC/PP to define the required context information including relationships and dependencies, the limitations of CC/PP shown in this paper do not make CC/PP very suitable for future complex context models required for pervasive systems. The goal of this exercise was to show the gap between the currently proposed standard for context description and the needs of context models for pervasive systems. Many research teams address the issue of context modelling but their research will need to be followed by standardisation of context models to allow some cooperation between heterogeneous domains in pervasive systems.

References

[1] Dey, A., Salber, D., Abowd, G.: A context-based infrastructure for smart environments. In: 1st International Workshop on Managing Interactions in Smart Environments (MANSE'99). (1999)

[2] Schmidt, A., et al.: Advanced interaction in context. In: 1st International Symposium on Handheld and Ubiquitous Computing (HUC'99), Karlsruhe (1999)

[3] Kindberg, T., et al.: People, places, things: Web presence for the real world. Technical Report HPL-2000-16, Hewlett-Packard Labs (2000)

[4] Harter, A., Hopper, A., Steggles, P., Ward, A., Webster, P.: The anatomy of a context-aware application. In: Mobile Computing and Networking. (1999) 59–68

[5] Gray, P., Salber, D.: Modelling and using sensed context in the design of interactive applications. In: 8th IFIP Conference on Engineering for Human-Computer Interaction, Toronto (2001)

[6] Ebling, M., Hunt, G.D.H., Lei, H.: Issues for context services for pervasive computing. In: Middleware 2001 Workshop on Middleware for Mobile Computing, Heidelberg (2001)

[7] Henricksen, K., Indulska, J., Rakotonirainy, A.: Modeling context information in pervasive computing systems. In: Proceedings of The First International Conference on Pervasive Computing, Pervasive 2002. Volume 2414 of Lecture Notes in Computer Science., Zurich, Switzerland, Springer (2002) 169–180

[8] Reynolds, F., Woodrow, C., Ohto, H., Klyne, G.: Composite Capability/Preference Profiles (CC/PP): Structure and vocabularies. W3C Working Draft (2001)

[9] Saryanarayana, L., Hjelm, J.: Cc/pp for context negotiation and contextualisation. In: Second International Conference on Mobile Data Management, Hong Kong, Springer (2001) 239–245

[10] Brickley, D., (eds), R.G.: Resource description framework (RDF) schema specification 1.0. W3C Candidate Recommendation (2000)

[11] Lassila, O., (eds), R.R.S.: Resource description framework (RDF) model and syntax specification. W3C Recommendation (2000)

[12] Bond, A., Gallagher, M., Indulska, J.: An information model for nomadic environments. In: Proceedings of the Ninth International Workshop on Database and Expert Systems Applications, Vienna, IEEE (1998) 400–405

[13] Efstratiou, C., Cheverst, K., Davies, N., Friday, A.: An architecture for the effective support of adaptive context-aware applications. In: Mobile Data Management - MDM, Hong Kong, China, Springer (2001) 15–26

[14] Sousa, J., Garlan, D.: Aura: An architectural framework for user mobility in ubiquitous computing environments. In: Proceedings of 3rd IEEE/IFIP Conference on Software Architecture, Montreal (2002)

[15] WAPForum: User agent profile specification. WAP Specification Information Note (2000)

[16] Segall, B., Arnold, D.: Elvin has Left the Building: a Publish/subscribe Notification Service with Quenching . In: Proceedings of AUUG97, Brisbane, Australia (1997) Available at <http://www.dstc.edu.au/Elvin/doc/papers/auug97/AUUG97.html>.

[17] Melnik, S.: The RDF API Homepage. http://www-db.stanford.edu/~melnik/rdf/db.html, last accessed June 2002 (2001)

[18] Schilit, B., Theimer, M., Welch, B.: Customising mobile applications. In: USENIX Symposium on Mobile and Location-Independent Computing. (1993)

Document Visualization on Small Displays*

Ka Kit Hoi, Dik Lun Lee, and Jianliang Xu**

Hong Kong University of Science and Technology, Clear Water Bay, HK
dlee@cs.ust.hk

Abstract. Limitation in display size and resolution on mobile devices is one of the main obstacles for wide-spread use of web applications in a wireless environment. Web pages are often too large for a PDA (Personal Digital Assistant) screen to present. The same problem exists for devices with low resolution such as WebTV. Manual reconstruction of web pages for these devices would ease the problem; however, the large variation of display capabilities will greatly increase the burden of web page designers since they have to customize a web page for each possible display device. In this paper, we propose a document segmentation and presentation system. The system automatically divides a web document into a number of logical segments based on the screen size and the structure and content of the document. Additional information such as overviews and summaries is also extracted to facilitate navigation. The system presents the segments and structural information of a web document to make full use of the screen for information finding.

1 Introduction

Mobile computing has drawn much attention in these few years due to the technology advances. However, viewing documents on a mobile device is not an easy task because of the limited display size, CPU power, and bandwidth. This is an obstacle for wireless web browsing as there is no automatic optimal transformation of HTML [12] documents designed for display on large screens to small PDA screens.

The same problem appears in devices for web access (or web-like access) on television. Although the size of TVs generally is larger than most of the ordinary computer monitors, the relatively low resolution and long viewing distance reduce the amount of information they can display. In this paper, we use the term "small displays" to refer to devices which can only display a small amount of information for comfortable user viewing, either because the screens are physically small or have very low resolution.

Small displays increase the burden of web page authors and application designers. They need to consider all the possibilities of the display ability of user's devices when the web pages are created. One of the solutions to the problem

* This work was supported in part by grants from the Research Grant Council of Hong Kong (Grant No. HKUST 6079/01E).

** The author is now with Hong Kong Baptist University, Kowloon Tong, HK.

M.-S. Chen et al. (Eds.): MDM 2003, LNCS 2574, pp. 262–278, 2003.

is to use a scroll bar to navigate a web page when the page is too large for the screen. Obviously, this approach requires many user interactions before the desired information can be reached. Another approach is to break down the web page into small segments in a top-down left-right order and display them one by one. However, users would still have difficulties in locating relevant information if no prior knowledge of the web page is known.

In this paper, an automatic document segmentation and presentation system (DSPS) is presented to solve the problem. The system has three primary functions. Firstly, it automatically divides a web document into different logical segments based on the display size of the devices, and the hierarchical structure and content of the documents. Secondly, it extracts the summary and overview information from the logical segments to help users locate relevant information. Thirdly, an interface for clear and user-friendly presentation of the segments is created for rapid access to the desired information.

The rest of the paper is organized as follows: Section 2 discusses previous work which was done related to the problem; Section 3 describes the algorithms used in the DSPS system; Sections 4 and 5 present the description of the DSPS system and the case studies; lastly, the conclusion is presented in Section 6.

2 Previous Work

2.1 Existing Solutions

The most obvious solution to present information on small display devices is to tailor it for displaying on these devices. There are several information providers who manually create documents for specific mobile devices so that users can download them into their devices [4, 9, 10]. Portability is the major problem for tailor-made information. The same design of a document is difficult to convert from one device to the others. Different formatted versions of the same document are necessary for different kinds of users.

Besides tailor-made methods, there are several web browsers running on different kinds of hand-held devices and webTV devices [1, 6, 11]. The usual technique of these web browsers in presenting web pages is to fill up the screen line by line. This approach provides quite a nice solution for simple web pages, which do not use tables and frames as formatting tools. When a web page contains many tables and frames, which are heavily used for information formatting, the layout of the web page in those web browsers becomes messy and almost impossible to read. The main problem is that they do not perform any kind of content transformation to make the information suitable for display on small screens.

2.2 Studies on Small Display and User Interaction

A number of studies on the effect of reading and understanding of using small displays rather than ordinary displays were carried out. The studies on the effect on reading and comprehension in a small display environment by Duchnicky and

Kolers [3], Dillon et al [2], and Shneiderman [14], concluded that a small display did not dramatically impact the reading speed or the comprehension level of the readers. However, heavy interaction like scrolling would annoy many users. Users had to take a lot more time and key strokes to get their desired information.

An experiment on the impact of small displays which was carried out at the Interaction Design Center, Middlesex University, UK by Matt, Gary, Norliza and Kevin [5] revealed an interesting behavior of users. Twenty computer science staff and student volunteers were asked to view a web site with a common design that is similar to many commercial web sites. In conclusion, large screen users showed a greater tendency to follow navigation paths while small screen users had much shorter navigation paths.

Such finding illustrated some principles in designing or creating web pages for small display devices. Direct access to the information items and search functions play very important roles. The layout of the information should reduce navigation operations like scrolling. Key information of the overall web sites should be presented before the detailed contents of different information items are shown.

2.3 Structured Data Tree, Content Tree, and Logical Tree

To process HTML documents, a document model is needed in DSPS. Lim and Ng [7, 8] proposed two types of tree representations of HTML documents, the Structured Data Tree (SDT) and the Content Tree (CT), to represent the hierarchical information of the data contents of the HTML documents. SDT is one kind of parse trees for an HTML document based on the HTML grammar. Fig. 1(a) shows an example of SDT. A CT captures the hierarchical relationships among the contents of an HTML document. Fig. 1(b) shows the CT which is constructed from the SDT in Fig. 1(a).

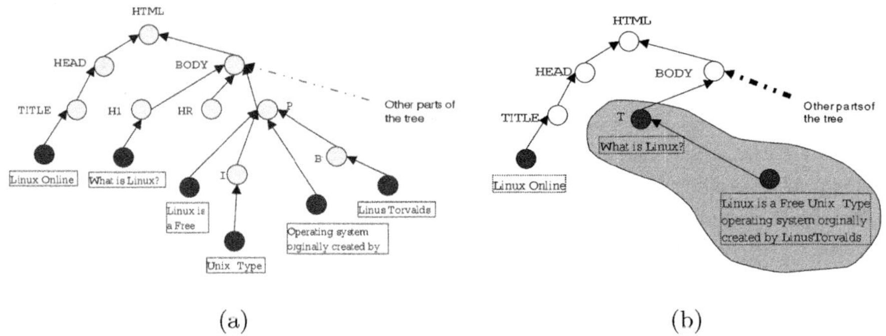

Fig. 1. (a) Structured Data Tree (SDT); (b) Content Tree (CT)

Shinagawa and Kitagawa also proposed a structure called Logical Tree which was similar to CT [13]. The construction of the Logical Tree has used context-free grammar which is, in principle, the same as the push down automata approach, which has been proposed by Lim and Ng.

However, there is a serious limitation in both Lim and Ng's algorithm and Shinagawa and Kitagawa's algorithm when they are applied to real-world web pages. Their algorithms only consider some HTML tags like $H1$, $TITLE$ and $BODY$. Tables and frames, which are heavily used in formatting web pages, are excluded.

3 Converting HTML Document to Content Tree

An HTML document is semi-structured in nature since the high-level structural meaning of the document itself is not explicitly expressed. Also, the HTML grammar provides a high flexibility for controlling the appearance of an HTML document. Such flexibility complicates the process of obtaining useful structural meaning for document processing. In order to obtain high-level structural meaning from an HTML document, this section proposes a series of algorithms and heuristic methods to extract the document structure (expressed in a Content Tree) from an HTML document.

3.1 Overview

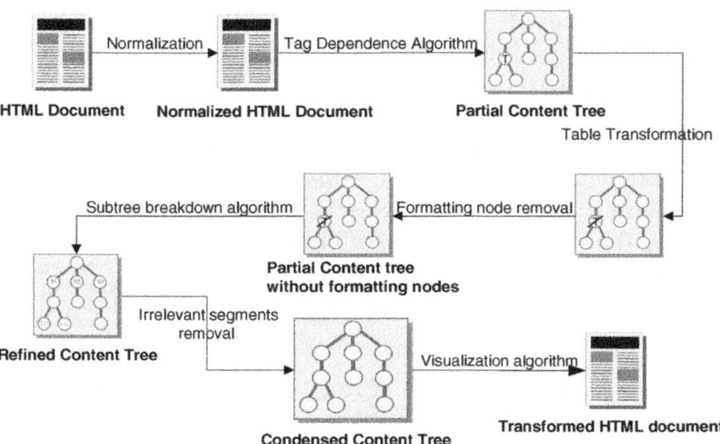

Fig. 2. Overview of the heuristic conversion from HTML document to Content Tree

The objective of all the algorithms that are going to be discussed is to transform an HTML document D into a structured document D_s which can be represented by a CT, CT_{D_s}. The CT_{D_s} should be able to facilitate the visualization of D on a small display screen $Scr(w, l)$. which can only display l lines and each line consists of up to w characters.

An HTML document is first converted and normalized into an intermediate Content Tree by an HTML parser. The intermediate CT is then converted into an initial CT using the *Extended Linear Dependence* algorithm. Then the table refinement algorithm transforms table structures in the HTML document into different subtrees. The *Segment Breakdown* algorithm further refines the CT structure by analyzing the formatting style of each node. Some irrelevant subtree segments are removed from the CT by the *Irrelevant Segment Removal* algorithm. The resulting CT is then transformed into a form that can be used for visualization on the screen. Fig. 2 shows the overview of the sequence of algorithms used to approximate an HTML document into a CT.

3.2 Classification of HTML Tags

The HTML tags are classified into three types, structural tags, block tags and formatting tags in the algorithms.

- **Structural Tags**: Structural tags contain different structural meanings in the HTML Specification. Among these tags, we can assign the order of dependence. For example, `H1, H2, H3 ...` are structural tags. Based on their semantic meanings, linear dependence order can be assigned to them. (e.g., `H3` is dependent on `H2`, and `H2` is dependent on `H1`).
- **Block Tags**: Block tags should be treated as individual entities. For example, `TABLE`, `TR` and `TD` is a set of tags in which the whole entity is defined between the start and the end tag `TABLE`. The set of block tags can further be classified into block begin tags and block element tags. Block begin tags are the tags that indicate the beginning of the block entity construct. `TABLE` is an example of block begin tag. Block element tags are the building elements of the block entity. `TR` and `TD` are examples of block element tags.
- **Formatting Tags**: Formatting tags used in formatting the text rather than indicating the structural meaning. These tags can be considered as an attribute of that piece of text rather than a node in the CT. These kinds of tags provide heuristic information in the heuristic transformation process described in section 3.6.

Table 1 shows the classification of tags. HTML tags which are not included in the table are discarded in the DSPS to simplify implementation.

3.3 Normalization of HTML Document

In order to simplify the algorithms and avoid handling too many special cases, an HTML document needs to be preprocessed and normalized into a simple and clean form.

According to the HTML grammar, formatting tags have very little restrictions on where they can appear. They can enclose a whole table or exist within

Tag Type	HTML Tags
Structural Tag	`Title, H1, H2, H3, H4, H5, H6, P, BR`
Block Begin Tag	`TABLE, UL, OL`
Block Element Tag	`TH, TR, TD, LI`
Formatting Tag	`TT, I, B, U, BIG, SMALL, EM, STRONG, FONT, A`

Table 1. Classification of HTML tags

the table cells. In order to reduce the complexity of the structural analysis, every HTML document should be normalized such that for every path from the root node of the intermediate CT to any leaf node, no formatting tag appears before any structural or block tag on that path.

To accomplish the above condition, all the formatting tags should be shifted down within the intermediate CT. The *Push-down Formatting Tags* algorithm traverses from the root of the intermediate CT to all the leaf nodes. When a formatting tag is encountered, the tag is saved in a stack and then removed from the intermediate CT. All the saved formatting tags will be released when a text leaf node is reached by the algorithms.

The operation does not affect the original structure of an HTML document. The algorithm just deletes a formatting tag and duplicates it to each element of an HTML structure if the formatting tag encloses the structure.

3.4 Extended Linear Dependence Algorithm

The *Extended Linear Dependence* algorithm is based on the construction of CT proposed by Lim and Ng [7, 8]. The original construction of CT is based on the tag object dependence. The linear dependence defines a sequence of the tag objects (`H1 > H2 > H3 .....`) which indicates the conceptual, structural and scope meaning of the tags. Any access path from the root node to any leaf node of the CT should follow the linear dependence.

Structural Tags		Formatting Tag
`TITLE, H1, H2, H3, H4, H5, H6`	`P`	`(TT, I, B, BIG ,SMALL, EM, STRONG, FONT, A)`
	`BR`	
	`UL, LI`	(all the formatting tags
	`OL, LI`	are of the same dependence value)
Lower dependence value		Higher dependence value

Table 2. Dependence order of HTML tags

In order to cope with the problem, we extend Lim and Ng's algorithm to create an initial CT for a real-world web page. The main element of the extension is to use different strategies for different types of tags. Different algorithms are applied to the structural, block and formatting tags. First, let us define the linear dependence values of the structural and formatting tags. Table 2 shows the linear dependence among the HTML tags.

The algorithm of creating an initial CT from an intermediate CT, CT_{SDT}, is defined as follows. Define a Block Elements Restricted In-order Traversal on a node n as the in-order traversal that begins at node n but excludes the subtrees rooted at a block begin tag. The block begin tags are included in the traversal.

Define Linear Dependence Transformation Function $\mathtt{ldf}(\text{node}n_r)$ as follows.

1. Create a list of nodes $L_{CT_{SDT}}$ with Block Elements Restricted In-order Traversal on node n_r.
2. Create a content subtree root node $sroot = (n_r.tag, \text{``''})$.
3. Create an empty stack BS.
4. Set *currnode* to *sroot*.
5. For each node cn_i in $L_{CT_{SDT}}$
   (a) create a node $n_i(cn_i.tag, \text{``''})$
   (b) if cn_i is a leaf node, set $n_i.text = cn_i.body$
   (c) if $cn_i.tag$ is a block begin tag, attach the node n_i to *currnode* as a child, push the (cn_i, n_i) pair into BS
   (d) if $cn_i.tag$ is a structural tag or formatting tag and the dependence value of $cn_i.tag$ is larger than the dependence value of *currnode.tag*, i.e. $cn_i.tag$ depends on *currnode.tag*, attach n_i to *currnode* as a child, set $currnode = n_i$
   (e) otherwise, set *currnode* = the parent node of *currnode*, repeat step d
6. For each pair (cn, n) in BS
   (a) attach the return subtree of $\mathtt{ldf}(cn)$ to n
7. Return the subtree rooted at *sroot*.

The transformation of the intermediate CT to an initial CT is done by applying the `ldf` function on the root node of the intermediate CT. The node returned by the algorithm is the root node of the initial CT. Fig. 3 shows an example of an initial CT.

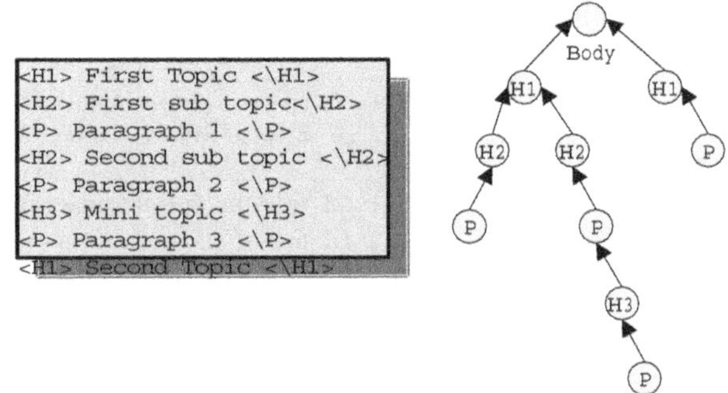

Fig. 3. Example of Extended Linear Dependence algorithm

3.5 Table Analysis

Table is heavily used in formatting web pages due to the fact that normal HTML tags cannot provide high formatting power. However, this makes the extraction of the actual logical structure from a web page much more complicated. This subsection proposes a heuristic approach to analyze the table structures in the HTML documents. In the following, we first describe several assumptions.

We assume that tables contain some textual content. Tables containing only figures can be identified at the parsing stage using some heuristic measures such as cell size. Thus, each table can be transformed into a single node in the construction of the initial CT. We also assume that there is no column or row span in the table. If column and row spans are used, the table is transformed into a normal table by splitting the spanning columns or rows and duplicating the content of the spanning cell accordingly.

Another assumption for the table analysis is that the text segments belong to the same level of the logical hierarchy do not span across different columns. For example, if all the text segments belong to a section, all of them will be located within one column. As such, the logical hierarchy will not break when the table analysis is separating two adjacent columns. The assumption is valid for most of the web pages because HTML pages do not have limits on their length. It is not necessary to split a logical paragraph into two columns.

Based on the above assumptions, we can transform a table as follows. Let $T(m,n)$ be a table of m columns by n rows, and each cell of the table represented as $c(i,j)$ where $0 < i < m$ and $0 < j < n$. Let $S_{(i,j)}$ be the sectional subtree on $c(i,j)$ produced by the *Extended Linear Dependence* algorithm. An initial CT of $T(m,n)$ is constructed by the following *Table Transformation* algorithm:

1. Create a root node *root*(`TABLE`, “”).
2. Set *currnode* to *root*.
3. For each column i of $T(m,n)$
   (a) create a node *cell*(`C`, “”)
   (b) set *cell* to *root* and set *currnode* to *cell*
   (c) for each row j of column i
      i. create a node *para*(`P`, “”)
      ii. append the root node of $S_{(i,j)}$ to *para* node
      iii. append the node *para* to *currnode*

The algorithm classifies each column of the table structure as a separated logical unit. It also assumes that cells in the same column are highly related but not related to the cells in other columns.

3.6 Heuristic Transformation

After performing the *Extended Linear Dependence* algorithm and *Table Transformation* algorithm, the resulting CT captures most of the structural information provided by the structural tags and table structures. However, not every web page authors or web page authoring tools use structural tags or table structures to indicate the structural information of the web pages.

If the HTML fragment does not contain structural tags or tables, heuristic can be applied to fill up the gap. In the following section, the use of formatting styles of the web pages to extract hidden structural meanings from the web pages is discussed. The idea behind the heuristic is that if a small piece of text contains a lot of formatting, it is more important and structurally meaningful than a plain text is.

Transformation of Formatting Nodes A node is defined as a formatting node if the tag of the node is a formatting tag. Formatting nodes complicate the structural analysis, since a heavily formatted text segment would become a large subtree. In order to facilitate the heuristic breakdown algorithm, this kind of nodes need to be removed. We introduce the concept of formatting value which can simplify a complicated subtree structure into a single text node together with a quantified measure of the formatting style.

Different formatting tags have different degrees of emphasizing effects. For example, `B` has a stronger emphasizing effect than `I` and `SMALL`. Each formatting tag is assigned with a value, a higher value means a higher degree of emphasizing effect. Table 3 shows the formatting value of each formatting tag in our DSPS implementation. The value is derived by estimating the emphasizing effect of each formatting tag.

Formatting Tag	Formatting Value	Formatting Tag	Formatting Value
STRONG	3	I	1
BOLD	2	EM	1
BIG	2	U	1
FONT with size inc.	3	TT	0
FONT with size dec.	-1	SMALL	-1
A	2		

Table 3. Formatting value of formatting tags

The *Formatting Nodes Removal* algorithm removes all the formatting nodes by using a numeric value to replace a collection of formatting nodes. The numeric value reflects how heavy the text is formatted. The numeric value is calculated by summing up all the formatting values of the formatting tags applied to the text. Each nodes in the CT will be assigned a formatting value after the *Formatting Nodes Removal* algorithm. Fig. 4 shows an example of the *Formatting Nodes Removal* algorithm.

Segment Breakdown Algorithm The basic idea of the *Segment Breakdown* algorithm is to segment an HTML segment by identifying titles within the segment and using them as the cutting points to create sub-segments. The *Segment Breakdown* algorithm can be applied to each HTML segment until the size of a sub-segment is small enough for display.

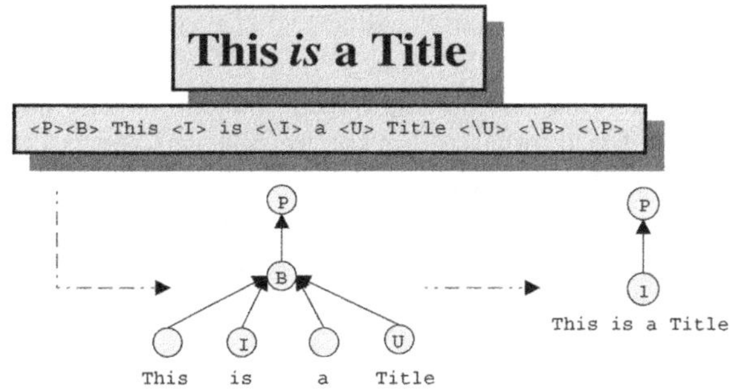

Fig. 4. Example of Formatting Nodes Removal algorithm

The *Segment Breakdown* algorithm identifies the highly formatted text nodes and promote them to a higher hierarchical level of the CT. Less formatted texts then follow the highly formatted texts as dependent nodes. The algorithm can break a flat tree structure into a hierarchical one by analyzing the formatting styles. This operation is essential in DSPS for visualizing the document on small display devices as the document fragment represented by the flat tree structure can be replaced by a number of logically separated segments. Also, an overview, which consists of highly formatted texts, of these segments is also generated.

3.7 Content Analysis

In order to provide more relevant and concise information to the users, analysis on the textual content of the CT is needed.

Irrelevant Segment Removal Algorithm It is common that there are some irrelevant segments in HTML documents, especially for commercial web pages. Irrelevant segments can be eliminated so that the HTML document can be visualized on small display device effectively. In the design of the algorithm, two assumptions are made for irrelevant segments: 1) irrelevant segments are usually small in size; 2) irrelevant segments have no or few keywords in common with the relevant segments within the same logical level.

For a node n in a CT, define Seg_n as the text of the segment represented by the subtree rooted at n. Let $SIM(t1, t2)$ be the similarity scores of text $t1$ and $t2$. The algorithm for calculating the relevance scores RS_{n_i} of a node n_i is given:

1. Let $n_1, n_2 ... n_{i-1}, n_{i+1} ... n_m$ be the sibling of n_i.
2. Set text $t1 = \cup(Seg_{n_1}, Seg_{n_2} ... Seg_{n_{i-1}}, Seg_{n_{i+1}} ... Seg_{n_m})$.
3. Set text $t2 = Seg_{n_i}$.
4. $RS_{n_i} = SIM(t1, t2)$.

Subtrees with relevance scores which are lower than threshold tr are considered irrelevant and can be purged from the original CT to provide more precise information. In practice, the relevance scores of all sibling subtrees at the same level of the logical hierarchy are normalized to the range of 0 to 1. Thus a single threshold value can be used in different levels of logical hierarchy.

The conceptual meaning of the algorithm is based on the fact that a segment should be irrelevant if it is not similar to the rest of the text that is within the same level of the logical hierarchy.

Similarity Measure The Boolean vector space model is used to calculate the similarity scores in our implementation of the DSPS.

Let the keyword space be $(w_1, w_2, ...w_n)$. Let $v_t(b_1, b_2, ...b_n)$ be the bit vector of text t, such that

$$b_i = \begin{cases} 1 & \text{if } t \text{ contains } w_i \\ 0 & \text{otherwise} \end{cases} \tag{1}$$

Given two document vectors v_1, v_2, the scores $SIM(v_1, v_2) = \frac{|v_1 \cdot v_2|}{\sqrt{|v_1|+|v_2|}}$. The reason for using the size of the vector as the divisor is to minimize effects of the long paragraph, especially in intra-document analysis. This is because most of the entities in a document are titles, subtitles and short paragraphs which will be unfairly treated if normalization based on segment size is not used.

Special Case The *Irrelevant Segment Removal* algorithm has an interesting behavior when we apply it to subtrees with a very small number of siblings. The scores for all sibling would be very low if they have too few or even no common keywords. For example, if there are two children of a node and they have no common keywords, the scores of the first sibling and the second one are both zero. The algorithm cannot conclude which child is irrelevant. Thus the DSPS will not remove any children if all the children have scores lower than a minimum threshold.

4 Document Segmentation and Presentation System

4.1 System Overview

The Document Segmentation and Presentation System was developed using the Java2 Platform [15]. The DSPS is a Java application which contains four internal windows within a main window. These four windows are the original HTML document, the complete CT, the CT with irrelevant segments removed and the card view of the transformed HTML documents. Both CTs are presented using the Swing tree component of the Java 2 Swing package.

The Java2 Platform contains a robust HTML parser and supports both the Extended Markup Language (XML) and the Document Object Model (DOM). As most of the operations in the DSPS interact with the tree structures like SDT and CT, the XML and DOM support is essential. There is a clear mapping from a tree structure to an XML document or DOM structure.

Internally, DSPS consists of two main sub-systems, the HTML document transformation sub-system and the document visualization sub-system. DSPS takes two inputs, the URL of the target HTML document which is passed in as

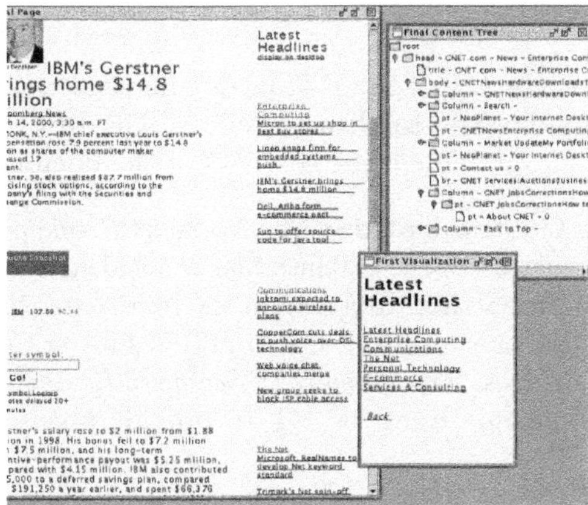

Fig. 5. Document Segmentation and Presentation System screen shot

a command line argument, and the information of the display which is stored in a property file. A transformed HTML document is generated after a series of operations.

4.2 HTML Document Transformation

The HTML document transformation sub-system is the critical part of the DSPS. It takes an HTML document as input and transforms the unstructured HTML document into a meaningful and structural data which is represented as a CT structure, following three steps:

1. **HTML Document Parsing:** DSPS first converts the HTML document into an intermediate CT using the Hot-Java HTML parser, which was built inside the Java 2 Platform, by implementing the call back functions. The call back functions are called when appropriate tokens are encountered. The Hot-Java HTML parser provides error handling and recovery on some common HTML errors like inserting missing end tags and rearranging wrongly ordered HTML tags. DOM has been used as the underlying data structure to represent the CT structure.
2. **Content Tree Construction:** The normalized intermediate CT is transformed into an initial CT using the *Extended Linear Dependence* algorithm. All the tags information including the type and the dependence information have been stored in different property files. The initial CT structure has also been represented by DOM.
3. **Content Tree Refinement:** The *Formatting Tags Removal* algorithm has been applied to the initial CT. The formatting value of each node in the CT has been stored in the attribute of each node. All the formatting values of the formatting tags have been stored in a property file.

4.3 Document Visualization

For a specific device, the CT of the HTML document needs to be customized in order to obtain an adaptive visualization effect. The DSPS needs to transform the current CT first to provide suitable views of the document for the device.

1. **Content Tree Customization:** The *Segment Breakdown* algorithm breaks down the flat structure of the CT into hierarchical structure by incorporating two factors, the formatting values of all the text segments and the size of the display. Therefore, the resulting CT contains subtrees that can fit into the corresponding devices. The *Irrelevant Segment Removal* algorithm removes the irrelevant segments from the CT to provide precise information to the users.
2. **Card-View Presentation:** After the device-specified CT has been obtained, the display engine visualizes the CT using the card-view methodology. The display engine converts the CT into a deck of cards. The content of each card consists of three elements, the text of the node as the header, the list of the texts of all its children and a backward hyperlink. Each text in the list contains a hyperlink to the card of the corresponding child node. The backward hyperlink is used to access the parent of the current card. The first card of the document is the overview which is generated by performing the overview extraction operation on the CT. The depth of the CT to be extracted is adjusted to an optimal level such that the overview can fit into the display of the device. Users can access the information by first looking at the overview of the document and then selecting the header to access different logical sections of the document. The card-view method is suitable for small displays since the size of each card is small but users can obtain the overall picture of the document by accessing the top level overview.

5 Case Studies

This section demonstrates how the DSPS converts some important components of the real-world web pages into different Content Tree structures and thus shows the strengths and weaknesses of the DSPS. The effect of each algorithm is highlighted and explained for critical parts of the documents.

5.1 Case 1 : Wired.com - Front Page

Let us look at the front page of Wired.com first. As we see in Fig. 6, the page contains links to other places in the site on both sides of the page while the center part contains the collection of news items.

Regarding to the three main segments in the page, let us define *seg*1 as the left column represented by the subtree which is rooted at "`SEARCH HEADLINES`", *seg*2 as the center column represented by the subtree which is rooted at "`Not Just for Kids`" and *seg*3 as the right column represented by the subtree which is rooted at "`Quite Marks`" (See Fig. 6). *seg*1 and *seg*3 contain lists of links to

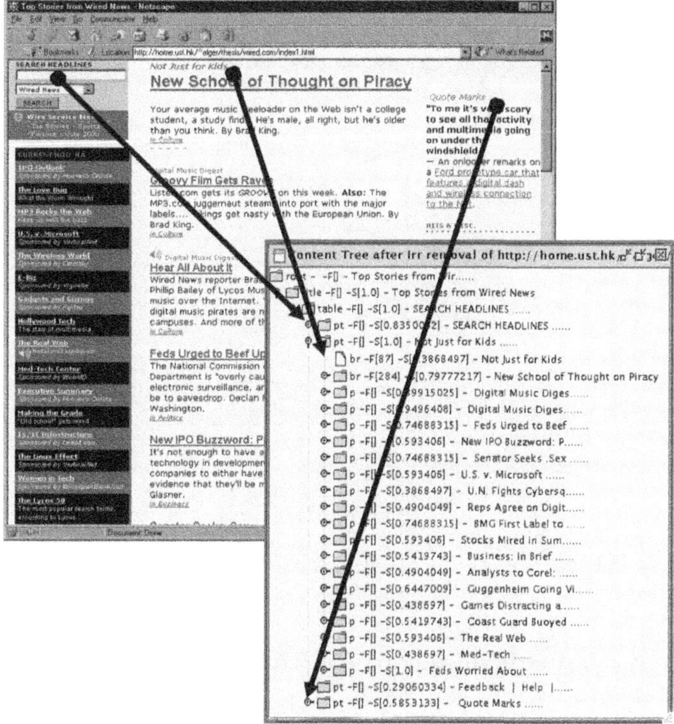

Fig. 6. HTML document of the front page from Wired.com and its CT

other places in the web site which could be considered as irrelevant segments. The scores for *seg*1 and *seg*3 are 0.8350 and 0.5853 respectively. If we set the irrelevant threshold to 0.9, *seg*2 which has score of 1.0 will not be purged. *Seg*1 and *seg*3 will be considered irrelevant and removed from the CT. The result of irrelevant segment purging is satisfactory in this case.

5.2 Case 2: Wired.com - Article

The news article in Fig. 7 of the Wired.com shows the pitfall of the *Irrelevant Segments Removal* algorithm. Considering two segments in the page, let us define *seg*4 as the left column represented by the subtree which is rooted at "`POLITICS` ⋯" and *seg*5 as the center column represented by the subtree that is rooted at "`Read ongoing US vs` ⋯."

In the CT, there exists only two siblings at the level of *seg*4 and *seg*5 (indicated by the arrow in Fig. 7). The scores for these two segments are the same, i.e., 1.0. It cannot be concluded that which one is more relevant than the other. One way to solve this problem is to use the size of the segment as a heuristic to decide which segment can be purged.

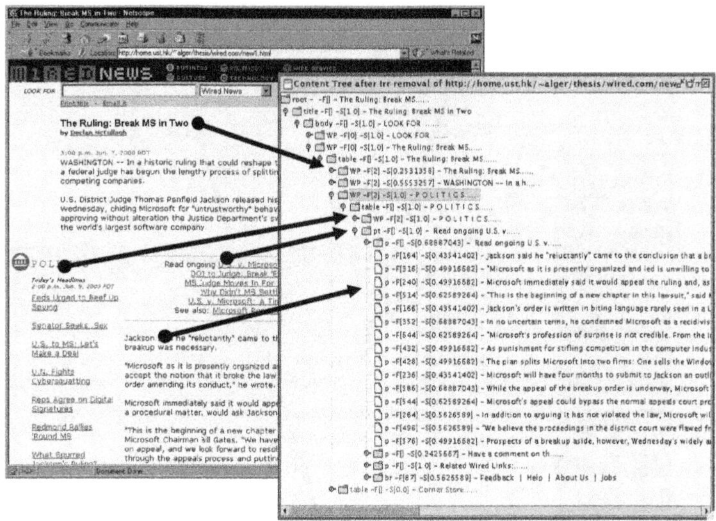

Fig. 7. HTML document of the article from Wired.com and its CT

5.3 Case 3 : Linuxdevices.com - Article

We use an article from Linuxdevices.com to illustrate how DPSP processes an HTML document which contains a long passage composed of a collection of paragraphs. Fig. 8 shows the snapshot of the sample article. The formatting style of this article is similar to that of the front page of Wired.com. Banners are on the top of the pages. A two-column table forms the main body of the article. The long passage locates in the first column while a listing of articles locates in the second column.

Unlike the front page of Wired.com, the long passage at the center column is composed of a collection of paragraphs with `BR` as the newline delimiter. The long passage can be viewed as a collection of logical segments and each logical segment is in the form of a header followed by a number of paragraphs. The logical header is highlighted by using formatting tags such as `B` and `FONT`, instead of using structural tags such as `H1` and `H2`.

As the structure of the article is similar to the front page of Wired.com, we focus on the long passage and describe the effect of the heuristic breakdown algorithm. During the first stage of parsing, a flat tree is obtained. Each node is a text segment which is separated by tags `<P>` or `<BR>`. As no structural tag exists to help breaking down the flat tree into a more structural one, Segment Breakdown algorithm plays an important role in solving the problem.

It is obvious that the formatting value of the header in each logical unit is higher than those of the paragraphs. By calculating the formatting value of each node and applying the Segment Breakdown algorithm, the flat tree becomes a more structural one. Each header becomes the root node of its associated

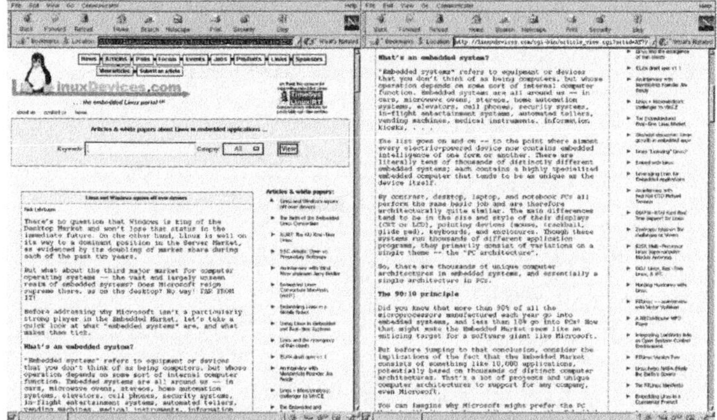

Fig. 8. Article from LinuxDevices.com

paragraphs and all the headers becomes siblings. The resulting CT is shown in Fig. 9.

6 Conclusions

In this paper, we proposed a document segmentation and presentation system (DSPS) to transform an HTML document into a data structure called a "Content Tree", which represents the logical structure of the HTML document. The Content Tree structure can be used to provide various views that are suitable for display in various small displays. One of interesting ideas of the paper is the introduction of the notion of formatting value, which helps to simplify formatting structures and hence ease structural analysis. The DSPS is targeted for real world commercial HTML documents that contain many screen-oriented formatting styles like heavily nested tables and complicated text formatting structures.

References

[1] F. G.-Castano et al. Hierarchical atomic navigation for small display devices. In *Proc. WWW10*, 2001.

[2] A. Dillon, J. Richardson, and C. McKnight. The effect of display size and text splitting on reading lengthy text from the screen. *Behaviour and Information Technology*, 9(3):215–227, 1990.

[3] R.L. Duchnicky and P.A. Kolers. Readability of the text scrolled on visual display terminals as a function of windows size. *Human Factores*, 25:683–692, 1983.

[4] InfoRover. *InfoRover*. Commercial company delivers transformed web contents to Palm Pilot handheld devices. (http://www.inforover.com).

[5] Matt Jones, Gary Marsden, Norliza Mohd-Nasir, and Kevin Bonne. Improving web interaction on small displays. Technical report, Interaction Design Centre, School of Computing Science, Middlesex University, 1997.

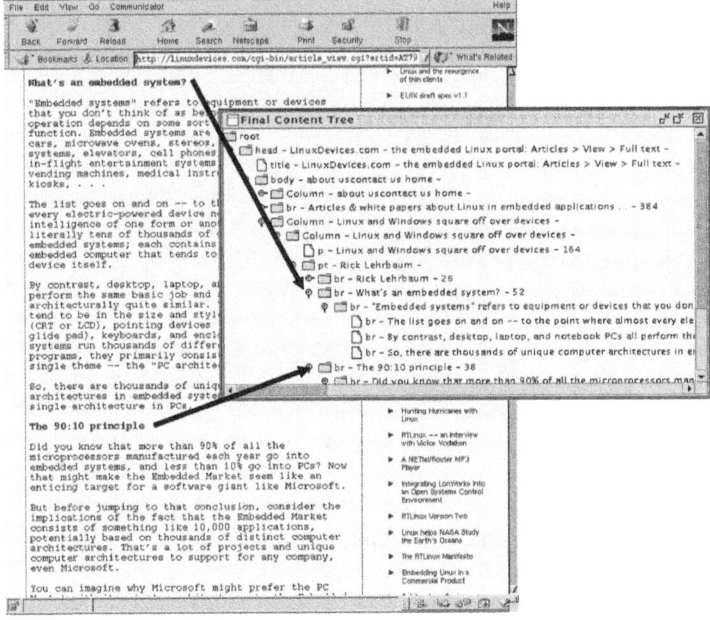

Fig. 9. CT of the article from LinuxDevices.com

[6] Oku Kazuho. *Palmscape PR 5.0a5.* A free web browser running on Palm Pilot handheld devices. (http://www.muchy.com/review/palmscapeloc.html).

[7] Seung Jin Lim and Yiu Kai Ng. Constructing hierarchical information structures of sub-page level html documents. In *Proceedings of the 5th International Conference of Foundations of Data Organization (FODO'98), Kobe, Japan*, pages 66–75, 1998.

[8] Seung Jin Lim and Yiu Kai Ng. Extracting structures of html documents. In *Proceedings of the 12th International Conference on Information Networking (ICOIN-12), Tokyo, Japan*, pages 420–426, 1998.

[9] Microsoft. *Microsoft WebTV.* Commercial company provides web-based access services through TVs. (http://www.webtv.com).

[10] Pocket Info. *Pocket Info.* A public domain initiative in 1998 to provide information for the users of handheld computers, palmtops and organizers. (http://www.pocketinfo.org).

[11] ProxiNet. *ProxiWeb.* A web browser running on Palm Pilot handheld devices. (http://www.proxinet.com).

[12] D. Raggett. *HTML 3.2 Reference Specification.* World-Wide Web Consortium, 1997. http://www.w3c.org/TR/REC-html32.

[13] Norihide Shinagawa and Hiroyuki Kitagawa. Dynamic generation and browsing of virtual www space based on user profiles. In *Proceedings of the Fifth International Computer Science Conference, Hong Kong*, pages 93–108, 1999.

[14] B. Shneiderman. User interface design and evaluation for an electronic encyclopedia. *Cognitive Engineering in the Design of Human-Computer Interaction and Expert Systems*, pages 207–223, 1987.

[15] Sun Microsystem. *Java 2 Platform*, 1998. http://java.sun.com.

Towards Autonomous Services for Smart Mobile Devices

Thomas Strang

German Aerospace Center (DLR)
Institute for Communications and Navigation
D-82230 Wessling/Oberpfaffenhofen, Germany
thomas.strang@dlr.de

Abstract. In this paper a framework is presented which allows the discovery and execution of services on connected and partially autonomous mobile devices. Discovery and execution procedures are sensitive to the user's context (current location, personal preferences, current network situation etc.). We present a description language for service offers which is used to provide the necessary information for a service registry running on the mobile device itself. Services are executed in an abstract manner (in the sense of a non-specific implementation) from the user's point of view, getting an optimal result with respect to the current context out of a set of parallel invoked service implementations.

1 Introduction

In the past few years mobile computing has become very popular, the penetration of mobile devices like mobile phones or PDAs is growing fast. By using technologies implemented today, mobile devices enable the user to access Web-based data from nearly any place in the world when online by utilizing a Web/WAP-browser in their device. In contrast, when offline (e.g. in a plane, where the usage of online connections is prohibited), the user is unable to find new interesting services and restricted to execute only a small set of offline-applications like a calculator. It is a challenge to design an architecture which provides the user of a mobile device with personalized, situation-related services for discovery and execution, online and offline in a best effort sense.

Upcoming mobile devices are capable of using multiple access networks, and are partially programmable for third parties.The architecture presented in this paper is designed with such new features in mind. Even if the architecture does not restrict the mobile device to be a client only, this paper does not further investigate scenarios, where the mobile device is the server for clients in the wired part of the network.

In section 2 of this paper we describe our definition of the service terminology subsequently used. We introduce our architecture used to manage mobile services. A new kind of service registry, residing on the mobile device itself, enables autonomous service discovery and service execution in times of bad or no network coverage. The concept of context sensors, presented in section 3, shows

M.-S. Chen et al. (Eds.): MDM 2003, LNCS 2574, pp. 279–293, 2003.

how to integrate context-awareness during discovery and execution. The service offer description language proposed in section 4 will be used to create context-dependent service offers, and also provide a way to exchange mobile services for partial autonomous execution on the mobile device. A new type of abstract service invocation is discussed next (section 5), which delivers an optimal result with respect to the current situation. Finally, we discuss some implementation aspects (section 6), and draw our conclusion (section 7).

2 Services in Distributed Systems

Within the context of computer networks, the terminology of a "service" is used do describe a wide variety of different things. Thus, to avoid misinterpretation of what a service is and how they are used, a detailed specification is required.

We define a service as a *namable* entity being *responsible* for providing information or performing actions with specific *characteristics*. We distinguish between *MicroServices* and *MacroServices*. MicroServices encapsulate "atomic" operations, meaning that no MicroService engages any other MicroService. In contrast, MacroServices engage at least one other MicroService to perform its operation.

The entry point of a service, be it a MicroService or MacroService, is represented by an addressable instance in a sense of WSDL's "service ports" [4]. The entity representing a service needs to be namable to identify its operation in relation to a given namespace to distinguish services across different namespaces. Each service is engaged through its addressable instance, which is instantiated and controlled by a service provider. The service provider is required to manage the lifecycle (start/stop, pause/resume, etc.) of a service instance and to coordinate service interaction (input/output, invocation/result delivery, etc.) by adopting to a common protocol like JINI, CORBA, DCOM or SOAP for information exchange in distributed environments.

For the time of the duration of a service's operation, the service providing entity bears the responsibility to act in a sense the calling entity envisioned. This is modeled by a caretaker entity, which resides at the service provider and acts e.g. as a life-cycle-manager as well as a representative of the calling entity when the execution of the service is asynchronous. Thus, a service provider in our definition has similarities to a platform for software agents [8], and the caretaker entity can be seen as an extended life-cycle-manager like a *dock* [5].

2.1 Service Discovery and Interaction on Mobile Devices

In a distributed environment, a service has to be found by a client entity (human and/or program) before it can be used. Therefore, a potential service client contacts any form of service registry (e.g. UDDI registry, Bluetooth SDP Database, JINI Lookup-Service or Corba Trader), and requests the availability of providers which have been registered with the service registry to offer services with specific characteristics. These characteristics are usually classifications of a common ontology well known to both parties. An overview of typical protocols in use for service discovery can be found in [2] or [3].

Most architectures for distributed systems silently imply the existence of a reliable, wired network which interconnects all hosts involved in distributed operations. In the wireless world, important additional aspects like limited bandwidth, cost of bandwidth, sometimes unavailable network connections etc. have to be taken into account. Whereas the network layers are usually designed to handle these differences to the wired world very well, the consideration of them in the service layers is often totally insufficient.

In fast-changing contexts [6, 17] such approaches are not sufficient. Thus, to handle at least some of these dynamic context issues and to enable the most possible seamless operation of services on devices with occasionally unavailable network connection of any type and intermittent failures, we suppose an architecture element shown in fig. 1, where an instance of a service registry is running on a mobile device. In [11, 19] the authors showed the several advantages of having a service registry instance on a mobile device, whether in short range or wide range networks. This makes our approach different from other mobile service architectures like the *Capeus* [16], the *Centaurus* [10] or the *Ninja* [9] system, where the service registry (sometimes also named *Trader* or *Broker*) respectively the offer-demand-matcher resides in the wired part of the network.

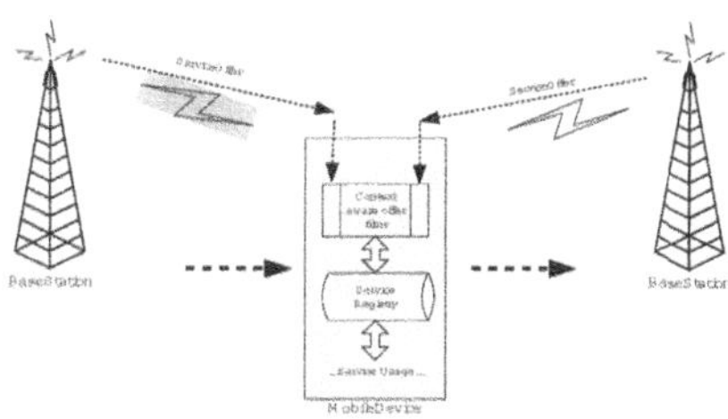

Fig. 1. Service Registry on the Mobile Device

Mobile devices like mobile phones or personal digital assistants (PDAs) are predominantly limited devices with respect to computational power and both volatile and persistent storage (memory). These limitations make it impossible to follow "heavyweight" approaches like CORBA or JINI for distributed service discovery and distributed service usage, even if they have attractive solutions in some areas. One architecture [21] tries to handle dynamic environments by *virtual services* which are composed using some flow language and bound to some physical services at runtime to reflect reconfigurations of the current situation. The scenario behind the architecture of [21] is not mobile device centralized, wherefore the proposed protocols and other architecture elements are not designed to be used on usually very resource-limited mobile devices. The MOCA architecture [1], whose design has the most similarities to ours, claims to be a service framework for small mobile computing devices. But having the service registry on the device is not sufficient. So is MOCA's invocation model for remote services based on dynamically downloaded bytecode, RMI or IIOP and custom Java classloaders, all together not available on typical wireless information devices. As the authors state in [15], MOCA addresses mainly back-end application logic, and a typical MOCA application is either implementing "traditional Java technologies such as the AWT" or "on the use of servlets that are hosted under MOCA", which are both too heavyweight for small mobile devices. The aspect of context-awareness (see section 3) is not handled within MOCA.

Instead of this, we propose an architecture in the following sections based on *Web Services*, a technique incorporating HTTP as the default transport protocol and XML as the representation format for data of any type. XML offers a platform-neutral view of data and allows hierarchical relationships to be described in a natural way. Compared to other architecture models, HTTP and XML are less resource consuming and, therefore, supported by upcoming smart mobile devices (see section 6).

The discovery of web services is usually performed by contacting a UDDI registry located somewhere in the network, but typically not on the same host as the client performing the search query. As explained, to be able to answer search queries even in times where no network connection is available, it is necessary to have a service registry running on the mobile device itself. Just running an UDDI registry on a mobile device is insufficient for several reasons, mainly because of its missing volume control features and its context-unawareness. Instead, we propose an *Advanced Mobile Service Registry* (AMSR) according to fig. 2 running on the mobile device.

The design of this architecture leans towards a CORBA object trader, e.g. allowing one to apply constraints, as well as a JINI lookup server by enabling code deposition at the registry, but also enables service discovery and execution on mobile devices even in cases of poor or no network connection. The adaptation to the various different types of (usually wireless) network interfaces is done by the *Network Communication Manager*. This module handles network specific issues like adressing in ad-hoc networks or interfacing to SDPoverIP when peering to Bluetooth devices.

Service offers (see section 4) are announced to the AMSR from the network (or the mobile application suite itself, if it also implements corresponding services) through a *Register Interface*, which routes such offers to a context aware offer filter (see section 3). Any search for a service with specific characteristics is performed using the *Search Interface*, and the demand is tried to match against any entry in the registration, where all the offers reside which passed the offer filter constraints. If the search is successful, an instance of a generic *Intermediate Service Provider* (IMSP) is created, which can be accessed using the *Invocation Interface* to perform an *Abstract Service Invocation* (see section 5).

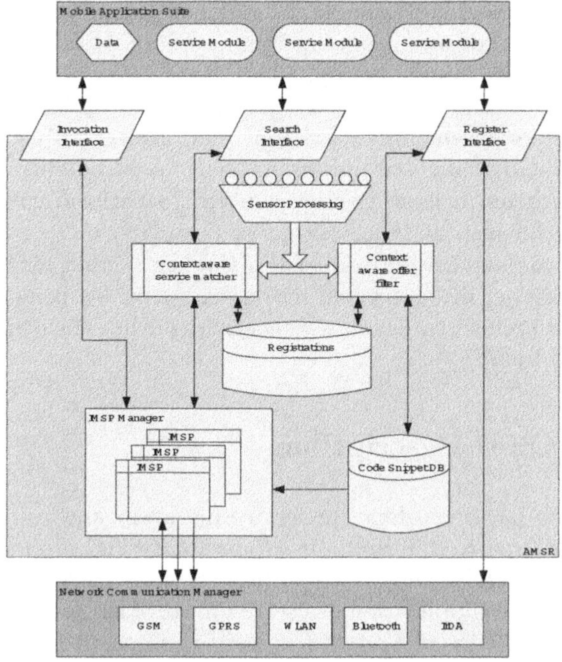

Fig. 2. Advanced Mobile Service Registry (AMSR)

3 Context-Awareness through Context Sensors

The effectiveness of discovery is heavily affected by the quality of information mutually exchanged to describe the requirements on and the capabilities of a service. In a human conversation, the listener uses context information as an implicitly given additional information channel to increase the conversational bandwidth. By enabling access to context information during but not restricted to service discovery, we improve the amount of transinformation from service provider to service user (human or application) and thus the quality of offer-demand-matching. Changes in context during service execution may directly affect a service's output.

We rely on the term context as defined in [6], which is "any information that can be used to characterize the situation of an entity. An entity is a person, place, or object that is considered relevant [..]". Context information is obtained by *observing* relevant entities with *context sensors* (CS). Each CS is responsible for acquiring a certain type of context information and encapsulates how the information is actually sensed. It represents the state of a specific context type against user- and system-services.

Even if most publications like [14] rely on the current geographical position of a mobile device as the most important context information, there is more to context than location [17]. A substantial amount of entities may be observed by CS which base on hardware or pure software sensors. Most of the characteristics of a CS (high degree of individuality, adaptation-ability and autonomy) are also known as the main characteristics of software agents [8]. Thus part of our work is to analyze the requirements of an agent platform on mobile devices [20], and how a CS may be implemented following the agent paradigm. We assume agent-based CS to be a much more flexible concept to perform context sensing and processing than for instance the concept of *Cues* [18].

The AMSR architecture introduced in the last section enables access to the context during service discovery and service execution by using a bank of context sensors, thus our architecture is *context-aware* regarding the definition of context-awareness given in [6].

4 Service Offer Descriptions

A lot of research has been done on service discovery and associated protocols. They all can be used to determine interoperability like supported interfaces or valid value ranges. Only some of these protocols try to consider the context of the requesting entity (application or user). This is of key interest in mobile usage scenarios, where the current context has a strong impact to the needs of a mobile user. Different aspects of context like the position, time, personal preferences or costs should be considered in order to discover a service which matches the user's needs best.

As an instrument for describing a service demand containing context dependent constraints one can use *demand descriptions* like Content Aware Packets (CAPs) [16]. CAPs may be "used as communication units between service requester and service provider", and are predominantly designed to "describe a demand from the consumer perspective" [13].

Even less research has been done in context-dependent service *offer descriptions.* Service availability may be restricted to certain places or times. The result quality of a service may depend on network conditions, type of contract with the provider, availability of sub-services and so on. A concrete example is a printer which announces by an offer description to be available for print jobs in principle, but if a context sensor indicates a "toner low" state, this offer is a "non-preferable" one.

Any demand description must be analyzed and matched against the service offers known to the AMSR. To become a service candidate, the service provider or a representative entity called *registrar* is required to announce the service's availability, it's requirements and capabilities to the AMSR. For this purpose we propose the usage of offer descriptions called "Rich Service Offer" (RSO) according to fig. 3.

It is important to notice that each MicroService/MacroService must be announced by at least one RSO to ensure its "visibility" to the AMSR.

Offer Description
Software Interface Signature
List of Addressable Instances
Property Value Ranges (optional)
Dependency Declarations (optional)
Cost Indications (optional)
Code Snippets (optional)

Fig. 3. Rich Service Offer

4.1 Commit on Software Interfaces and Addressable Instance

During service discovery the first thing all participating entities have to commit on are the signatures of the software interfaces in use (*signature interoperability*). Thus a primary element of any RSO is a software interface signature, expressed as a WSDL *port type* identifier [4]. WSDL port types together with a namespace identifier are perfectly suitable to express in an open and extendable way, how to invoke a service, which input parameters are required and which output parameters can be expected.

If the service instance applying to be a candidate for being discovered is running somewhere in the network, but not on the mobile device (standard case for web services), this is the location identifier of the running instance, e.g. the content of the <soap:address ...> tag of the port definition of a WSDL document [4]. If the issuer of the RSO knows about the existence of an AMSR-local service implementation (e.g. by sending a code snippet along with the RSO), this may also be a location identifier pointing to a local implementation.

4.2 Classification Ontology

Beneath information about implemented interfaces and addressable interfaces, it is essential to characterize the offered service with properties any interested entity is able to evaluate (*semantic interoperability*), meaning those properties share a code space known to both parties (often called *Ontology*). Each property is expressed as a valid code out of a common taxonomy or identifier system. A taxonomy is defined as a set of categories, whereas a identifier system is defined as a set of identifiers, according to

$$taxonomy_A = \{cat_1, cat_2, cat_3, ..cat_n\} \tag{1}$$

$$identifiersystem_B = \{id_1, id_2, id_3, ..id_m\} \tag{2}$$

Samples for (1) are industry codes like NAICS, product and service classifications like UNSPSC or geographic location codes like ISO 3166. Samples for (2) are unique identifiers like Dun & Bradstreet D-U-N-S or phone numbers.

The most simple implementation of properties are key-value pairs, attached to an offer. A typical extension is to group a set of key-value pairs together, building a *template* of parameters (see UDDI `tModels` or JINI LUS `entry` attributes).

Another extension are *dynamic* properties, where the value of the property is not held within the service registry itself, but is obtained on-demand from an entity nominated by the registrar (see CORBA trading object service or Bluetooth SDP). Known variations of key-value properties are models which take into account the hierarchical relationship of typical classifications. A sample for this is the CORBA service object `ServiceType` model or the Bluetooth SDP `ServiceClassIDList`.

If two or more service offers are evaluated by their properties to be a service candidate, this means those services fulfill the minimum requirements regarding technical (e.g. implemented software interfaces) and quality level (e.g. minute-precision departure times) constraints from the client's perspective.

The minimum requirements itself may be expressed as threshold values for each evaluated property. Book [12] gives samples how to classify different service characteristics, how to project general deterministic, probabilistic and statistical properties to a common metrical base, and how to calculate the distance between any combination of $\{request(property_i), offer_j(property_i)\}$ pairs, which allows one to specify an overall graduation of service offers with respect to a specific service request. This book gives also a good overview of different metric models with the interesting observation that a complex metric model is not necessarily a guarantee for the best result.

4.3 Costs of Usage

An important part of any service offer is the indication of the costs of usage. [7] postulates any service to have a nominal price, which is defined as a charge for the service being provided. We would refine this requirement by a distinction between the price indication for service usage (*cost of usage*) and a price indication for the content (*cost of content*). A sample for the first one is a reservation-processing fee, and the latter one is a ticket price for the same train ticket reservation service.

For any non free-of-charge service, the service offer must contain information about all applicable payment systems (e.g. eCash, credit card), payment channels (e.g. network, phone) and beneficiary, which is the entity to which the payment is addressed. It should be noted that we concentrate here on the cost of usage. The cost of content may base on the same debit system, but is negotiated on the service level.

Most payment systems require online validation of any account data (e.g. remaining credit units on a prepaid account, or credit-standing of a visa account). Service implementations described by offers containing cost indications for these payment systems thus require an established online connection at the time of validation, which is usually at the time of execution of the service, and hence these services cannot be executed offline.

For MacroServices cost indications are accumulative, meaning that cost indications of the composed MicroServices are merged. The service user typically requires a valid account at least at one payment system announced in the offer of any engaged MicroService.

4.4 Dependency Declarations

It is important to notice that in cases where services are offered which internally make use of other services - "visible" to the AMSR or not - this dependency has to be indicated to the AMSR. For instance if a service announced by the current RSO is able to perform its task without any network connection, but makes use of other services via the AMSR, those other services may require a network connection.

If those sub-services indicate to be non-free in the cost-sense, any budget which has been granted during service discovery for the service represented by the current RSO must be splitted and shared with all non-free sub-services.

4.5 Code Snippet Deposition

In cases where a service implementation does not require an existing online connection to perform its operation (e.g. a credit card number validation service), a code fragment implementing the operation can be deposited at the AMSR for usage. These snippets must follow some common code exchange format, like interpretable code (e.g. Java Bytecode, JavaScript) or abstract rendering directives (e.g. XML/XSLT files defining a GUI as a service front-end [15]) Note that it must be ensured that a proper execution handler for the code snippet exists on the mobile device (e.g. Java VM, Scripting handler or Renderer). One may apply an additional protocol to negotiate a proper format. In our case this information can be obtained from the `User-Agent` HTTP header field.

In principle and in comparison to JINI full service proxies, code snippets may, but need not be, a marshaled form of an already instantiated object. The AMSR is required to instantiate the local implementation if a service demand matches with this offer and the offer is not marked as a static service. Some devices may have a Virtual Machine which does for security reasons not allow the execution of bytecode which has not been present at the time of installation of the calling application (*closed late binding*) [20]. In this case, the usage of code snippets is limited to non-Java scripting.

4.6 XML Encoding of RSOs

RSOs make use of XML respectively XML schemas as its canonical base to describe service offers with the characteristics shown in the previous sections because of XML's platform and application independent, structure preserving and extensible capabilities. It is reputable to use XML as a base for service or content descriptions, as one can compare with e.g. CAPs [16] or Centaurus' CCML [10]. The example in fig. 4 may illustrate the content of a RSO.

5 Abstract Service Invocation

Traditional mediators like the CORBA trading service mediates services in a distributed environment either according to the operation *select* or the operation

```
<?xml version="1.0" encoding="UTF-8"?>
<rso xmlns="urn:schema-RSO.xsl">
  <interface name="urn:position-service#getPosition" portType="PositionPortType"
    wsdl="http://foo.com/wsdl/position.wsdl"/>
  <instance port="http://foo.com/soap/servlet/rpcrouter"/>
    <property codesys="DUNS" code="12-345-6789" title="WGS84">
    <restriction class="region" type="country">DE</restriction>
    <precision min="1m" max="30km" shape="sector"/>
  </property>
  <dependency interface="urn:math#gaussKruegerToWGS84"/>
  <cost type="perRequest">
    <option system="LHMilesAndMore" beneficiary="foo.com" price="3" unit="miles"/>
    <option system="Visa" beneficiary="foo.com" price="5" unit="cents"/>
  </cost>
</rso>
```

Fig. 4. RSO sample

search. The first one selects one service out of the list of offers which meets the requirements best, the latter one list all service offers meeting the minimum requirements to the requesting entity for further decisions (which is in fact a delegation of a selection decision). The mediator itself is not involved in the service invocation procedure after either operation has been performed.

5.1 A New Operation: Spread & Join

We would like to introduce a third mediator operation which we called *spread& join*. This operation engages a group up to all services (best n) meeting the minimum requirements of a given request. The services are executed in parallel (*spreaded*, see fig. 5), and after a set of conditions (e.g. quality threshold passed) are reached, all results are collected and reworked (*joined*, see fig. 5) in a result object, to which the service requestor is holding a reference.

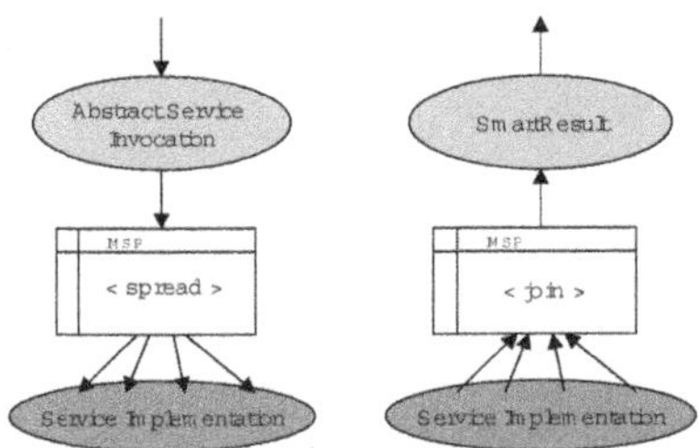

Fig. 5. Spread and Join operations

In this constellation the mediator becomes an active part of the service invocation procedure as well: It acts as a single service provider against the service requesting entity, and as a service requestor against any addressable instance. Thus, we call this entity a generic *intermediate service provider (IMSP)*. Inspired by the concept of virtual services in [21], each IMSP instance can be seen as an *abstract service* invoked by the client. At least one implementation of the abstract service is contributed by the service implementation registered for the corresponding matching offer.

Spreading a service operation over a group of n service providers according to their service offers means invocation of the same operation on n addressable instances in parallel, which leads to $m \leq n$ results reported to the IMSP. Spreading a service operation requires a budget splitting according to a distribution reasonable to the IMSP, if the operation has been restricted by a budget granted to the IMSP.

One problem is the increased the amount of exchanged messages caused by this kind of operation, especially when using expensive wireless network links. Thus an interaction between the IMSP and the Network Communication Manager shown in the AMSR (see fig. 2) is required to handle the trade-off between minimizing the costs of network connections and maximizing the gain of joining multiple results. Additionally it makes sense to hierarchically cascade the IMSP facilities over the time-variant network infrastructure, which enables the spreading of an operation partly in the wired part of the network, which is much less expensive. Cascading AMSRs and its integrated IMSP facilities guarantees a store-and-forward functionality optimal for a variable network situation. Although any results are optimal in the local context of the reporting node, they are re-evaluated after reported to the next level of the hierarchy. Any AMSR/IMSP in the hierarchy tree is responsible to act in a sense as the calling entity envisioned, e.g. not exceeding any budget granted for a specific operation [5]: If necessary, budgets have to be splitted between service providers (cost sharing).

For any service invocation request received from a client the IMSP Manager manages an instance of a control structure named *SmartResult object* containing the elements *RequestProperties*, *ListOfAddressableInstances*, *Budgets*, *CurrentBestResult* and *AllResults*. The control structure is updated when any new information is available with relevance to the particular operation. This may be the receipt of a new result, or a new context situation (e.g. transition from connected to not connected), or something else. So the spread&join operation relates very much to context-aware service discovery and execution.

An interesting option is enabled by the AMSR/IMSP's autonomy in combination with object references. If the situation, which has been the base for distance vector calculations, changes (e.g. the user moves into a new spatial context or a network link is broken), the metric for any related SmartResult object has to be re-calculated. And even if a result has already been reported to the calling entity, a "better" result stated by the re-calculated metric can be signaled to the calling entity if this entity is still interested. This affects result updates as well as result revocations.

5.2 Selecting or Merging

If more than one result from a spreaded service operation is available, it is the IMSP's task to join them into a single result for the client. Figure 6 gives an overview of the available options.

client expects	join options
single primitive data type (e.g. *integer*)	select one sub-result
single object data type (e.g. *java.lang.String*)	select one sub-result (may be updated when reference given)
array of primitive data type (e.g. *integer[]*) or array of objects (e.g. *java.lang.Boolean[]*)	merge sub-arrays or select one sub-array
collection object of primitive data type (e.g. *java.lang.Vector* containing *integer* elements) or collection object of object data type (e.g. *java.lang.Vector* containing *java.lang.String* elements)	merge sub-collections or select one sub-collection (may be updated when reference given)

Fig. 6. IMSP join options

If the return value expected by the client is a single primitive data type, the IMSP must extract one result out of the received results. If the return value expected by the client is a single object(-reference), the task of the IMSP is the same, extended by the ability to adopt values from a modified result object upon following sub-results with a postponed distributed object access. If the return value expected by the client is an array of primitive data types, it may be desirable to merge the sub-results, or to select a sub-result along certain criteria (e.g. longest array). If the return value expected by the client is a collection object like the Java Vector, the facilities of merging or selecting sub-results and distributed object access are combined.

Different strategies may be followed to perform this join, depending on the termination condition specified by the client as well as the type of return value and the desired quality. If the client specified to be interested in any result in a given period of time (e.g. the first result the IMSP receives within the next 2 seconds), the join strategy is pretty simple. But much more often a client claims results according to certain quality parameters. This requires the IMSP to re-order all sub-results according to a given metric.

Any parameter which is required during the calculation of distance vectors (which is the base operation for any metric) must be given by the client together with the call itself, either implicit (e.g. input parameter of a function) or explicit (e.g. vector distance procedure identifier or timeout value). Because the quality of the operation which a service performs may differ from the quality level promised during registration, additional taxonomy or identifier system codes attached to the sub-results may help the IMSP to adapt better to the calling entity's desired result.

6 Implementation Prototype

A prototype of the AMSR is currently under development, based on Java2 Micro Edition (J2ME), whose Virtual Machine (CVM/KVM) is designed to run on resource constraint devices from SetTopBoxes down to mobile phones or PDA's.

Devices with the most restrictive limitations are covered by the *Mobile Information Device Profile* (MIDP) on top of the *Connected Limited Device Configuration* (CLDC), which define the subset of Java classlibraries a CLDC/MIDP compliant device must support.

To not overstrain the computational power, network and memory resources of small mobile devices running CLDC/MIDP, a lot of functionality known from full Java (Standard or even Enterprise Editions) have been omitted. For instance the late binding concept of Java has been restricted to be able only to load classes at runtime, which have been downloaded and installed at the same time as the application itself (*closed late binding*). Thus it is impossible to use a service on a mobile device, whose implementation is loaded from the network at runtime, which is the standard procedure e.g. in a JINI network. But even if full late binding would be possible on CLDC/MIDP compliant devices, JINI like approaches would not work because of the lack of standard network access via sockets: The only available protocol is HTTP in client mode. Additional profiles may contribute additional functionality, but devices targeted by J2ME will always be more restricted than devices in the fixed wired world.

Hence, our AMSR implementation considers this situation and is built using an XML parser on top of HTTP as transfer protocol. An XML parser is a valuable application type for Java equipped mobile devices, and there exist several different implementations (kXML, NanoXML etc.). As described in section 1, our approach employs the web service technology of SOAP and WSDL, both based on XML and HTTP, and the footprint of the current prototype (still missing the Network Communication Manager block) is about 75 KB.

7 Summary, Conclusions, and Further Work

We have presented a new concept for service discovery and execution on resource constraint programmable mobile devices. Discovery and execution of services is provided if connected, but also if not connected to a backbone system. Discovery and execution procedures are sensitive to context changes. This is achieved by running a service registry called AMSR on the mobile device itself, which is updated with service announcements called RSO. Multiple service providers are, if available, employed to fulfill a service request concurrently. This parallel execution of services is hidden to the user by the abstract service interface, which is invoked by the client, and delivers the optimal result from the AMSR after execution.

First results with the prototype implementation showed that the concept is valuable and outputs the desired results.

Further work has to be done in the area of RSO schema definition. Several strategies to prioritize service offerings have to be worked out and compared to decide which fits best to the limits in storage space and processing power of the target devices. Scalability and inter-system tests have to be performed, and a wider range of sample services must show if all aspects are covered and reasonable. Further investigation has to be done if results from agent systems can be applied to our system (e.g. when comparing spread&join with an agent federation architecture), and if approaches for automated service chaining can be utilized in our approach.

References

[1] Beck, J., Gefflaut, A., and Islam, N. MOCA: A service framework for mobile computing devices. In *Proceedings of the ACM International Workshop on Data Engineering for Wireless and Mobile Access, August 20, 1999, Seattle, WA, USA* (1999), ACM, pp. 62–68.

[2] Bettstetter, C., and Renner, C. A comparison of service discovery protocols and implementation of the service location protocol. In *In Proceedings of Sixth EUNICE Open European Summer School - EUNICE 2000* (Twente, Netherlands, September 2000).

[3] Chakraborty, D., and Chen, H. Service discovery in the future for mobile commerce. http://www.cs.umbc.edu/~dchakr1/papers/mcommerce.html, 2000.

[4] Christensen, E., Curbera, F., Meredith, G., and Weerawarana, S. Web Services Description Language (WSDL). http://www.w3.org/TR/wsdl, 2001.

[5] Dalmeijer, M., Hammer, D., and Aerts, A. Mobile software agents. http://wwwis.win.tue.nl/~wsinatma/Agents/MSA.ps, 1997.

[6] Dey, A. K. Understanding and using context. *Personal and Ubiquitous Computing, Special issue on Situated Interaction and Ubiquitous Computing 5*, 1 (2001).

[7] Dumas, M., O'Sullivan, J., Heravizadeh, M., Edmond, D., and ter Hofstede, A. Towards a semantic framework for service description. http://sky.fit.qut.edu.au/~dumas/publications.html, April 2001.

[8] Franklin, S., and Graesser, A. Is it an agent, or just a program?: A taxonomy for autonomous agents. In *Proceedings of the Third International Workshop on Agent Theories, Architectures, and Languages* (1996), Springer-Verlag.

[9] Gribble, S. D., Welsh, M., von Behren, R., Brewer, E. A., Culler, D. E., Borisov, N., Czerwinski, S. E., Gummadi, R., Hill, J. R., Joseph, A. D., Katz, R. H., Mao, Z. M., Ross, S., and Zhao, B. Y. The ninja architecture for robust internet-scale systems and services. *Computer Networks, Special Issue on Pervasive Computing 35*, 4 (March 2001), 473–497.

[10] Kagal, L., Korolev, V., Chen, H., Joshi, A., and Finin, T. Project centaurus: A framework for indoor mobile services. http://www.cs.umbc.edu/~finin/papers/centaurus/.

[11] Kammann, J., Strang, T., and Wendlandt, K. Mobile services over short range communication. In *Workshop Commercial Radio Sensors and Communication Techniques - CRSCT 2001* (Linz/Austria, August 2001).

[12] Linnhoff-Popien, C. *CORBA - Communications and Management*. Springer, September 1998.

[13] Michahelles, F. Designing an architecture for context-aware service selection and execution. Master's thesis, University of Munich, 2001.

[14] Nord, J., Synnes, K., and Parnes, P. An architecture for location aware applications. In *Proceedings of the Hawai'i International Conference on System Sciences, Big Island, Hawaii* (January 2002), IEEE.

[15] Romn, M., Beck, J., and Gefflaut, A. A device-independent representation for services.

[16] Samulowitz, M., Michahelles, F., and Linnhoff-Popien, C. Capeus: An architecture for context-aware selection and execution of services. In *New developments in distributed applications and interoperable systems* (Krakow, Poland, September 17-19 2001), Kluwer Academic Publishers, pp. 23–39.

[17] Schmidt, A., Beigl, M., and Gellersen, H.-W. There is more to context than location. *Computers and Graphics 23*, 6 (1999), 893–901.

[18] Schmidt, A., and Laerhoven, K. V. How to build smart appliances. *IEEE Personal Communications* (August 2001).

[19] Steingass, A., Angermann, M., and Robertson, P. Integration of navigation and communication services for personal travel assistance using a jini and java based architecture. In *Proc. GNSS '99* (Genova, Italy, October 1999).

[20] Strang, T., and Meyer, M. Agent-environment for small mobile devices. In *Proceedings of the 9th HP OpenView University Workshop (HPOVUA)* (June 2002), HP.

[21] Wang, Z., and Garlan, D. Task-driven computing, May 2000.

Nomad: Application Participation in a Global Location Service

Kris Bubendorfer and John H. Hine

School of Mathematical and Computing Sciences, Victoria University of Wellington, P.O.Box 600, Wellington 6001, New Zealand
Kris.Bubendorfer@vuw.ac.nz, John.Hine@vuw.ac.nz

Abstract. The rapid growth of the world wide web has led to the wider appreciation of the potential for Internet-based global services and applications. Currently services are provided in an *ad hoc* fashion in fixed locations, which users locate and access manually. Mobility adds a new dimension to the location of such services in a global environment. Most systems supporting mobility rely on some form of home-based redirection which results in unacceptable residual dependencies. Nomad is a middleware platform for highly mobile applications. A significant contribution of the Nomad platform is a novel global object location service that involves the participation of both applications and global structures.

1 Introduction

Mobile code is a powerful paradigm for the creation of distributed software [1]. Applications constructed of mobile code are able to execute more efficiently, exploit heterogeneity and dynamically adapt to changes in their environment. The Nomad (Negotiated Object Mobility, Access and Deployment) [2] architecture is a platform that provides a distributed systems infrastructure to support applications constructed of mobile code. The challenges faced by such applications include: the discovery of resources required for execution, the establishment of rights to access these resources, and the ability of an application to locate and coordinate its component parts. These issues are addressed by three complementary platform elements. Firstly Nomad embodies an electronic Marketplace, through which applications locate and obtain the resources they require. Secondly, applications form contracts with the virtual machines on which they execute, specifying the quality of service to be provided. Lastly, Nomad provides a global location service, to track the migration of code throughout the system, and to enable the various components of an application to coordinate, in order to satisfy their collective goals. The ability to rapidly and reliably locate a mobile object or other resource is a critical and limiting factor in the growth of any large scale distributed system.

This paper presents a brief overview of the Nomad architecture in Sec. 2, followed by Sec. 3 which details the focus of this paper — the Nomad location service. The novel design of the location service involves the participation of both applications and global infrastructure in the task of locating mobile objects. Finally, Sect. 4 introduces the Nomad prototype.

M.-S. Chen et al. (Eds.): MDM 2003, LNCS 2574, pp. 294–306, 2003.

2 An Overview of the Nomad Architecture

An application within Nomad is a logical grouping of mobile code that performs one or more services (functions) for some client (another application, user etc.), employing the resources of one or more Depots, see Fig. 1. An application can range from a single mobile agent, implemented within a single piece of code and offering a single service, through to a collection of many pieces of mobile code offering many different services. The distinction is that an *application* is an autonomous unit of administration and identity, made up of cohesive parts that are managed collectively.

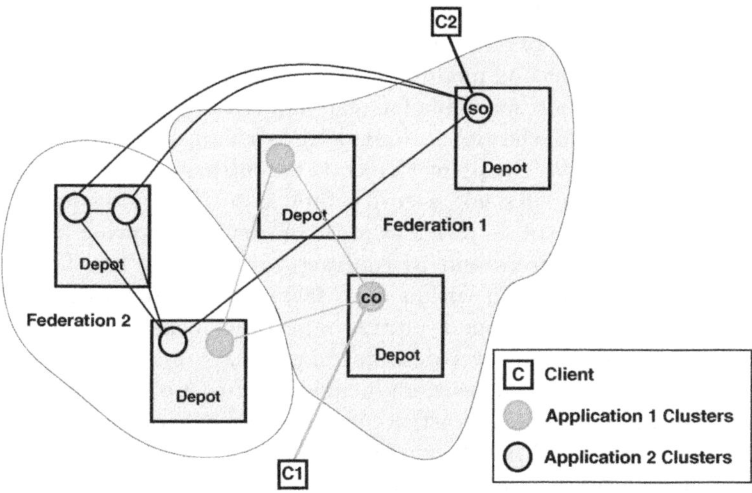

Fig. 1. *Interactions.*

The individual objects of an application are collected into Clusters, that are the basic unit of mobility in Nomad. The Cluster is a sophisticated wrapper that provides mechanisms for mobility and the isolation of namespaces. The Cluster groups language objects, such as integers and strings, which are too small to be independently mobile. References outside the Cluster take place as interprocess communications via proxies, while those within a Cluster are handled by standard run-time invocation mechanisms.

An application is responsible for: initiating negotiation for resources, directing the distribution of its *own* clusters, handling unrecoverable failures, and maintaining part of the location service. Each application controls its degree of distribution by placing and replicating its clusters to achieve QoS goals such as reduced latency to clients, redundancy and concurrent processing. The application negotiates with Depots via the Nomad Marketplace for the resources that it requires.

Each Nomad application is required to have at least one: Contact Object (CO), Service Object (SO) and Location Table (LT). A client seeking a service from an application must first obtain a reference to a CO. It then negotiates with the CO to obtain the desired service from the application. After a successful negotiation, the CO presents the client with a reference to a SO that will provide the client with the negotiated service. This is illustrated in Fig. 1 where client C1 is bound to the CO of Application 1, with which it is negotiating. Client C2 has completed negotiations, and is receiving service from a SO of Application 2.

Each Depot [3] consists of a Depot Manager and a set of managed virtual hosts (vHosts). Management policies [4] dictate how each Depot will react to specific circumstances: how it will ensure levels of service (QoS) and how it interacts with Depots within and outside of its federation. A federation of Depots (as shown in Fig. 1) reflects common ownership or administration of a set of Depots, and is used to organise such things as the balance of work within the federation, or the prices and quality of service ranges that the Depots offer. The Depot Manager is also responsible for responding to negotiation, charging, and arranging the hosting of the global Nomad infrastructure components. A vHost is a virtual machine on which application code executes. A vHost is responsible for managing physical resource use, security, fault detection, and communication.

The Nomad Marketplace provides a set of services allowing applications to discover available resources and to contract with Depots for the provision of those resources. Negotiation within the Marketplace utilises the Vickrey auction [5] protocol. Applications requiring a contract construct a description of their requirements in a resource description graph.

The combination of the consistent local infrastructure provided by the Depot architecture, and the global infrastructure provided by Nomad services, combine to form a consistent and interlocking environment for distributed mobile applications. The location service provides dynamic binding to mobile objects and the initial discovery of services.

3 The Nomad Location Service

Location service research has previously concentrated on architectures with a geographical or organisational basis [6, 7, 8, 9, 10]. Most middleware location services, such as those offered by CORBA and Java RMI, do not address the question of mobility on a global scale, nor solve problems with residual dependencies[1]. Other technologies such as Mobile IP [11] do not address location independence, transparency or residual dependencies.

A location service needs to track the location of each mobile object throughout that object's lifetime. In a large scale distributed system the location service may have to deal with billions of objects, some of which will move frequently. Scaling a system to this degree requires wide distribution, load balancing and a

[1] A residual dependency arises when an object *relies* upon state remaining on a machine on which it is no longer resident. This includes both the machine on which the object was instantiated, and any machines on to which it subsequently migrated.

minimum of coordinating network traffic. The general capabilities required for the Nomad location service are that: a client should be able to locate and bind to a target application for which it holds only a name, the relocation of an object should be supported in a manner that is transparent to the users of that object, an application should be able to locate and bind to all of its element objects, and the service should scale.

3.1 Naming

There are many naming systems that exist side by side and will continue to do so. The Internet Engineering Task Force uniform resource naming scheme (URN) provides a meta naming service that enables a client to locate the appropriate name resolution service. We take advantage of this scheme and integrate Nomad within it, so as to utilise this seamless and general naming strategy.

The URN [12, 13, 14] is a human readable name which is translated into a physical location (possibly a URL) by a URN resolver.

> *The purpose or function of a URN is to provide a globally unique, persistent identifier used for recognition, for access to characteristics of the resource or for access to the resource itself [12].*

URNs are intended to make it easy to map other namespaces (legacy and future support) into the URN-space. Therefore the URN syntax must provide a means to encode character data in a form that can be sent using existing protocols and transcribed using most keyboards [15].

Each URN has the structure `<URN> ::= "urn:"<NID>":"<NSS>`. The `NID` (namespace identifier) distinguishes between different namespaces, or contexts, which may be administered by different authorities. Such authorities could include ISO, ISBN, etc. The `NSS` (namespace specific string) is only valid within its particular `NID` context, and has no predefined structure. As pointed out in [13], the significance of this syntax is in what is missing, as it is these omissions which make the URN persistent. That is, there is no communications protocol specified, no location information, no file system structure and no resource type information.

URN resolution can be broken down into two steps; finding a resolver for a particular URN, and then resolving that URN. Finding a specific resolver requires a Resolver Discovery Service (RDS) [16, 14] as shown in Fig. 2. It is important to emphasise that the RDS need not cope with the high degrees of mobility, as found in Nomad, but rather must cope with the lesser mobility of NID resolvers.

The Nomad location service fulfils the role of URN Resolver for the Nomad context. However, the location problem is not addressed by the URN scheme, and solutions to the location problem are domain specific. We now turn our attention to the design of the Nomad location service.

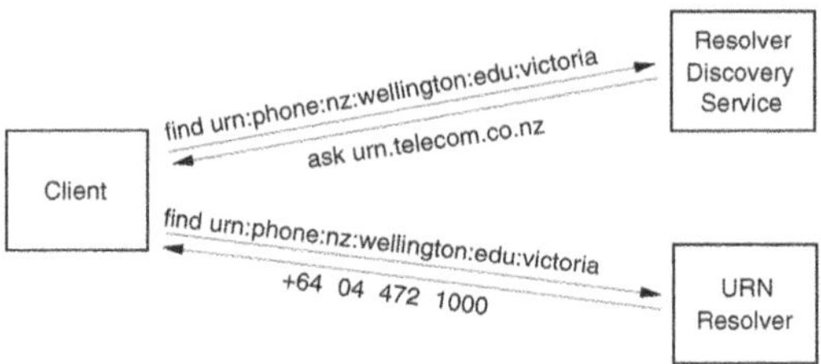

Fig. 2. *Resolver Discovery Service.*

3.2 Exploiting Application Locality

Existing systems, such as those cited in Sect. 1, have one thing in common: a reliance on geography — either based partly on country codes as in the DNS, or totally as with the quad tree mapping in Globe. The major argument behind Globe's use of a geographical structure is to provide locality of reference. In highly mobile systems such as Nomad this no longer holds true, as application objects are envisioned as using their mobility to move closer to clients and resources. Using an organisational basis is also problematic for the same reasons.

Instead there is another form of locality which can be exploited — the locality of reference within an application. If you interact with an application object once, chances are that you are quite likely to interact with it, or a related object from the same application, again.

The Nomad location service endeavours to take advantage of this and builds the basis of its location service around the concept of the application. That is, each application is wholly responsible for tracking and locating its objects on behalf of itself, the Nomad system and its clients. In order to do this, the application maintains a set of location tables (LTs) that hold the current locations of all of the application's Clusters, and consequently their objects. In addition to providing locality of reference, an application's location tables exhibit a high degree of inherent load distribution (they are as distributed as the applications themselves), and permit the application to tailor its location table to suit its needs and reference characteristics.

3.3 Design Overview

The Nomad location service must meet two distinct requirements; the discovery of services and the resolution of out-of-date bindings. Discovery provides an external interface used to establish an initial binding, while rebinding occurs internally and transparently. These two systems act independently but synergistically and are illustrated in Fig. 3. In addition, this figure highlights the separation between the Nomad level services above the dividing line, and the application level ones below.

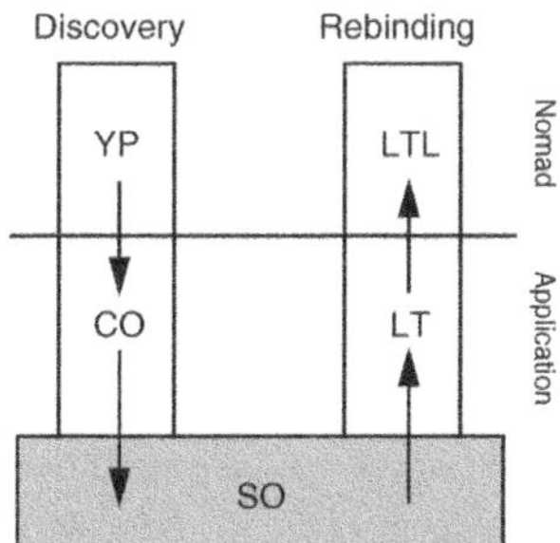

Fig. 3. *Discovery and rebinding are complementary parts of the location service.*

The following list is a brief overview of the various location service components which feature in Fig. 3:

- **Yellow Pages (YP):** The Yellow Pages is the Nomad URN resolver, returned by the URN RDS from Fig. 2. The Yellow Pages takes a textual URN string and returns a reference to a contact object representing the service named. This is a Nomad level service.
- **Contact Object (CO):** Part of the application responsible for negotiating with the client to provide needed services. Once negotiation is successfully completed, the contact object returns a service object interface to the client.
- **Service Object (SO):** This is any object within the application which provides a contracted service to a client. It is to this object that the final binding is made.
- **Location Table Object (LT):** A required part of each application within Nomad. A location table contains references to all of the Clusters composing an application and is responsible for tracking their movement.
- **Location Table Locator (LTL):** The location table locator is a global Nomad service. It is responsible for locating and tracking all the location tables of all of the applications executing within Nomad. Each Depot is responsible for maintaining a cached reference to the LTL. As with all services, this reference is initially obtained from the Yellow Pages.
- **Local Interface Repository (LIR):** This is not part of the location service as such (not shown in Fig. 3), rather it is part of the Depot architecture. It is maintained by the local Depot and acts as a source of useful references, such as the LTL. The LIR may be accessed by both applications and vHosts resident on a particular Depot.

Each mobile Cluster has a reference to a location table object, which is shared by all mobile Clusters belonging to the same application. The location table object, which is in its own mobile Cluster, has in turn a reference to the global location table locator. All objects created in a mobile Cluster share the same location table.

As an example, consider a client which holds a textual URN for a service which it needs to use. This requires the discovery facet of the location service and the first step is to find a resolver for the URN using the RDS. Once the RDS system identifies the Yellow Pages as the Nomad URN resolver, the name is resolved returning a binding to an application specific contact object. The service is then provided to the client via a service object returned through the contact object.

When an object moves, all existing bindings held on that object become outdated. Resolving these bindings needs the rebinding hierarchy of the location service. Once a binding is outdated, subsequent invocations on that object will fail. The local proxy representing the binding will transparently rebind via a cached reference to the object's Cluster's location table. As the location table is also a mobile object, it may also have moved, requiring the rebinding to resort the next level — the location table locator (LTL).

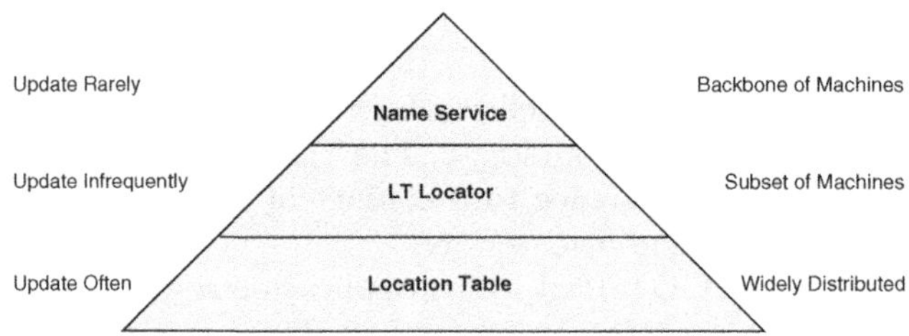

Fig. 4. *Division of Labour.*

Figure 4 illustrates the division of labour between the components of the naming and location service resulting from this location service architecture. As the load increases towards the bottom of the pyramid, so does the degree of distribution.

The remainder of this paper is dedicated to discussing in greater detail, the individual location service components. Sections 3.4 through 3.7 are arranged following the different phases of an application's life within the system, that is, its creation, registration, discovery, mobility and rebinding.

3.4 Creating an Application

Let's first consider creating an application from outside Nomad. To do this, a launcher negotiates for an initial vHost on which the application can be started.

The launcher then invokes the createApplication method on the vHost, triggering the events detailed in Fig. 5. The vHost starts by creating a Cluster for holding the new application's location table, and sets the resolver[2] for the location table Cluster to be the location table locator. The location table object is then instantiated inside the new Cluster.

The vHost next creates a Cluster for the application object and sets its resolver to the newly created location table object. It then instantiates an object

[2] Each mobile Cluster has a cached reference to its resolver.

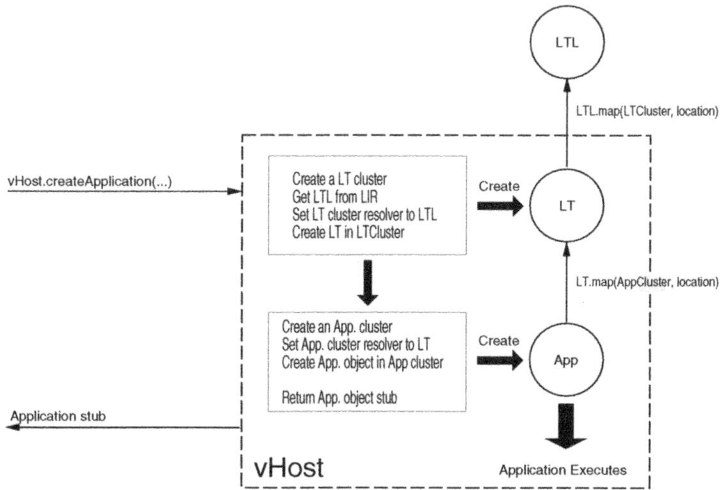

Fig. 5. *Creating an application in Nomad*

of the specified application class in the application Cluster and returns to the launcher the interface (if any) on the application object. From this point the application will start to execute and construct itself independently of the launcher.

The steps for one Nomad application to create another application are similar.

3.5 Registering a Service

The creation of the application will not automatically create a corresponding entry in the Yellow Pages. It is now up to the application to register a URN via the steps illustrated in Fig. 6. Otherwise it can simply remain anonymous to the outside world and respond only to the holder of the interface returned upon its creation.

The application starts by obtaining a reference to a contact object (CO) representing itself, usually by creating it. The next step is to find the Yellow Pages with which the application needs to register, by using the Depot's LIR to acquire a reference to the Yellow Pages. With the Yellow Pages stub, it then registers the URN::CO association.

There is no dictate over what mappings are registered, combinations could include; one URN to one contact object, multiple URNs to one contact object, or multiple URNs to multiple contact objects.

As an example of what may be registered, consider the following URN which an application wishes to register: `urn:nomad:apps/currencyconverter`. The associated contact object may offer different levels of service e.g. gold, silver and bronze, and will return the appropriate service object for each level. Alternatively the application could register three different URNs, each with a unique contact object for the individual service levels.

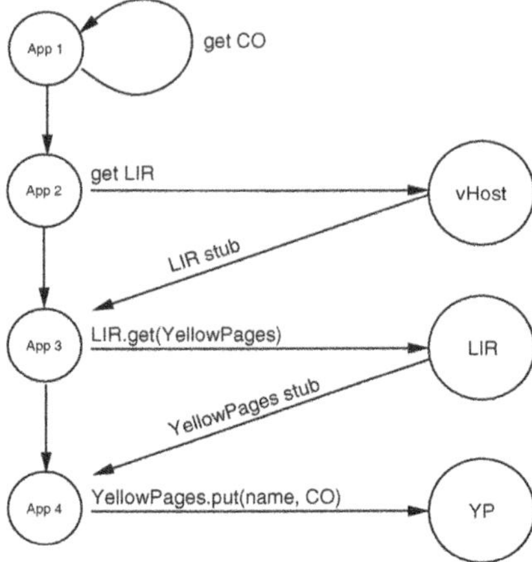

Fig. 6. *Registering a service*

3.6 Obtaining a Service

Figure 7 illustrates the four steps and seven messages which are needed to obtain a service from an application, starting from a textual service name. The example given here shows the initial binding to a Whiteboard application — from outside the Nomad system. From within the system, the RDS step would be unnecessary as the reference to the Yellow Pages is available from the LIR, the external access is shown here for completeness.

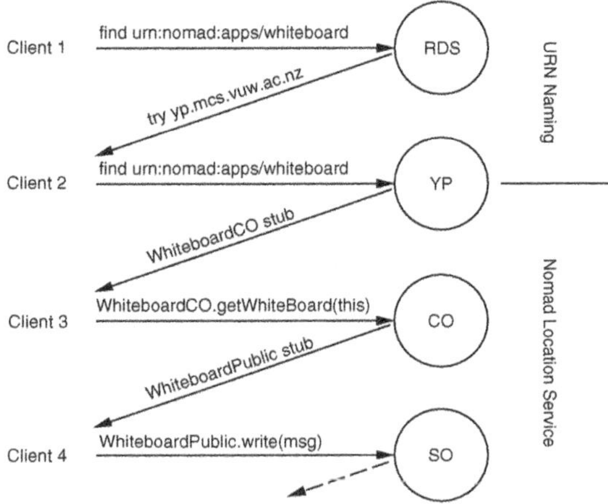

Fig. 7. *Locating an Object.*

The first step involves the discovery of the URN resolver that resolves names for the domain `urn:nomad`. In Nomad the URN name resolver is the Yellow Pages, to which a reference is returned by the RDS. The client then queries the Yellow Pages with the same URN, and in reply gets a reference to the Whiteboard contact object. The methods which a client can invoke on a contact object depend upon the application — in this case it is a request for a shared Whiteboard. The Contact object returns a WhiteboardPublic stub to the client which is an interface on the service object and is used to write to the Whiteboard.

All four steps need only be performed once during interaction with a service. However, when dealing with mobile objects it is possible that an object will have moved since it was last referenced. This then requires rebinding to occur transparently, as detailed in Sect. 3.7.

3.7 Moving and Rebinding an Object

From the perspective of the location service, moving an existing object is similar to creating a new object. The migration process transparently invokes the map method on the location table responsible for tracking the migrating object's Cluster. This ensures the correct binding of future references, but does not resolve the issue of outstanding references which no longer specify the correct location.

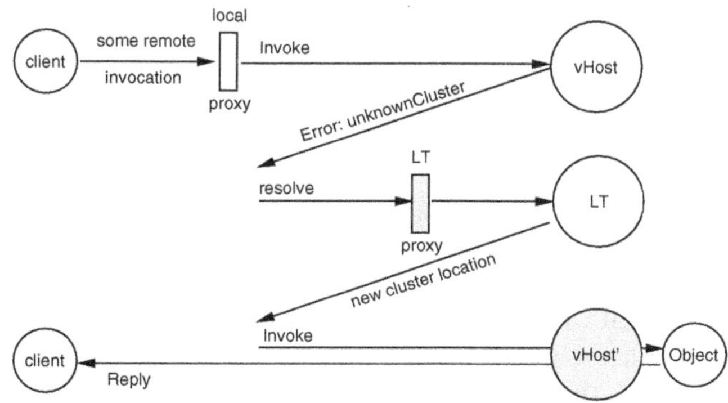

Fig. 8. *Rebinding to an Object.*

Fig. 8 illustrates what happens when one such outdated reference is used to invoke a method on a remote object. The invocation by the client causes the local proxy to use the object reference it held from the previous invocation. The proxy contacts a vHost which no longer hosts the target object. When the vHost's communications stack attempts to invoke the remote method on the object an exception occurs and an error is returned to the invoking proxy.

The local proxy next uses a cached reference to the location table responsible for tracking the target object's Cluster to request a new binding. The location table replies with the new Cluster location and the proxy reissues the original invocation to the object on its new vHost and caches the new location. Note that the call to the location table is simply another remote invocation via a proxy. Since the location table is also mobile, this may result in a cascaded rebinding, as shown in Fig. 9.

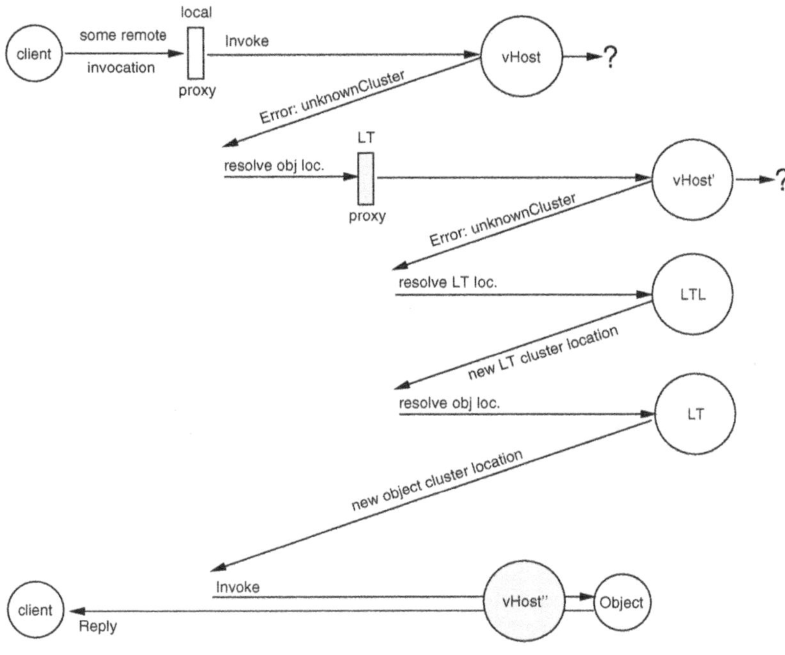

Fig. 9. *Cascaded location table rebinding.*

In Fig. 9, the call to the location table by the location table proxy fails. As in Fig. 8, this results in an error being returned to the location table proxy. The location table proxy then acts in exactly the same way as the local proxy and uses its cached reference to contact the location table locator responsible for tracking the target location table Cluster. The new Cluster location is returned, the location table's location is updated within the proxy, and the original query reissued to the location table. The location table returns the new location of the target object's Cluster to the original proxy, and the original invocation is reissued.

In the two previous examples all rebinding took place dynamically and transparently to the application. It is worth summarising a few cases where transparency can no longer be maintained:

– In the case where an **object no longer exists**, the same steps as shown in Fig. 8 take place, except that the rebinding fails. At this point transparency

can no longer be maintained and the failure is handed back to the client. How the application recovers is application dependent.
- If the **object no longer exists but was one of a set of replicas**, the default behaviour is to fail as previously. This behaviour supports strong replication in which replicas are distinguished and state may have been maintained by the lost object. However, the application can specialise the Location Table class to implement weak replication, in which case rebinding would return a suitable replica.
- Where there is a **complete failure of the application**, and the rebind through the location table locator fails to find a valid location table, then assuming the application can restart, all steps from Fig. 7 need to be repeated.

A significant advantage of the separate rebinding hierarchy is now evident — it does not matter if the contact object references held by the Yellow Pages are outdated, as they will be automatically rebound the first time they are invoked and found to be incorrect. This means that unless the application changes its contact object, the Yellow Pages does not need to be immediately updated to reflect changes due to mobility of the contact objects. Instead, the Yellow Pages can utilise a lazy, best effort update as omissions, delays and inconsistencies will be caught and then corrected by the rebinding system.

4 Implementation

The location service has been implemented in a Java prototype of the Nomad architecture. Location tables are created and managed automatically as part of the Nomad mobile Cluster package. The system accommodates transparent replication of location tables, consistency and fault tolerance, details of which appear in [2].

This paper has concentrated on the overall architecture of the twin discovery and resolution hierarchies, and the novel application level location tables. Development of a specific location table location service would be redundant, as other research efforts, in particular Globe [8, 9, 10], are readily applicable solutions for this component of the system.

5 Summary

This paper presents a novel solution to the design of a distributed location service for large scale mobile object systems. The use of the application to optimise the distribution of the location tables, limits the impact of the majority of updates to these small and infrequently mobile data structures. The remaining global workload only involves resolving the locations of the location tables themselves, ensuring an implicit ability to scale. Higher level bindings are stable within the discovery hierarchy.

The application architecture, which separates the roles of contact and service objects, along with the use of application specific location tables, encourages an application to intelligently distribute itself to meet negotiated client QoS requirements.

References

[1] Lange, D.B., Oshima, M.: Seven Good Reasons for Mobile Agents. Communications of the ACM **42** (1999) 88–89

[2] Bubendorfer, K.: NOMAD: Towards an Architecture for Mobility in Large Scale Distributed Systems. PhD thesis, Victoria University of Wellington (2001)

[3] Bubendorfer, K., Hine, J.H.: DepotNet: Support for Distributed Applications. In: Proceedings of INET'99, Internet Society's 9th Annual Networking Conference. (1999)

[4] Damianou, N., Dulay, N., Lupu, E., Sloman, M.: PONDER: A Language for Specifying Security and Management Policies for Distributed Systems. Technical Report DoC 2000/1, Imperial College of Science Technology and Medicine (2000)

[5] Vickrey, W.: Counterspeculation, Auctions, and Competitive Sealed Tenders. The Journal of Finance **16** (1961) 8–37

[6] Mockapetris, P.: Domain Names: Concepts and Facilities. Network Working Group, Request for Comments (RFC) 1034 (1987)

[7] Comer, D.E.: Computer Networks and Internets. 1st edn. Prentice-Hall (1997)

[8] Ballintijn, G., van Steen, M., Tanenbaum, A.S.: Exploiting Location Awareness for Scalable Location-Independent Object IDs. In: Proceedings of the Fifth Annual ASCI Conference, Heijen, The Netherlands, Delft University of Technology (1999) 321–328

[9] van Steen, M., Homburg, P., Tanenbaum, A.S.: Globe: A Wide-Area Distributed System. IEEE Concurrency **7** (1999) 70–78

[10] van Steen, M., Hauck, F.J., Homburg, P., Tanenbaum, A.S.: Locating Objects in Wide-Area Systems. IEEE Communications Magazine **36** (1998) 104–109

[11] Tanenbaum, A.S.: Computer Networks. 3rd edn. Prentice-Hall (1996)

[12] Sollins, K., Masinter, L.: Functional Recommendations for Internet Resource Names. Network Working Group, Request for Comments (RFC) 1737 (1994)

[13] Sollins, K.R.: Requirements and a Framework for URN Resolution Systems. Internet Engineering Task Force (IETF) Internet-Draft (1997)

[14] Slottow, E.C.: Engineering a Global Resolution Service. Master's thesis, Laboratory for Computer Science, Massachusetts Institute of Technology (1997)

[15] Moats, R.: URN Syntax. Network Working Group, Request for Comments (RFC) 2141 (1997)

[16] Iannella, R., Sue, H., Leong, D.: BURNS: Basic URN Service Resolution for the Internet. In: Proceedings of the Asia-Pacific World Wide Web Conference, Beijing and Hong Kong (1996)

Mobiscope: A Scalable Spatial Discovery Service for Mobile Network Resources*

Matthew Denny[1], Michael J. Franklin[1], Paul Castro[2], and Apratim Purakayastha[2]

[1] U.C. Berkeley, Berkeley, CA, USA
{mdenny,franklin}@cs.berkeley.edu
[2] IBM T.J. Watson Research Center, Hawthorne, NY, USA
{castrop,apu}@us.ibm.com

Abstract. Mobiscope is a discovery service where clients submit long-running queries to continually find sets of moving network resources within a specified area. As in moving object databases (MODBMSs), moving resources advertise their positions as functions over time. When Mobiscope receives a query, it runs the query statically against advertisements (ads) that are currently cached by Mobiscope, and then continuously over ads that Mobiscope receives subsequently. For scalability, Mobiscope distributes the workload to multiple nodes by spatial coordinates. Application-level routing protocols based on geography ensure that all queries find all matching resources on any node without significant processing overhead. Simulation results indicate that Mobiscope scales well, even under workloads that stress Mobiscope's distribution model.

1 Introduction

With the growth in wireless connectivity, we are increasingly surrounded by objects that are connected to the network. Many of these items are highly mobile, such as cars, cell phones, PDAs, and even mobile sensor stations floating in rivers used to forecast weather (e.g. [19]). Many new applications require the discovery of mobile network resources keyed on position. Location-based service providers may want to find cars within certain geographic regions, and monitoring applications for mobile sensor stations may want to know sensor station density within some area in order to adjust sampling rates.

Mobiscope is a service for discovering moving network resources. Mobiscope accepts advertisements (ads) from network resources containing position information, and queries for network resources within a specified spatial region. For each query, Mobiscope returns a set of resources that are inside the region, and keeps this set current as resources move over time. Ads represent position as

* Franklin and Denny have been partially supported by the NSF under grants IIS-0086057 and SI-0122599, by an IBM Faculty Partnership Award, and by research funds from Intel, Microsoft, and the UC Micro program.

M.-S. Chen et al. (Eds.): MDM 2003, LNCS 2574, pp. 307–324, 2003.

functions over time. Functions allow a system to interpolate the position of a resource between successive ads. A resource does not have to update its advertised position function until the function no longer reflects the resource's position.

Position functions originate in the moving objects database (MODBMS) literature [17, 15, 21, 12]. MODBMSs can use position functions to predict which objects will be within a given spatial region for some time interval. Although this previous work is relevant to Mobiscope, current MODBMSs have two major limitations when position functions are rapidly changing. First, MODBMSs do not provide the correct semantics for Mobiscope queries because MODBMSs derive future positions based only on advertisements that are currently stored by the MODBMS. If resources re-advertise new position functions after the MODBMS processes a query, the answer may become incorrect. Second, MODBMSs store position functions in a static, centralized database. These databases may not scale to a large number of moving objects if the objects rapidly update their position functions. Mobiscope must not suffer from these limitations if it is to be deployed in environments where many resources change position functions rapidly (e.g. a service tracking cars that start, stop, and change direction often in a large city).

Mobiscope addresses both the semantics problem and the scalability problem of MODBMSs. In Mobiscope semantics, the service first runs each new query statically over cached ads, and then continuously over new incoming ads. To ensure that clients maintain correct result sets for each query, Mobiscope employs a unique model that combines results from the static and continuous portion of each query.

To achieve scalability, ad and query processing in Mobiscope is distributed to a set of *directories* in a network. Each directory has a spatial *service region*, which dictates the ads and queries for which the directory is responsible. The Mobiscope network design accommodates large numbers of moving resources that communicate using wide-area wireless infrastructure. Mobiscope directories can be deployed alongside wireless infrastructure, such as basestations, basestation controllers, or gateways. The service regions of each directory can be defined by the range of the corresponding piece of infrastructure. If we assume that the density of wireless infrastructure per spatial region reflects the density of resources advertising in the region, such a deployment is likely to partition the workload efficiently. Mobiscope directories function at the application level, and can work over any underlying wireless infrastructure. In our environment, we assume that directories do not come online, go offline, or change service regions unless a) the total area covered by a Mobiscope network is changed, or b) a directory fails.

Mobiscope ensures *full reachability* such that, in the absence of failures, the response set for each query returns every matching ad in the system. Mobiscope directories route ads and queries among themselves to maintain full reachability. The routing protocols are designed to conserve processing overhead, which can become significant if the protocols are poorly designed.

To cope with failures, Mobiscope stores both ads and queries as *soft-state*. That is, Mobiscope directories do not require any expensive logging and recovery mechanisms because resources and clients periodically re-transmit ads and queries, respectively. If a directory fails and restarts or becomes temporarily disconnected from the network, it will eventually be repopulated with soft-state ads and queries, and every response set eventually contains every matching ad. Previous service discovery work (e.g. [11, 1, 2, 22, 9]), stores ads as soft-state. However, Mobiscope also stores continuous queries as soft-state.

To evaluate Mobiscope, we implemented a simulator and workload generator. We generate and run workloads that model a large number of rapidly updating resources and stress Mobiscope's distribution model. In these experiments, Mobiscope performs quite well, even under workloads that are less than ideal for its distribution model.

The rest of the paper is structured as follows. Section 2 describes the components of Mobiscope, as well as the query semantics. Section 3 presents the distributed processing algorithms used by Mobiscope. Section 4 presents the performance study, Section 5 discusses related work, and Section 6 concludes the paper.

2 Mobiscope Overview

In this section, we present the structure and semantics of the Mobiscope system. In Section 2.1, we introduce the major components of the system, as well as the ad and query formats. In Section 2.2, we describe the query semantics.

2.1 Mobiscope Components

In Mobiscope, there are three types of components: directories, resources that submit ads, and clients that submit queries.

Directories Mobiscope directories are located throughout the network and accept both ads and queries. For a given network of Mobiscope directories, we define $R_{totalArea}$ to be the 2-dimensional rectangle that represents the area for which the network provides service [1].

Each directory has a service region (SR), which is a rectangle that covers some subset of $R_{totalArea}$. Mobiscope is optimized for the case where the entire area of $R_{totalArea}$ is completely covered by directory service regions. If $R_{totalArea}$ is not totally covered, however, Mobiscope will still function. If a Mobiscope network is deployed alongside of wireless infrastructure, coverage holes in the infrastructure can be subsumed by directories with adjacent service regions. The Mobiscope network tolerates overlap between service regions. Overlap may occur, for example, if service regions are determined by the coverage area of wireless infrastructure components, and these coverage areas overlap.

[1] Although we limit our discussion to the discovery of mobile resources in 2 dimensions, Mobiscope could be easily modified to discover resources in 3 dimensions.

Mobile Resources Resources notify Mobiscope of their position functions by sending advertisements to a directory in the network. Ads contain a unique identifier for a resource (UID), an expiration time ($tExp$), the resource's network address, and a position function ($\overline{pos}$). Resources send ads to a directory such that the resource resides in the directory's service region. In Section 3, we explain how resources find the correct directory as they move throughout $R_{totalArea}$.

The function $\overline{pos}$ takes time as a parameter and returns an upper and lower bound for each dimension. Readers can find a detailed explanation of position function representations in [20]. Figure 1 shows an example of position functions for 3 mobile resources (mr1-3). For simplicity, this diagram shows functions in only one dimension plotted against time. These functions are shown as parallelograms, which represent the value range that each resource may take at a given time. For a resource mr that sends an ad a at time $tXmit_a$, mr sends a successive ad with a new position function when mr detects that it has moved outside of $a.\overline{pos}(t)$ at any time $t : tXmit_a \leq t \leq a.tExp$. Discontinuities in the parallelograms in Figure 1 represent changes in the position functions. If the position function does not change, mr will send a new ad at time $a.tExp - \Delta$, where Δ is enough time for the ad to arrive at a directory. At least one directory in Mobiscope will cache each ad until it either expires or is replaced by an ad with the same UID. In the example in Figure 1, we assume that each ad a transmitted at time $tXmit_a$ has $a.tExp = tXmit_a + 2$ min. $+ \Delta$. In this case, resources must transmit new ads every other minute.

Clients To query Mobiscope, clients send queries to any directory in the Mobiscope network. Each query Q has a rectangle, *rect*, the client's network address, an expiration time $tExp$, and an *isRefresh* flag. Clients periodically send *query refreshes* for each query with a new expiration time to maintain the query's soft-state. Figure 1 shows a query $Q1$ in one dimension, where the rectangle is only a range of values for a time interval. $Q1$ expires at $t = 14$ min. The queries will cease running after $t = 14$ min. if the $Q1$ client does not send a query refresh. In the following section, we describe the semantics of the query output.

2.2 Query Semantics

For a client that submits Q to a directory at time $tSubmit_Q$, the directories work together to find a set of UID/network address pairs that indicate the resources currently in $Q.rect$. This set continually changes as resources move in and out of $Q.rect$. For example, consider $Q1$ in Figure 1, where $tSubmit_{Q1} = 4$ min. Initially, $Q1$'s result set should contain query responses for $mr1$ and $mr2$. At time $t = 6$ min., it contains responses for $mr1$ and $mr3$. At time $t = 12$ min., the result set again contains responses for $mr1$ and $mr2$.

When a directory receives a new query Q at time $tSubmit_Q$, directories find all cached ads a such that $a.\overline{pos}(t)$ intersects $Q.rect$ for some time $t : tSubmit_Q \leq t \leq a.tExp$. The query and ad routing policies we present in Section 3 ensure that Q finds all matching ads.

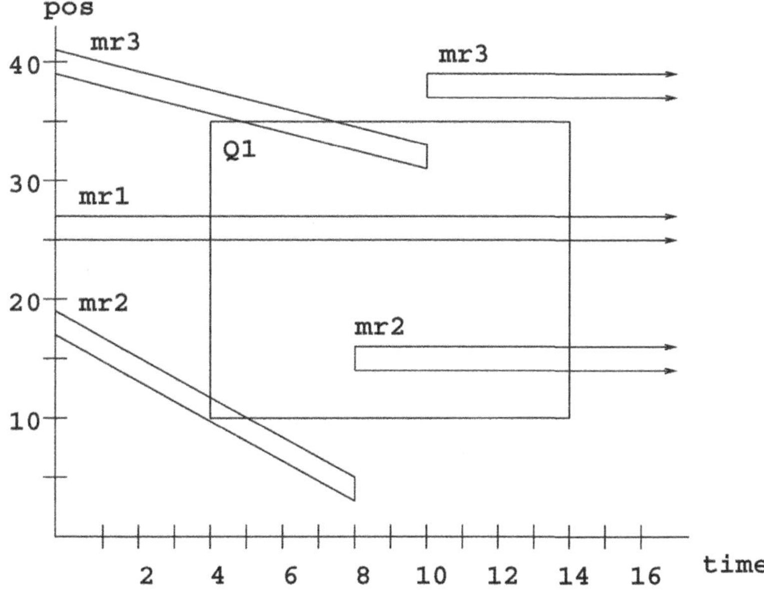

Fig. 1. Moving resources moving along one dimension, plotted against time.

For each ad a found, a *query response* (qr_a) is generated for the client. A query response qr_a contains the UID and network address from a, as well as a time interval ($tStart$, $tExp$). For query response qr_a, $qr_a.tStart$ is the smallest time $t \geq tSubmit_Q$ such that $a.\overline{pos}(t)$ intersects $Q.rect$, and $qr_a.tExp = min(a.tExp, t_{leaveregion})$, where $t_{leaveregion}$ is the smallest time $t \geq tSubmit_Q$ such that $a.\overline{pos}(t)$ no longer intersects $Q.rect$.

The directories send initial query responses to the client and run the query continuously over subsequent incoming ads. Consider an ad a_{new} that arrives at a directory where Q is running continuously. If a_{new}'s position function intersects $Q.rect$ within the lifetime of the ad, the directory creates a query response $qr_{a_{new}}$ for Q's client. $tStart$ and $tExp$ for continuous query responses are defined in a similar manner to the initial query responses. Again, routing ensures that all continuous queries find all matching ads.

To demonstrate these semantics, we again refer to our example in Figure 1. To keep our example simple, we again assume that any ad a transmitted at time $tXmit_a$ arriving at a directory at time t has an expiration time of $tXmit_a + 2\ min. + \Delta$, where $tXmit_a + \Delta \approx t$ (i.e. Δ is approximately the network transit time for a). We also assume that all ads are transmitted on even numbered minutes. Consider query $Q1$, with $tSubmit_{Q1} = 4$ min. and $Q1.tExp = 14$ min. Table 1 shows the query responses generated for $Q1$ at different times, along with their ($tStart, tExp$) intervals. If the client does not send a refresh for $Q1$ before 14 min., the query expires and ceases to run continuously.

Time	Responses (tStart, tExp)
4	$\{mr1(4,6+\Delta)$, $mr2(4,5+\Delta)$, $mr3(5,6+\Delta)$
6	$\{mr1(6,8+\Delta)$, $mr3(6,8+\Delta)\}$
8	$\{mr1(8,10+\Delta)$, $mr2(8,10+\Delta)$, $mr3(8,10+\Delta)\}$
10	$\{mr1(10,12+\Delta)$, $mr2(10,12+\Delta)\}$
12	$\{mr1(12,14+\Delta)$, $mr2(12,14+\Delta)\}$

Table 1. Query responses returned to client for Q1.

If a directory receives a refresh for a query that is not currently running, the directory processes the query as a new query to ensure that the client has current query responses. This situation can occur if a directory fails and restarts. Although a client may receive duplicate query responses, duplicates will not affect the correctness because all query responses contain UIDs.

The clients cache query responses that they receive from the directories. For each active query Q running at a client, the client maintains two data structures: a cache and a *future response queue* (FRQ). When a client receives a query response qr_a at time t, the client places qr_a in the cache if $t \geq qr_a.tStart$. Otherwise, it places qr_a in the FRQ. When $Q1$ is issued in our example, the client puts responses from $mr1$ and $mr2$ in the cache, and the response from $mr3$ in the FRQ. If there is a query response in the FRQ or cache for the query with the same UID, the client deletes the old response. The FRQ keeps an index on the $tStart$ field. When the current time equals the $tStart$ field of the query response at the head of the queue, the client moves the query response to the cache. If a query response expires in the cache, the client deletes it. Thus, the cache for a query represents the continuously updated result set of the query.

3 Mobiscope Distributed Processing

In this section, we show how the Mobiscope architecture supports the above semantics in a distributed environment. As described above, a directory processes an ad by caching it, replacing any ad with the same UID, and finding any continuous queries that intersect the ad's position function for some portion of the ad's lifetime. Similarly, a directory processes a query by finding any cached ads that match the query, and then running the query continuously over any new incoming ads. Within a single directory, this processing is similar to the symmetric treatment of queries and ads in PSOUP [3]; however, Mobiscope must perform this processing in a distributed environment. A directory processes an ad it receives if and only if the ad's position function intersects the directory's service region for some time in the ad's lifetime. Similarly, a directory only processes queries that intersect its service region. Directories route ads and queries among themselves to ensure full reachability. The routing protocols must be efficient in their use of processing, or they could introduce unneeded overhead into the directories.

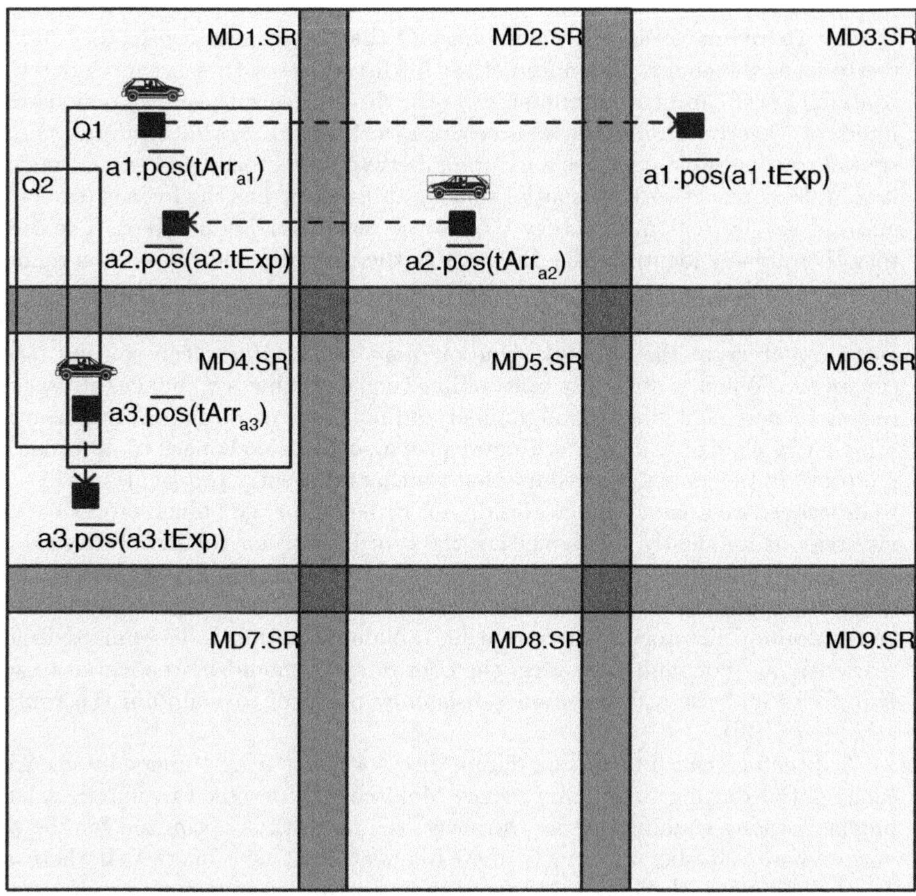

Fig. 2. A sample Mobiscope network, with ads, queries and service regions. The white regions indicate area covered by one service region, where the gray regions indicate service region overlap.

The ad and query routing protocols ensure that a) all ads are routed to directories that can process them, and b) each query is routed to a set of Mobiscope directories $mdset$ such that $\bigcup_{md \in mdset} md.SR$ contains the query rectangle. An example Mobiscope network in Figure 2 demonstrates why ad and query routing is needed. In this example, we assume that a resource sends ads to a directory such that the resource resides in the directory's service region at the time it sends its ad. We also assume that clients can submit queries to any directory in the network. Without routing, query $Q1$ will not find ad $a2$, and $Q1$ and $Q2$ will not find $a3$ unless their clients submit them directly to $md4$. The following two sections explain the query and ad routing algorithms, respectively.

Query Routing Consider a new query Q that comes into a directory md directly from the client. The minimal set of directories with service regions that contain $Q.rect$ can be approximated by the directories with service regions that intersect $Q.rect$. To find these directories, md uses a *spatial routing table.* A spatial routing table provides a mapping between directories and service regions in a Mobiscope network. A spatial routing table entry has the format (directory address, service region, directory ID, service region version number). The directory ID uniquely identifies the directory in the network, and the version number increases each time a directory the service region changes.

Each directory in a Mobiscope network has a spatial routing table entry for every directory in the network. One or more *coordinators* keep routing tables consistent. When a directory goes offline, comes online, or changes its service region, it increments its version number and notifies a coordinator of the routing table entry change [2]. The coordinator propagates these changes to all other directories. In this paper, we assume that routing table entry propagation will scale well because we assume directories do not come online, go offline, or change service regions frequently. Thus, neither the coordinator nor the directories should experience scalability problems in processing the changes. For scenarios where these assumptions do not hold, our technical report describes a hierarchical organization for directories where routing table information is disseminated more efficiently [6]. For fault tolerance, the coordinators should be replicated to several nodes and use a distributed consistency protocol to maintain the routing table (e.g. [13]).

A directory can find the matching directories for a new query by checking each spatial routing table entry. Since Mobiscope is designed to handle a large number of long-running queries, however, the network is optimized for the scenario where refreshes are much more frequent than new queries. If there are many directories, checking all routing table entries for frequent refreshes could result in significant overhead. For each query received directly from a client that is routed to other directories, a directory creates an entry in a query hashtable, keyed on a concatenation of the client address and rectangle [3]. It attaches to this entry a list of directories to which it sent the query, along with the version numbers from their routing table entries. When a directory receives a refresh for the query directly from the client, it checks that each of the version numbers listed in the query entry is equal to the version number currently in the corresponding routing table entry. If all of these checks succeed, the directory routes the refresh to the directories listed in query entry. Only if a version check fails

[2] Directory failures can be detected by heartbeat messages from neighboring directories, which update the coordinator accordingly.

[3] Although a client can issue multiple queries with the same rectangle, Mobiscope can effectively treat them as the same query

does the directory need to consult the routing table. Since we assume a client will generally send the initial query request and all refreshes for a query to the same directory, directories normally avoid checking the routing table to process a refresh, which can significantly reduce overhead.

Ad Routing Consider a resource that submits an ad a to a directory md. To route a, md could check the entire spatial routing table for matching service regions as it does for queries. This solution could create significant overhead if md needs to process ads rapidly and the number of directories is large. Instead, directories usually use use *intersection* and *adjacency* entries to more scalably route ads through the network. For a given directory, these are the entries with service regions that are either adjacent to or intersect the directory's own service region. In our example in Figure 2, $md1$ has intersection entries for directories $md2$, $md4$, and $md5$, but has no adjacency entries. Each directory finds its intersection and adjacency entries by consulting its routing table.

Using intersection and adjacency entries will provide more scalable routing than checking each spatial routing table entry. Given reasonable service region overlap, there are usually few intersection and adjacency entries per directory. Also, the number of intersection and adjacency entries is not usually dependent on the size of the routing table. In Figure 2, any directory in the network has a maximum of 8 intersecting entries and no adjacency entries. As long as service region overlap does not grow, this number does not change if more directories are added in a similar configuration.

Directories can route most ads using only intersection and adjacency entries. Only under exceptional conditions, which we explain at the end of this section, does a directory default to checking the entire spatial routing table. Mobiscope ensures that each resource sends ads to a directory md such that the resource resides in $md.SR$. When a resource comes online, it sends its position to a coordinator [4] or a well-known directory. The coordinator or well-known directory finds a directory with a service region that currently contains the resource, and sends the resource the corresponding routing table entry. The resource then transmits an ad to the directory associated with this entry.

Unlike queries, a directory attempts to route any ad it receives to other directories, regardless of whether the ad was sent directly from a resource or from another directory. A directory usually checks only adjacency and intersection entries to find directories that should process the ad. Before forwarding an ad to other directories, it adds itself to the ad's *visitedDirList* (VDL). The VDL is a list of directories that have already processed the ad. A directory only sends an ad to a remote directory if a) the directory has an adjacency or intersection entry for the remote directory, b) the directory's service region intersects the ad's position function for some time in the ad's lifetime, and c) the remote directory is not already on the ad's VDL.

To illustrate ad routing, we refer to our example in Figure 2. In this example, the ad $a1$ is received by $MD1$ because the ad's position function intersects

[4] We assume the coordinator addresses are well known to all resources in the system

$MD1$'s service region at time $tArr_{a1}$. $MD1$ puts its ID on $a1$'s VDL, and sends $a1$ to $MD2$. Similarly, $MD2$ puts its ID on $a1$'s VDL, and sends $a1$ to $MD3$. $MD1$ is in $a1$'s VDL, so $MD2$ does not send $a1$ back to $MD1$. Because $MD2$ is on $a1$'s VDL when it reaches $MD3$, $MD3$ does not send the ad to any other directories. Using this routing algorithm, a is routed to all directories that process a, given that a's position function predicts that the resource will be in some area covered by a service region until a expires.

When a directory is processing an ad, it sends its own routing table entry to the resource if it did not receive the ad directly from the resource. This way, the resource will have the routing table entries for all directories that process the ad. If the ad's position function predicts that the resource will always be in some area covered by a service region, then the resource always knows the appropriate directory to send a subsequent ad to. Thus, a coordinator or a directory only has to check all routing table entries when a resource comes online. In the above example, the resource that sends $a1$ will have routing table entries for directories $MD1$, $MD2$, and $MD3$. If the resource changes position functions at any time during the lifetime of $a1$, the resource will send a new ad to $MD1$, $MD2$, or $MD3$, depending on which service region the resource is in at the time it sends the ad.

In some cases, an ad's position function may predict that the resource will be in an area that is not covered by any service region. This condition may occur, for instance, when a directory fails. A directory routing a can recognize this case if i) the directory finds no other directories to route a to from its adjacency and intersection entries, and ii) a's position function does not intersect the directory's service region at time $a.tExp$. If this situation occurs, the resource exits the service region before it's ad's expiration time, but does not enter another service region. In our example, either $MD1$ or $MD2$ would find this situation with $a1$ if either $MD2$ or $MD3$ fails, respectively. In these cases, the directory defaults to checking the entire routing table.

4 Performance

In this section, we report on a performance study of the Mobiscope distributed processing algorithms. For this study, we implemented a workload generator and simulator to test the scalability of Mobiscope and Mobiscope's behavior under workloads that stress Mobiscope's distribution model. In Section 4.1, we describe the workload generator and simulator. In Section 4.2, we describe the parameters varied in the study, and we present results in Section 4.3.

4.1 Simulation Environment

Our simulation environment consists of a workload generator and Mobiscope simulator, both written in C++. Our workload generator creates workloads that emulate queries over resources moving in a congested area with rapid position function updates. Although other moving object data generators are available (e.g. [17]), we wrote our own so that we could introduce parameters to stress our distribution model. The ad workload reflects resources that move in a 10 x 10 mi. area, traveling along paths in a grid that are .1 mi. apart. Resources traveling in such a grid could reflect cars moving around a city with square blocks. By default, the resources are initially evenly distributed. However, the initial placement of resources can be skewed to a "hot" region using a simulation parameter.

Resources are either stopped or moving along one of the paths. Every minute, a resource chooses a new direction or stops with equal probability. Resources move at a speeds between 35 and 55 mi./hr, chosen randomly. A resource transmits an ad after choosing a new trajectory. The ad represents the resource either stopped or moving along a grid line, with the upper and lower bound functions providing a 1/5 mi. error bound on both dimensions. All ads have the same lifetime, determined by a parameter.

The query workloads consist of long running queries with square query regions that cover a subset of the simulation area. Since queries are long running, most queries are already running when the simulation starts, but the workload generator also generates 10 new queries per minute. The query rectangle centers are distributed evenly throughout the area by default, but can also be skewed to the hot region with a simulation parameter. The length of a square side is also a parameter. All queries are refreshed once a minute.

All the workloads presented here have 1000 continuous queries initially running and 100000 mobile resources. For each set of parameters, the workload generator generates a set of ads and queries initially in the Mobiscope network, along with new ads, queries, and refreshes for a given period of simulated time. For each simulation, we generate ads, queries, and refreshes for 2 minutes of simulated time[5].

The simulator runs the workloads over different Mobiscope network configurations. Each configuration consists of n directories, each with a unique service region that covers 1/nth of the total area plus overlap with other service regions. The directories are placed in a $\sqrt{n}$ by $\sqrt{n}$ matrix, and overlap is controlled by a parameter. Figure 2 shows an example configuration where $n = 9$.

The simulator determines the ads and queries that are processed by each directory, and runs the workload against each directory in sequence. The directory implementation functions as described in Section 3, but it receives ads and queries via procedure calls instead of from real resources and clients. The implementation processes ads and queries by checking them against each continuous query and cached ad, respectively. Directory processing might be improved by indexing the queries and ads with spatial indexes (see [8] for a survey) and spatio-temporal indexes (e.g. [17, 12]), respectively. Unfortunately, finding a spatial index that will improve performance is highly dependent on the workload.

[5] We ran additional experiments where we varied the number of moving resources from 10000 to 200000, the queries from 1000 to 10000, and the simulated time from 1 to 5 min. Although the magnitudes were different, the overall trends were essentially the same as the reported 100000 resources/1000 initial continuous query case

Parameter Description	Values Range (default)
Number of directories	1-256
% of each service region that overlaps other service regions	5-100 (5)
Ad expiration time (min.)	1-7.5 (1)
Length of the side of each query (mi.)	.25-5 (.25)
Percentage of resources initially in hot region	6.25-100 (6.25)
Percentage of query rectangle centers in hot region	6.25-100 (6.25)

Table 2. Simulation Parameters

As stated in [10] , there is currently no thorough, unbiased experimental analysis to guide systems designers in choosing a spatial index for a given workload. The design space of spatio-temporal indexes is even less well understood. Although a performance study of various indexes under Mobiscope workloads provides interesting future work, this paper concentrates on studying the performance of the distribution model.

For each run, the initial queries and ads are loaded into the directory before the simulation begins. For each configuration and workload, we report the "wall clock" time for one directory to process all ads and queries in its entire workload for the given simulated time period. The directory is usually chosen from the middle of the directory matrix. For the experiments varying ad and query skew, however, we report the time for the directory covering the hot region. Experiments report a mean time for each set of parameters. Workloads for each set of parameters was generated and run repeatedly until the 95% confidence interval of the times was within 10% of the mean.

4.2 Experiment Parameters

Table 2 shows the parameters that we vary in our experiments, along with the ranges used and the default values. We tried to vary parameters that either a) test the scalability of the Mobiscope network, or b) reduce the effectiveness of Mobiscope's distribution model.

To test scalability, we vary the number of directories in a configuration while keeping the workload fixed. These tests allow us to measure the incremental performance benefit of extra directories. To test the routing protocol performance, we also ran experiments with "naive" routing. In naive routing, a directory that receives an ad, query, or refresh directly from a resource or a client routes it by checking the entire routing table. To isolate the routing overhead, we ran experiments both with the ad and query processing turned off as well as normal processing.

We vary the rest of the parameters in Table 2 to increase the overhead due to the Mobiscope distribution model. In our study, we run several experiments to observe the effects of increasing the proportion of ads and queries that are processed by more than one directory. A larger number of ads and queries processed by multiple directories creates more ad and query routing, as well as redundant processing costs at multiple directories.

We can vary several parameters to increase the number of these ads and queries. One such parameter is the amount of area a Mobiscope directory service region overlaps with neighboring service regions. An increase in overlap increases both the number of queries and ads processed by multiple directories. Increasing ad expiration time increases the average number of directories with intersecting service regions. Similarly, an increase in query side length increases the average number of directories that processes each query.

In addition to increasing the number of ads and queries processed by multiple directories, we also introduce ad and query skew into our workloads to stress the distribution model. As opposed to increasing the overhead at each directory by increasing the number of ads and queries processed by multiple directories, we reduce the parallelization in Mobiscope by increasing the amount of ad and query skew. As ads and queries are increasingly skewed toward a hot region, only one or a few directories will process most of the ads and queries. In our skew experiments, the hot region for both ads and queries is 1/16th of the total simulation area. The hot region is covered by one directory's service region, and we report this directory's time in the skew experiments.

4.3 Results

Experiment 1: Vary Number of Directories Table 3 shows the processing and routing times for different numbers of directories, using both naive and Mobiscope routing. The one directory time is given only for comparison. In this and all other experiments, all parameters except for those varied are set to the default values as listed in Table 2.

We expect to initially see large benefits to increasing the number of directories due to parallelization. At high numbers of directories, however, ads and queries match several service regions, and the performance increases should be diminished. At high numbers of directories, we expect to see much of the processing time from ad and query routing.

These experiments confirm the expected trends. The processing performance increase from using 4 directories instead of 1 is more than tenfold, but the performance increase from using 256 instead of 144 directories is less than double. At 16 directories and above, the routing time is more than 1/5th of the total processing time. In addition, we can see a large benefit to the Mobiscope routing policy in comparison to the naive routing policy when the number of directories is large. At 256 directories, for instance, the naive processing time is more than 3 times that of the Mobiscope processing time, and almost all of the time difference is due to routing costs.

Experiment 2: Vary Service Region Overlap In this experiment, we vary the percent of service region overlap and report the processing times. We define percent of overlap to be the percent of area that a given service region overlaps with any other service region. Table 4 shows the results. For the remaining experiments, we only run configurations with 1 (1dir), 4 (4dir), and 16 (16dir) directories. In this experiment, we increase service region all the way to 100% to create a workload where all ads and queries match multiple service regions. The processing times for 16 and 4 directories increase significantly with overlap. Even at 100% overlap, however, multiple directories still process their workloads much faster than a single directory.

Number of Directories (numDirs)	Total Processing Time (sec.)	Ad and Query Routing Time Only (sec.)	Total Processing Time (Naive , sec.)	Ad and Query Routing Time Only (Naive, sec.)
1	41.331	N/A	41.331	N/A
4	3.541	.060	3.482	.060
16	0.466	0.032	0.473	0.037
36	0.165	0.018	0.185	0.035
64	0.089	0.013	0.117	0.039
100	0.060	0.010	0.102	0.050
144	0.042	0.008	0.092	0.055
256	0.030	0.006	0.095	0.068

Table 3. Vary Number of Directories

% SR Overlap	Processing Time 1dir (sec.)	4dir (sec.)	16dir (sec.)
20	41.331	4.218	0.538
40	41.331	4.531	0.818
60	41.331	5.446	0.996
80	41.331	6.183	1.118
100	41.331	6.766	1.359

Table 4. Vary Percent of Service Region Overlap

Experiment 3: Vary Query Side Length In Figure 3, we graph the processing times for simulations with different query side lengths. Note that the y-axis in Figure 3 is on a logarithmic scale. At higher values of query side length, each incoming ad matches more queries on average, creating a higher cost for generating a query response. Thus, the single directory processing time rises with query side length and expiration time. However, ads and queries that are processed by multiple directories still add significant overhead, as the increase in 16 and 4 directory processing times is more than proportional to the single directory increase. 16 directories takes less than 1/50th the processing time of 1 directory when query side length = .25 mi. At query side length = 5 mi., this ratio is more than 1/6th.

Again, we use extreme parameter values to stress the Mobiscope distribution. At their maximum values shown here, query sides are half the length of the total simulation region. Even with these large query areas, 16 and 4 directories still significantly outperform the single directory. In addition to query side length, we ran similar experiments where we varied ad expiration time from 1 to 7.5 minutes. These experiments showed similar trends to the experiments varying query side length, and thus we do not show this graph here in the interest of conserving space.

Experiment 4: Vary Percent of Ads and Queries in the Hot Region To measure the effects of ad and query skew, we vary the percentage of queries and ads in the hot region. Figure 4 shows processing times for various percentages of query centers in the hot region when the resources are either initially evenly distributed or all placed in the hot region. Unlike Figure 3, the y-axis in Figure 4 is not on a logarithmic scale. First, consider the processing time growth as the queries in the hot region increase and resources are kept evenly distributed. As expected, the 16 and 4 directory processing times rise with query skew. However, 1 directory processing time decreases slightly. This decrease is caused the many ads that generate no query responses when most queries are in the hot region and ads are evenly distributed. When all mobile resources are initially in the hot region, the 1 directory processing time also increases with query skew. Because all ads are now all in the hot region, an increase in query skew increases the average number of matching queries per ad.

Note that when either ads or queries are evenly distributed, total skew in the other workload does not cause the 16 or 4 directory processing times to approach that of 1 directory. Not until all ads and most queries are in the hot region do the 16 and 4 directory processing times come close to the corresponding 1 directory processing time.

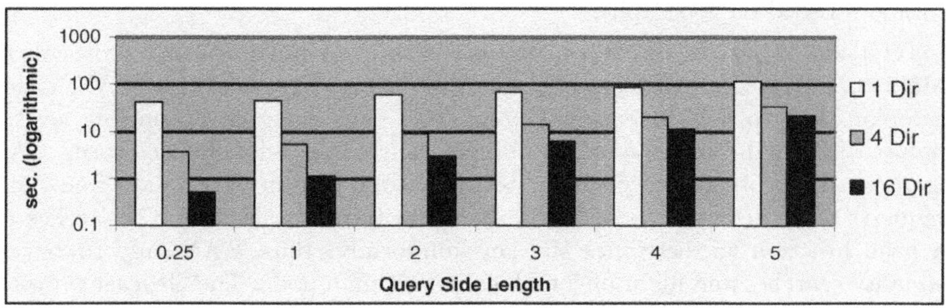

Fig. 3. Vary Query Side Length

5 Related Work

Besides the MODBMS work mentioned earlier, much of the service discovery literature is also related to Mobiscope, such as [11, 1, 2, 22, 9]. None of these services efficiently support discovery of moving network resources based on location. To our knowledge, none of these systems process advertisements with position functions. These services would face serious problems dealing with moving resources that advertise a continuously changing position.

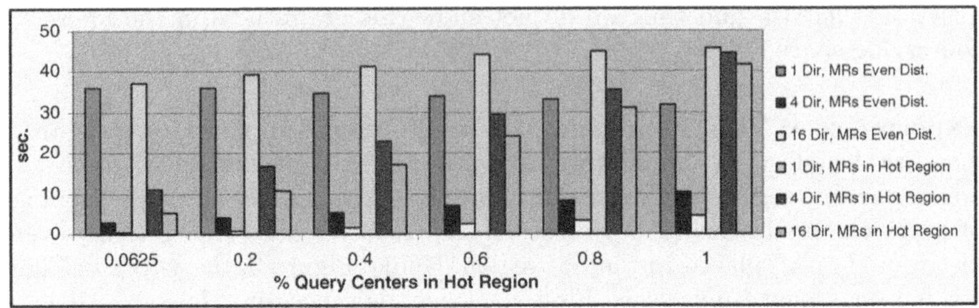

Fig. 4. Vary Percent of Queries and Moving Resources In Hot Region

Like Mobiscope, most service discovery services are distributed to multiple nodes. Service standards such as LDAP [22] and SLP [9] use very elementary distribution models that are not apropos to Mobiscope. Berkeley SDS [11] and VIA [2] attempt to route queries to only the relevant nodes by filtering queries as they travel through the system. Berkeley SDS filters queries through an administratively defined hierarchy of nodes to increase scalability. VIA nodes adaptively form clusters to determine query filtering policies based on query workloads. Berkeley SDS and VIA are both primarily built for queries with equality predicates, although [11] describes a design that allows Berkeley SDS to support range queries. None of these services have protocols that route queries and ads to nodes based on geography.

Outside of service discovery, two projects use similar routing protocols to Mobiscope: the CAN distributed hashtable [16] and Geocast [14]. CAN hashes a lookup key to a d-dimensional point, and each node is responsible for the points within a d-dimensional bounding rectangle in a d-dimensional torus. CAN uses routing protocols very similar to Mobiscope to route both data items and requests to the correct node. Like other distributed hashtables, CAN relies on a hash function to distribute the keys uniformly; thus, CAN only routes on equality searches and maintains no locality within a node. The Geocast protocol uses routing protocols similar to Mobiscope to route messages between hosts based on their position. Although Geocast does allow a form of addressing that specifies a region as a destination, it does not allow hosts to express positions as functions.

Most recently, the continuous query work in the database community shows some similarity to Mobiscope. Like Mobiscope, PSOUP presents a system where a query response can consist of data that arrives either before or after a query arrives at the system [3]. Although PSOUP is built on a single node, the work on Flux shows how continuous query systems can be parallelized across nodes in a cluster [18]. However, neither of these systems deal with the problems of queries over moving objects.

Both PSOUP and Flux were developed as part of U.C. Berkeley's TelegraphCQ project [7]. The Telegraph project, along with the IBM iQueue system for processing streaming data from pervasive data sources [4, 5], was the inspiration for the Mobiscope service. In the future, we hope to integrate Mobiscope with both these projects.

6 Conclusion

Mobiscope is a service where clients submit long running queries to continuously find the set of network resources currently in a given region. Like moving object DBMSs (MODBMSs), Mobiscope represents position with functions over time. Unlike MODBMSs, Mobiscope uses a new query model to first run queries over cached position functions, and then continuously over subsequent functions received after the query. For scalability, Mobiscope distributes the ad and query workload to multiple directories by geography. Directories use application-level routing protocols based on geography so that all queries find all matching resources. Our performance study of Mobiscope shows that the Mobiscope service scales well, even under workloads which stress Mobiscope's distribution model.

References

[1] W. Adjie-Winoto, E. Schwartz, H. Balakrishnan, and J Lilley. The design and implementation of an intentional naming system. *Operating Systems Review*, 35(5):186–201, December 1999.

[2] P. Castro, B. Greenstein, R. Muntz, C. Bisdikian, P. Kermani, and M. Papadopouli. Locating application data across service discovery domains. In *Mobicom*, 2001.

[3] S. Chandrasekaran and M. J. Franklin. Streaming queries over streaming data. In *VLDB*, 2002.

[4] N. H. Cohen, H. Lei, P. Castro, J. S. Davis II, and A. Purakayastha. Composing pervasive data using iql. In *4th IEEE Workshop on Mobile Computing Systems and Applications*, pages 94–104, 2002.

[5] N. H. Cohen, A. Purakayastha, L. Wong, and D. L. Yeh. iqueue: A pervasive data composition framework. In *MDM*, 2002.

[6] M. Denny. Mobiscope: A scalable spatial discovery service for mobile network resources. Master's thesis, U.C. Berkeley, 2002.

[7] S. Chandrasekaran et. al. Telegraphcq: Continuous dataflow processing for an uncertain world. In *The First Biennial Conference on Inoovative Data Systems Research*, 2003.

[8] V. Gaede and G. Gunther. Multidimensional access methods. In *Computing Surveys 30(2)*, 1998.

[9] E. Guttman, C. Perkins, J. Veizades, and M. Day. Service location protocol, version 2. RFC 2608.

[10] J. M. Hellerstein and M. Stonebraker, editors. *Readings in Database Systems, 3rd Ed.*, chapter 2: Relational Implementation Techniques (Chapter Notes), pages 77–82. Morgan Kaufmann Publishers, Inc., 1998.

[11] T. Hodes, S. Czerwinski, B. Zhao, A. Joseph, and R. Katz. An architecture for a secure service discovery service. In *ACM Baltzer Wireless Networks (to Appear)*.

[12] G. Kollios, D. Gunopulos, and V. J. Tsotras. On indexing mobile objects. In *PODS*, 1999.

[13] C. Mohan, B. Lindsay, and R. Obermarck. Transaction management in the r* distributed database management system. In *TODS 11(4)*, 1986.

[14] Julio C. Navas and Tomasz Immielinski. Geocast - geographic addressing and routing. In *Mobicom*, 1997.

[15] D. Pfoser, C. S. Jensen, and Y. Theodoridis. Novel approaches in query processing for moving object trajectories. In *VLDB*, 2000.

[16] S. Ratnasamy, P. Francis, M. Handley, and R. Karp. A scalable content-adressable network. In *SIGCOMM*, 2001.

[17] S. Saltenis, C. S. Jensen, S. T. Leutenegger, and M. A. Lopez. Indexing the positions of continuously moving objects. In *SIGMOD*, 2000.

[18] M. A. Shah, J. M. Hellerstein, S. Chandrasekaran, and M. J. Franklin. Flux: An adaptive repartitioning operator for continuous query systems. In *ICDE*, 2003.

[19] D. Steere, A. Baptista, D. Mcnamee, C. Pu, and J. Walpole. Research challenges in environmental observation and forecasting systems. In *MobiCom*, 2000.

[20] O. Wolfson, S. Chamberlain, S. Dao, S. Dao, L. Jiang, and G. Mendez. Cost and imprecision in modeling the position of moving objects. In *ICDE*, 1998.

[21] O. Wolfson, B. Xu, S. Chamberlain, and L. Jiang. Moving objects databases: Issues and solutions. In *Conference on Scientific and Statistical Database Management (SSDBM98)*, 1998.

[22] W. Yeong, T. Howes, and S. Kille. Lightweight directory access protocol. RFC 1777.

{Arjan.Peddemors, Marc.Lankhorst, Johan.deHeer}@telin.nl

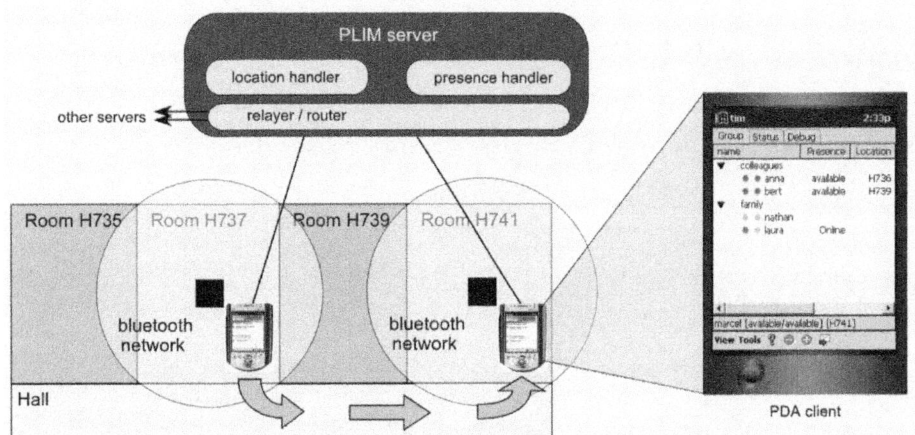
PLIM server
location handler
presence handler
relayer / router
other servers
Room H735
Room H737
Room H739
Room H741
bluetooth network
bluetooth network
Hall
PDA client

{yingch,raofy,yuxl,liudong}@cn.ibm.com

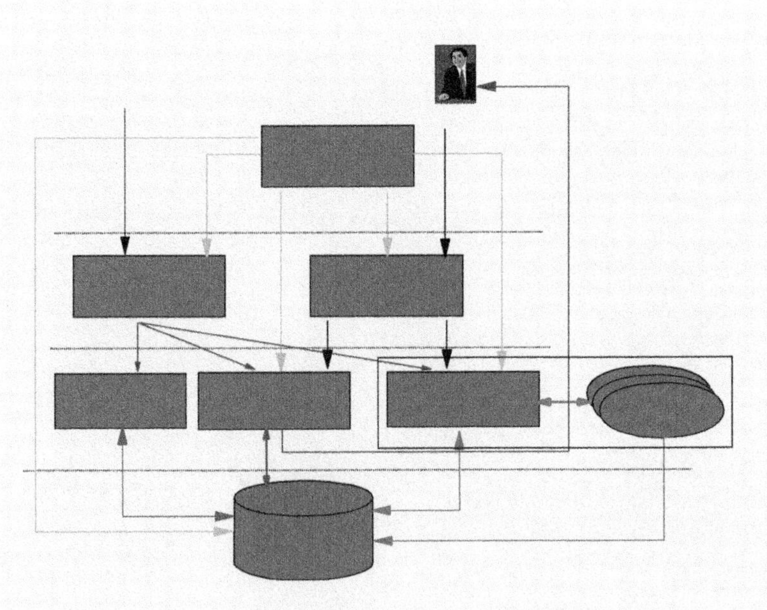

$$= \qquad <$$

$$= \qquad <$$

$$= \bigvee_{\in} \qquad <$$

Using Hashing and Caching for Location Management in Wireless Mobile Systems

Weiping He and Athman Bouguettaya

Department of Computer Science
Virginia Tech
{weiping, athman}@vt.edu

Abstract. In this paper, we propose a new technique for location management based on hashing and caching. This approach allows for the use of the location information cached at the local server to reduce the cost of call delivery. We also introduce replication as a mechanism to increase the reliability of Home Location Registers (HLRs). Additionally, we use hashing to achieve load balancing among HLRs. Preliminary results indicate that our proposed scheme is particularly suitable for large numbers of users and highly congested wireless networks.

1 Introduction

One of the challenging problems in mobile computing is location management. It enables systems to locate mobile hosts for call delivery and maintain connection with them. Location management aims at reducing the cost of querying and updating location information, and increasing the operating performance of location information databases.

Currently there are two standards for wireless communications: IS-41 and GSM [1]. They are very similar in location management. For simplicity, we would use IS-41 in this paper. In IS-41, the network coverage is divided into cells. Each cell has a base station. A few cells are grouped together to form a Registration Area (RA). All base stations in one RA are connected to a Mobile Switching Center (MSC). The MSC serves as an interface between the wireless network and the Public Switching Telephone Network (PSTN). The HLR is a centralized database. It contains the user profiles of its assigned subscribers. These user profiles record information such as quality of service requirement, billing information, and current location of the Mobile Host (MH). The Visitor Location Registers (VLRs) are distributed throughout the wireless network. Each one stores the information of current MHs residing in its location.

It has been shown that under heavy traffic conditions, the communications and processing overhead of location updates can result in quick blocking of resources [6]. To address the above problems, we propose a new method for location management based on hashing and caching. The rest of the paper is organized as follows: Section 2 provides an overview of recent research results. Section 3 describes our proposed schemes. Section 4 gives the analytical model of the proposed scheme. We provide a concluding remarks in section 5.

M.-S. Chen et al. (Eds.): MDM 2003, LNCS 2574, pp. 335–339, 2003.

2 Related Work

A review of location management strategies can be found in [1] [3] [4]. These strategies can be divided into two categories. The first category includes improvements on the current VLR/HLR scheme based on centralized database architectures. To reduce the lookup cost, the location of a mobile host can be maintained at additional sites. It may be cached at its caller's site or replicated at the sites from where the majority of calls originate [1]. The second category is based on a hierarchical organization of a distributed database architecture for locating mobile hosts. In [2], every level of the hierarchy represents a partition of geographic regions. The system records the location of every mobile user with a certain degree of accuracy in each level of the hierarchy. The degree of accuracy is increased as we go down the levels. The advantage of this scheme is that the upper bound of a search operation does not depend on the network size.

In [5], a load balancing protocol is proposed. Location updates and queries are multi-casted to subsets of location servers. These subsets change with time depending on the location of MH/querying MSC and load on the servers. They are chosen such that any pair of subsets have at least one common database. The construction of the subsets by using hashing function guarantees that the MSC is able to find a database with the desired location information. We extend this idea by using caching to reduce the cost of call delivery. Furthermore, we use hashing to choose only one location server. This process would simplify the load balancing protocol and reduce the database operation overhead.

3 Proposed Approach

Instead of using one HLR, we use a set of location servers. There is no difference between these location servers. Hashing is used to select one of the location servers dynamically. The location information cached at the MSC/VLR is checked before systems consult the location server.

An $\mathtt{MHID}$ consists of two parts: $\mathtt{MHID}_{\mathtt{AreaCode}}$ and $\mathtt{MHID}_{\mathtt{UserCode}}$. For example, if $\mathtt{MHID} = \mathtt{7035388379}$, the first three bits ($\mathtt{703}$) represent the $\mathtt{MHID}_{\mathtt{AreaCode}}$, the remaining 7 bits ($\mathtt{5388379}$) represent the $\mathtt{MHID}_{\mathtt{UserCode}}$. Each cell must know the $\mathtt{MSCID}$ of the MSC it connects to. The function $\mathtt{f(MHID, MSCID)} = \mathtt{MHID}_{\mathtt{AreaCode}} + \mathtt{h}()$ represents the location server in the cell represented by the $\mathtt{MSCID}$. The "+" here is the string concatenation operation. $\mathtt{h}()$ is a hashing function defined as $\mathtt{h}(\mathtt{MHID}_{\mathtt{UserCode}}, \mathtt{MSCID}) = (\mathtt{MHID}_{\mathtt{UserCode}} + \mathtt{MSCID})\mathtt{mod}\mathtt{m}$, where $[\mathtt{0}, \mathtt{m} - \mathtt{1}]$ is the range of the hashing function $\mathtt{h}()$. Given a $(\mathtt{MHID}_{\mathtt{UserCode}}, \mathtt{MSCID})$ pair, the hashing function $\mathtt{h}()$ is uniformly distributed over the range $[\mathtt{0}, \mathtt{m} - \mathtt{1}]$. Systems can dynamically change the value of $\mathtt{m}$ to get better performance and load distribution[1]. To reduce the lookup cost, the MH's location is cached at the VLR/MSC in the caller's zone each time a MH is called. It is checked before the location server is queried. If the location of the MH is found at the cache, then a query

[1] We assume that system engineer can decide how many servers to startup and shutdown according to the network traffic.

is sent to the indicated VLR without contacting the MH's location servers. We describe the algorithms of call delivery and location registration as follows.

- *Location Query Algorithm at Call Delivery*
  If a MH inside a cell belonging to a MSC wishes to call another MH, the actions to be performed are given as follows.

```
When (an MH call another MH) {
  Search current VLR's cache for MHID's location.
  if the cache has MHID's location information; {
    if the MH is still in the indicated VLR;
      {there is a cache hit; Return;}
    else {
      search the location server indicated by f( ) to
         locate MH with MHID and MSCID;
      Update cache with new location info and Return; }
  } // in the cache
  else {
    search the location server indicated by f( ) to
       locate MH with MHID and MSCID;
    Save location info in VLR cache and Return;
  }  // not in the cache
}
```

- *Location Update Algorithm at Location Registration*
  The operations in the following are performed to update the location information of a MH when it moves from the previous registration area to a new area. By using multicast, the location information of a MH can be updated in a single round of message exchange between servers.

```
When (An MH enters a new RA) {
 new serverID = f(MHID, new MSCID);
 The new MSC sends message containing MHID and the new MSCID
   to the location server indicated by the new serverID.
 On receiving the message, the location server indicated by
   new serverID multicasts the message to other m-1 location
   servers.}
```

4 Analytical Model

We assume the database access and signaling costs are measured by the delays required to complete the database query and signal transmission respectively. By using lazy caching, the cost of location update operations are almost the same as the basic scheme. We mainly concern the cost of call delivery.

A call in IS-41 scheme requires one query at the calling VLR (D_v), one query at HLR (D_h), and one query at the called VLR (D_v). The total database access cost is $2D_v + D_h$. In addition, delivering this call requires two message exchanges between the HLR and the calling VLR ($2C_h$), two message exchanges between the HLR and the called VLR ($2C_h$), and one message exchange between the calling VLR and the called VLR (C_v). The total signaling cost is $4C_h + C_v$. As a result, the total query cost of basic IS-41 scheme is: $C_b = 4 * C_h + C_v + 2 * D_v + D_h$.

According to our scheme, if the location information is in the cache and there is a cache hit, the total cost includes one query access in the cache and a message exchange between the calling VLR and called VLR (C_v). If the location information is in the cache but it is outdated (cache miss), the total cost includes one query access in the cache, a message exchange between the calling VLR and called VLR (C_v), the cost of the IS-41 steps, and an update in the cache. If the location information is not in the cache, the total cost includes one query access in the cache, the cost of the IS-41 steps, and an update in cache. For simplicity, we ignore the cost of querying or updating the cache. We define C_c is the average call delivery cost of the caching scheme, C_{ch} is The average call delivery cost if the location information is in the cache, p is Cache hit ratio, and q is the probability that the location information is in the cache. As a result, the average caching query cost of our model is: $C_c = q * C_{ch} + (1 - q) * C_b$, where $C_{ch} = p * C_v + (1 - p) * (C_v + C_b)$.

Assume that the call arrivals to MHs and MHs' mobility follow the Poisson distribution. The mean arrival rate is λ, the mean mobility rate is μ_m, $LCMR$ is the local call to mobility ratio. According to [4], $LCMR = \lambda / \mu_m$. The cache hit ratio $p = \lambda/(\lambda + \mu_m) = LCMR/(1 + LCMR)$. Figure 1 gives some preliminary results. When the LCMR increases, there is a decrease in the query cost.

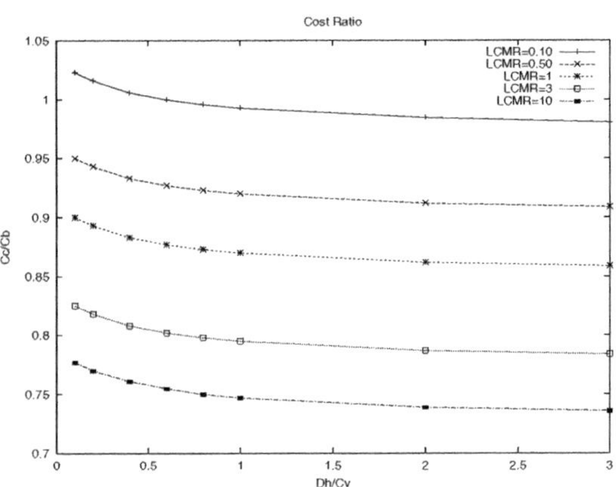

Fig. 1. Relation between LCMR and relative cost ratio

Throughput is used to evaluate location server's load performance. For the basic IS-41 scheme, the location server is the HLR. We define X is the throughput, λ is the call arrival rate, μ is the HLR query rate (service rate). We model the system as a M/M/1 system. It is a queue with call arrival times that are independent and exponentially distributed with rate λ. Service times are exponentially distributed with parameter μ and there is a single server in the system. The system throughput $X = \lambda$.

For our scheme, the MSC sends requests to $Server_i$, one of the location servers. We define Xc is one location server's throughput, λ_s is the call arrival rate at $Server_i$, μ_s is one location server's service rate, Pr_i is the probability to query $Server_i$, and m is the number of location servers. We assume that location servers use the same kind of computing power as HLR. The query service rate of one location server is $\mu_s = \mu$. Since we use hashing to decide which location server to query, the server is selected uniformly. As a result $Pr_i = 1/m$. The call arrival rate at $Server_i$ $\lambda_s = \lambda * Pr_i * (1-q) = \lambda * (1-q)/m$. where $(1-q)$ is the probability that the location information is not in the cache. We can model the system as m number of M/M/1 queues. The location server's throughput is $Xc = \lambda_s = \lambda * (1-q)/m$. if $q = 0.3, m = 5$, It reduces the server load dramatically by 86%.

The reliability of a system is its ability to function correctly over a specified period of time. There is only one location server(HLR) in the basic scheme. There are m servers for each area in our scheme. Assume one location server's reliability is r. Then the reliability of the basic scheme is: $R_{basic} = r$. The reliability of our scheme is: $R_c = 1 - (1-r)^m$. If $r = 0.8, m = 5$, the system reliability increases by whopping 24.96%.

5 Conclusion

In this paper, we proposed a combination of hashing and caching to improve the performance of location management in wireless networks. We show that it can significantly reduce the location management cost and improve location servers' reliability in wireless networks. The analytical model shows that the proposed scheme can dramatically reduce servers' load as well. We are currently investigating a better hashing function to query location server uniformly.

References

[1] I. Akyildiz, J. McNair, J. Ho, H. Uzunalioglu, and W. Wang. Mobility management in next-generation wireless systems. In *Proceedings of the IEEE*, volume 87, pages 1347–1384, August 1999.

[2] Y. Bejerano and I. Cidon. An efficient mobility management strategy for personal communication systems. In *The Fourth Annual ACM/IEEE International Conference on Mobile Computing and Networking*, October 1998.

[3] A. Bouguettaya. On the construction of mobile database management systems. In *The Australian Workshop on Mobile Computing & Databases & Applications*, Melbourne, February 1996.

[4] E. Pitoura and G. Samaras. *Data Management for Mobile Computing*, volume 10. Kluwer Academic Publishers, 1998.

[5] R. Prakash and M. Singhal. Dynamic hashing + quorum = efficient location management for mobile computing systems. In *Proceedings of the Sixteenth Annual ACM Symposium on Principles of Distributed Computing*, August 1997.

[6] S. Tabbane. Location management methods for third generation mobile systems. *IEEE Communications Magazine*, 35(8):72 –78, 83–4, Aug. 1997.

SEB-tree: An Approach to Index Continuously Moving Objects

Zhexuan Song and Nick Roussopoulos

Department of Computer Science
University of Maryland
College Park, Maryland 20742
zsong, nick@cs.umd.edu

Abstract. Recently, the requirement for storing the locations of continuously moving objects arises in many applications. The paper extends our previous work on zoning based index updating policy [1]. In the paper, we give the data format of object movement under the policy. Then, we propose the SEB-tree (Start/End time stamp B-tree). This index structure has fast insertion and query algorithm, and it outperforms the existing structures in the experimental evaluations.

1 Introduction

Recently, the demand for indexing and querying continuously moving data arises in many applications, including transportation and supply chain management, digital battlefield, air-traffic control, and mobile communication systems. The database that stores and indexes continuous moving objects is called the "Moving Object Database" (MOD).

In MOD, the motions of objects are represented by their trajectories. The positions of objects at some time stamps are sampled, and the location of an object between two consecutive sampled positions is estimated using interpolation (in most cases, linear interpolation). The movement of one object in a d-dimensional space is portrayed as a trajectory in a $(d+1)$-dimensional space after combining time into the same space [2]. Trajectory based methods include STR-tree, TB-tree [3] and MV3R-tree [4].

In [1], we proposed a zone-based policy: the space is partitioned into zones, and the locations of objects are represented by the zone id. But we did not give any methods to index the zoning data. In this paper, we extend the previous work and propose an index structure, which we called the "SEB-tree", to organize the history information of moving objects.

Before we discuss the SEB-tree, we briefly introduce the zone-based update policy [1]. First, the space is partitioned into zones. Objects always know which zone they are currently in. An object sends an update request to the database when it crosses the boundary of one zone and enters a new one, and the database stores the zoning information of objects. The benefit of this zone-based policy over other location-based policies (such as the dead-reckoning policy) is that the

M.-S. Chen et al. (Eds.): MDM 2003, LNCS 2574, pp. 340–344, 2003.

requirement for each object is low: instead of knowing the exact location all the time, in this policy, an object only has to know which zone it is current in. The zone information can be obtained with much less price. Therefore, the policy is practical in real-life applications.

The queries in which we are interested are range queries. The definition of the query is as follows: given a range R and two time stamps $t_1 \leq t_2 \leq now$, find all objects that were in R at any time between $[t_1, t_2]$. In a special case when $t_1 = t_2$, the query is called the time-slice query.

2 Index Structure

According to the definition of the zone-based policy, between two consecutive updates, objects stay at the same zone. The tuples have the following format (id, zid, t_s, t_e), where id uniquely identifies the object, zid indicates which zone the object stays during the time period $[t_s, t_e)$. If the tuple is about an object after the last update, $t_e = now$.

In the database, there are two kinds of records. The end time stamp of some records are *now*. We called these records the "current status records". For the other records, the end time stamp is a fixed value. We called them the "history records". There is one current status record for each moving object which shows the zone the object currently falls in and since when. Later, when the object leaves the zone at t, the end time stamp of the current status record is set to be t, (It becomes the history record now.) and a new current status record about the object is created. The start time stamp of the newly generated record is t.

Notice that zid are different from t_s and t_e. After the initial partition, zid is fixed but t_s and t_e increase with time; there is no order between two different zid but different t_s and t_e can be compared. We use an intuitive way to index the tuples: construct similar but independent structures for each zone. The structure we designed is called the SEB-tree (Start/End time stamp B-tree).

Suppose n is the population of objects. In zone z, after a period of time, N tuples, which include both the current status records and the history records, are generated. There is at most one current status record for each object, therefore, at most n among N records could be the current status record, where $n \ll N$. The format of each tuple is (id, z, t_s, t_e). Since all records have identical zone ids, we omit field z. Now each tuple has three fields (id, t_s, t_e), where $t_s < t_e$. The maximum number of records that can be stored in one disk page is B.

An interesting property about the history records is that the tuples are sorted by their end time stamps, i.e. if record $r_1 = (t_{s1}, t_{e1})$ are inserted before record $r_2 = (t_{s2}, t_{e2})$, then $t_{e1} \leq t_{e2}$. To index the history records, the first step is to construct a 2-dimensional space where the start time stamp and the end time stamp are the horizontal and the vertical axes. Each record are mapped to a point in the space. Since in each record $t_s < t_e$, the points are all in upper-left half of the plane.

Figure 1 shows the insertion procedure. Initially the tree contains only one node and points are inserted into that node (figure (a)). When the node is

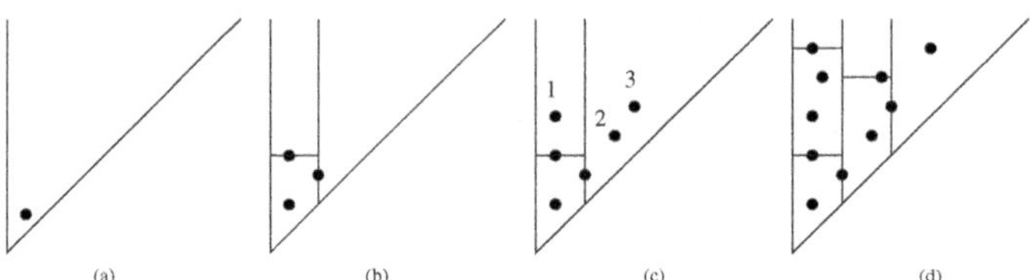

Fig. 1. The insertion procedure for $B = 3$

full (figure (b)), a vertical line and a horizontal line segments are marked. The vertical line is the maximum value of t_s in the node and the horizontal line segment shows the maximum value of t_e. Later, more points are inserted (figure (c)). For each point, there are two possibilities: if the point has the start time stamp no more than the vertical line, the point is inserted into the left side of the line (such as point 1 in figure (c)). Otherwise, the right side (such as point 2 and 3 in figure (c)). For the points that are inserted into the left side of the vertical line, because of the monotonous increasing of the end time stamps, the points must stay above the horizontal line segment. For points on the right side of the vertical line, the above procedure repeats. In figure (d), there are three vertical columns, and each column has 1 or 2 nodes.

Three steps are executed when we insert a new point: 1) check the start stamp list and find the column where the point is inserted; 2) check the end stamp list and find the last node in the column. If the node is full, create a new node in the column; 3) insert the point into the node. The cost for each step is $O(\log_B N)$, $O(1)$ and $O(1)$, and the overall cost for one insertion is $O(\log_B N)$.

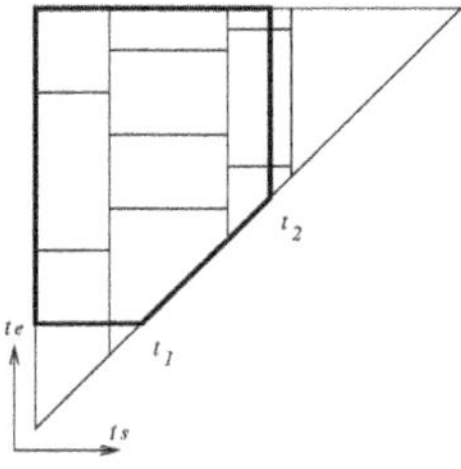

Fig. 2. Range query on the structure

Figure 2 shows a query on the transformed space. The query becomes to retrieve all points in the polygon. Observe that if a leaf node is inside the polygon, the points in the node must be in the query results; and if a leaf node is intersected with the polygon, the points in it require further check. The total query cost is $O((\log_B N) + ((n \log_B n)/B) + ((n + B)/B) + (t/B))$, where t is the size of the results. The first part $\log_B N$ is for searching the columns that contains t_1 and t_2; the second part $(n \log_B n)/B$ is for searching one node of each n/B

columns before column i; the third part $(n+B)/B$ is for searching column j and the fourth part t/B is for retrieving the nodes that fall into the query range.

3 Experiments

The uniform assumption of the moving patterns of objects seems to be unrealistic. We adopt the scenario that is used by other research. The scenario, which is similar to the data sets used in [5]. We use the Voronoi diagram on the destinations to partition the space into ND zones. The objects are able to find out which zone they are based on the distance to the destinations. The average number of history records that are created in one time interval is about 4,000. We observe the movement of objects for 1,000 time intervals. During the time period, the number of snapshots (i.e. update points) of each object is about 40.

Queries are generated as follows. First we create a set of rectangles. The area of each rectangle is 1% of the total area of the space. Those rectangles are randomly placed on the space. We define three kinds of range queries: short, medium and long. The corresponding time spans are 10, 50 and 100 time intervals. The time range are randomly placed between 1 and 1,000 and the query range is a rectangle randomly chosen from the above set. For time slice query, besides the range, the time stamp is arbitrarily selected between 1 and 1,000.

We are examining the index performance of the R-tree, the TB-tree and our method. The metric we used is the average number of nodes accessed when inserting a new record. The lower the better. The nodes we counted include both the leaf and non-leaf nodes. The number of objects varies from 20K to 100K and the size of the data sets for 1,000 time intervals varies from 800K to 4M. We count the total nodes accessed when all records are inserted and calculate the average values. Figure 3 shows the results.

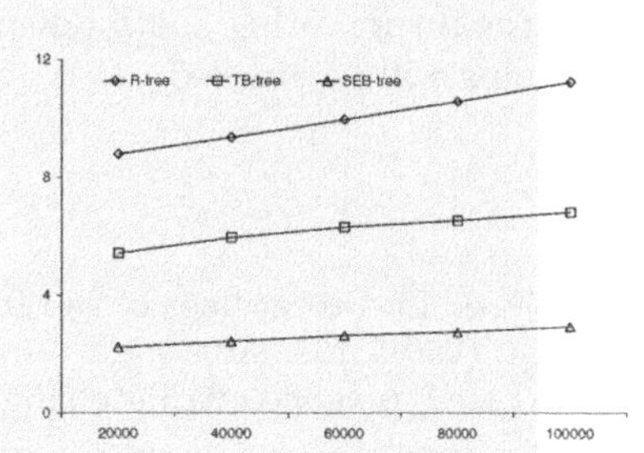

Fig. 3. Number of nodes accessed in one insertion

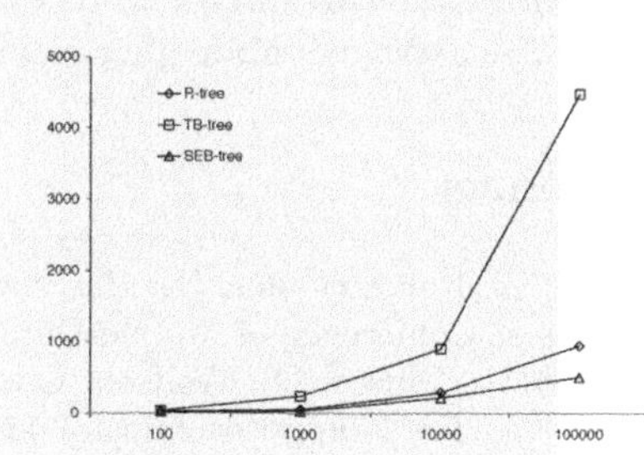

Fig. 4. Time slice query performance

In the figure, the horizontal axis is the number of objects and the vertical axis is the average nodes accessed in one insertion. In the R-tree, when a new record is insertion, to optimize the space proximity, a dedicated algorithm is used. If the tree is unbalanced, some adjustment functions are called. Thus, it has the worst performance for insertion. In the TB-tree, the new record about an object is always inserted into the end of the trajectory and the adjustment procedure is much easier. Thus, it is better than the R-tree. In the SEB-tree, the insertion position for new records can easily be figured out by checking the zone and the start time stamps and no adjustment needed. So the SEB-tree has the best performance.

Next, we check the query performance of all structures. We examine the time-slice query, short, medium and long range query and count the average nodes checked for each query. Due to the space limit, we only show the results about time-slice queries in figure 4. The results of the other types of queries are quite similar. In the figure, the horizontal axe is the number of objects. Notice that the axe is of logarithmic scale. The vertical axe is the average number of nodes checked for one query. The TB-tree has the worst query performance because the size of each trajectory is small and they are almost stored in one node. If the trajectory intersects with the query range, almost all line segments in the trajectory are checked. The R-tree is much better than the TB-tree because the records are clustered on time and space and each query only accesses part of the trajectories of objects. The SEB-tree uses totally different index method and has the best performance. It outperforms the R-tree by checking about 30% less nodes.

4 Conclusion and Future Work

The paper proposed an index structure to store the history data of moving objects. To use our structure, the records of objects are first transformed into points. And the records of one zone are stored together in a SEB-tree. The experiments show that the SEB-tree has fast insertion algorithm, and it beats the existing structures in all kinds of queries. In the future, other spatial queries, such as k-nearest neighbor queries and spatial joins will be studied.

References

[1] Song, Z., Roussopoulos, N.: Hashing moving objects. In: Proceedings of 2nd International Conference of Mobile Data Management. (2001) 161–172

[2] Tayeb, J., Ulusoy, O., Wolfson, O.: A quadtree based dynamic attribute indexing method. The Computer Journal **41** (1998)

[3] Pfoser, D., Jensen, C., Theodoridis, Y.: Novel approaches in query processing for moving objects. In: Proceedings of Very Large Database. (2000) 395–406

[4] Tao, Y., Papadias, D.: The mv3r-tree: A spatio-temporal access method for timestamp and interval queries. In: Proceedings of Very Large Database. (2001)

[5] Saltenis, S., Jensen, C.S., Leutenegger, S.T., Lopez, M.A.: Indexing the positions of continuously moving objecs. In: Proceedings of ACM SIGMOD International Conference on Management of Data. (2000)

Best Movement of Mobile Agent in Mobile Computing Systems

Chao-Chun Chen[1], Chiang Lee[2], and Chih-Horng Ke[3]

[1] Department of Computer Science and Information Engineering
National Cheng-Kung University, Tainan, Taiwan, R.O.C.
chencc@dblab.csie.ncku.edu.tw

[2] Department of Computer Science and Information Engineering
National Cheng-Kung University, Tainan, Taiwan, R.O.C.
leec@dblab.csie.ncku.edu.tw

[3] Department of Information Management
Chang Jung Christian University, Tainan, Taiwan, R.O.C.
kch@dblab.csie.ncku.edu.tw

Abstract. Retrieving data using mobile agents gains popularity due to the advance of mobile computing techniques. However, the transmission of wireless communication is still unreliable and bandwidth-limited. This paper investigates this issue and tries to find the most suitable location for a mobile agent in a vast network. We first propose a mechanism and operations for the support of this strategy. We then develop an analysis model to estimate the optimal location for a mobile agent.

1 Introduction

Mobile computing is now a reality owing to the rapid advances of wireless communication technologies and electronics technologies. Very often mobile users (MUs) need to interact with applications of very large and complex systems, such as mobile commerce applications, database management systems (DBMS), etc. When executing a task, e.g., a database transaction, a MU needs to frequently communicate with such a remote server. However, the transmission of wireless communication is now still unreliable (e.g., frequent disconnection) and bandwidth-limited. A large amount of waiting time would be incurred if the MU directly communicates with remote peers when needed. Thus, an important issue in this environment is to reduce the utilization of the wireless communication. A software approach to overcome the above problem is to place a special-purposed mobile agent (MA) in the wired network, and this MA executes communication-bound tasks for the MU when receiving a request message. In such a way, the MU only needs to connect to the wireless network twice while executing a communication-bound task: once for sending the request to the agent, the other for receiving the request result from the agent. The task performed by the MA could be time consuming and during this time the MU may keep moving and still the interaction between MU-MA and between MA-server may continue.

M.-S. Chen et al. (Eds.): MDM 2003, LNCS 2574, pp. 345–350, 2003.

Hence, the location of the MA may significantly affect the induced communication cost. In this paper, we propose a MA controlling mechanism and study the best suitable place for the MA to stay. We then develop an analysis model to represent and evaluate the effect of the distance between them.

2 Network Architecture and MA Controlling Mechanism

Figure 1 shows the system architecture based on which our mechanism is designed. Geographically, the regions are divided into cells. Each cell has a *base*

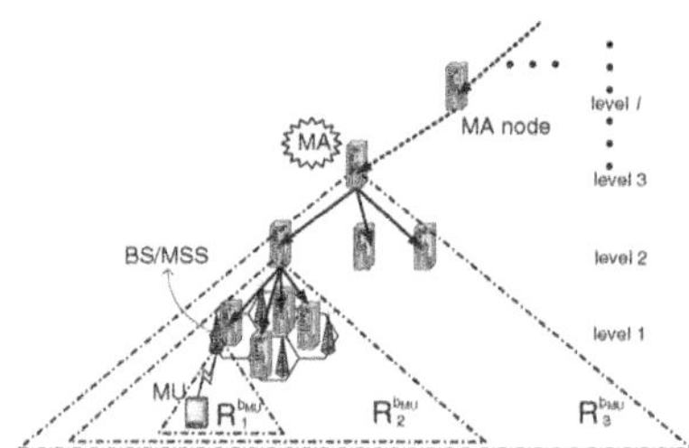

Fig. 1. The network architecture of the mobile environment

station (BS) and the management of the locations of MUs within a cell and the connection of these MUs with the outside world are all through the *mobile support station* (MSS) of the cell [IB94][PS01]. In the wired network, the fixed servers offering working places for MA's are called *MA nodes*. MA nodes are connected with each other through wired networks with different bandwidth, and the bandwidth of two connected MA nodes determines the distance between them. We assume the MA nodes form a hierarchy over the wired network. All BSs are inherently the leaf nodes. Level 1 of the hierarchy is at the leaves of the tree, and level 2 nodes are the parents of the level 1 nodes, and so on. As we mentioned above, a cell is a small geographical area overseen by a base station. Let the BS that a MU resides be b_{MU}. The ancestor MA node of b_{MU} therefore oversees a large region formed by the small areas of the sibling BSs of b_{MU} who have this same ancestor. If the ancestor is at level ℓ of the hierarchy, we denote its overseen region as $R_\ell^{b_{MU}}$ for convenience of presentation.

The MA controlling mechanism we proposed contains two operations: one is *request operation*; the other is *movement operation*.

Request Operation A request operation is issued when a MU is requesting a MA to execute some task. The network communication in a request operation includes (1) delivering the request from the MU to the MA, and (2) sending the result back from the MA to the MU. Figure 2 illustrates the scenario.

Movement Operation Figure 3(a) and Figure 3(b) are two cases that could occur while a MU moves. Figure 3(a) shows the case that a MU move inside the region of level $R_\ell^{b_{old}}$ (which is equal to $R_\ell^{b_{new}}$). In this case, the MA need

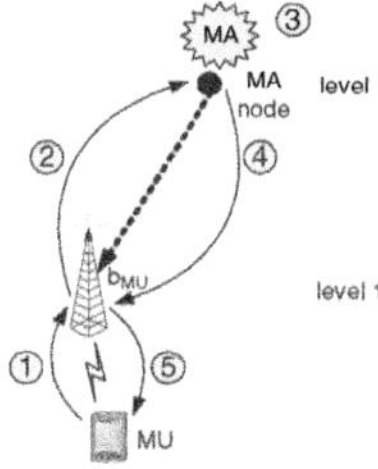

Fig. 2. An example of the MA's request operation.

not move. Figure 3(b) shows the case that a MU moves out of $R_\ell^{b_{old}}$ and the MA hence has to be moved to a new location. Notice that the MA is always the

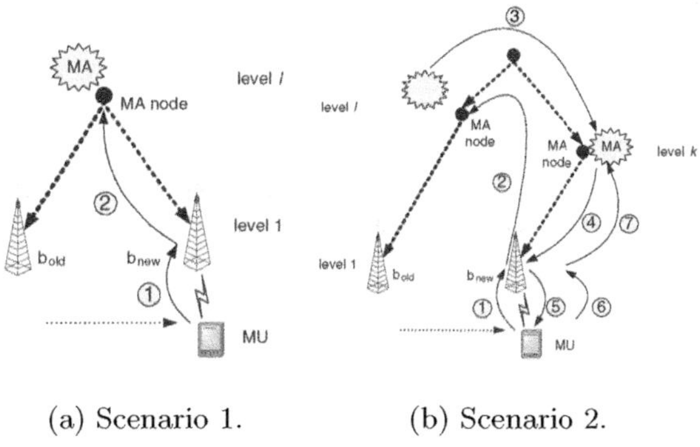

(a) Scenario 1. (b) Scenario 2.

Fig. 3. Two scenarios of the MA's movement operation.

ancestor of the MU in the network structure no matter where the MU moves to. In this scenario, the new MA location needs to be determined. The new MA location at a different level of the network incurs different costs for communicating with the MU and the server. Hence, the best level for the MA to stay should be carefully computed.

3 Analysis of the New Mechanism

The parameters used in our cost models are listed in Figure 4. In order to represent the MU's request and movement behavior, we design $\mathcal{MIR}$ as

$$\mathcal{MIR} = \frac{\text{frequency of movement operations}}{\text{frequency of request operations}} = \frac{\text{number of movement operations}}{\text{number of request operations}}$$

Parameter	Description
System parameters	
$\mathcal{MIR}$	The MU's mobility/request ratio.
$\mathcal{R}_x$	The size of x in the request operation, $x \in \{request, result\}$.
$\mathcal{M}_y$	The size of y in the movement operation, $y \in \{m, u, a\}$.
$\mathcal{A}$	The size of a MA.
α	The communication cost of transferring 1KB data through the wireless network.
β_ℓ	The communication cost of transferring 1KB data between the level ℓ and $\ell+1$ of the wired network.
P_ℓ	The probability that a MU moves inside the boundary of $R_\ell^{b_{MU}}$.
Q_ℓ	The probability that a MU moves out of the boundary of $R_\ell^{b_{MU}}$.
μ_ℓ	The probability that a MU stays within $R_\ell^{b_{MU}}$ when a MU moves out of $R_{\ell-1}^{b_{MU}}$.
Performance Metrics and Strategy-related parameters	
L	The farthest level that an MA can arrive.
C_R^ℓ	The request cost that a MA sends a request at the level ℓ.
C_m^ℓ	The movement cost that a MA moves inside the region of level ℓ.
$C_M^{(\ell,h)}$	The movement cost a MA moves out the region of level ℓ throuth level h.
C_{total}^ℓ	The total communication cost of a MA at the level ℓ.

Fig. 4. Parameters and their meanings.

We first derive the farthest level that a MA can reside. Due to the length limitation of the paper, we only present the result here. Interested readers can refer to [CLK02] for the details. Assume that L is the highest allowable level that the MU can moves. Then, we can infer that the following inequation should be true.

$$\sum_{i=1}^{L} \beta_i > \alpha \geq \sum_{i=1}^{L-1} \beta_i$$

From this inequation, L can be determined.

We then derive the communication cost of the MA. When a MA stays at a different level, it would incur a different cost of request operations and movement operations, and hence a different total cost. That is, the total cost is a function of a MA's resident level ℓ. From the view point of the request operations, it is beneficial if the MA is close to the MU; but from the view point of the movement operations, it is beneficial if the MA is far away from the MU. Thus, a tradeoff exists in this mechanism.

As there are only two types of operations: request and movement operations, the total cost can be represented on the per-request basis as follows.

$$C_{total}^\ell = (\text{cost of a request operation}) + \mathcal{MIR} \cdot (\text{cost of a movement operation})$$

The cost of a request operation that a MA is at level ℓ is denoted as C_R^ℓ. In terms of a movement operation, it can be divided into two subcases:

1. the movement of the MU is still within $R_\ell^{b_{MU}}$: the communication cost of the movement operation is denoted as C_m^ℓ. The probability of this case is P_ℓ.
2. the MU moves out of $R_\ell^{b_{MU}}$: the level of the least common ancestor of the newly arrived cell and the old cell affects the communication cost. We assume

the least common ancestor of MU's new cell and old cell is at level h, and denote such movement operation cost as $C_M^{(\ell,h)}$. The probability of this case is Q_ℓ.

Q_ℓ and P_ℓ can be easily expressed as follows.

$$Q_\ell = \begin{cases} 1 & \text{, if } \ell = 1. \\ 1 - P_\ell & \text{, if } \ell \geq 2. \end{cases}$$

$$P_\ell = \begin{cases} 0 & \text{, if } \ell = 1. \\ P_{\ell-1} + Q_{\ell-1} \cdot \mu_i = \sum_{i=2}^{\ell} Q_{i-1} \cdot \mu_i & \text{, if } \ell \geq 2. \end{cases}$$

where $\sum_{i=2}^{\ell} Q_{i-1} \cdot \mu_i$ is the result after $P_{\ell-1}$ is replaced by $P_{\ell-2} + Q_{\ell-2} \cdot \mu_{\ell-1}$, and recursively until all P_i is replaced, $1 \leq i \leq \ell$. Utilizing the above probability model, the total cost can be rewritten as follows.

$$C_{total}^{\ell} = C_R^{\ell} + \mathcal{MIR} \cdot \left(P_\ell \cdot C_m^{\ell}\right) + \mathcal{MIR} \cdot \left(\sum_{h=\ell+1}^{L} (Q_{h-1} \cdot \mu_h) \cdot C_M^{(\ell,h)}\right)$$

$$C_R^{\ell} = \mathcal{R}_{request} \cdot \alpha + \left(\sum_{i=1}^{\ell-1} \mathcal{R}_{request} \cdot \beta_i\right) + \left(\sum_{i=1}^{\ell-1} \mathcal{R}_{result} \cdot \beta_i\right) + \mathcal{R}_{result} \cdot \alpha$$

$$C_m^{\ell} = \mathcal{M}_m \cdot \alpha + \left(\sum_{i=1}^{\ell-1} \mathcal{M}_m \cdot \beta_i\right)$$

$$C_M^{(\ell,h)} = \mathcal{M}_m \cdot \alpha + \left(\sum_{i=1}^{h-1} \mathcal{M}_m \cdot \beta_i\right) + \left(\sum_{i=\ell}^{h-1} \mathcal{M}_m \cdot \beta_i\right) + \left(\sum_{i=\ell}^{h-1} 2 \cdot \mathcal{A} \cdot \beta_i\right)$$
$$+ \left(\sum_{i=1}^{\ell-1} \mathcal{M}_u \cdot \beta_i\right) + \mathcal{M}_u \cdot \alpha + \mathcal{M}_a \cdot \alpha + \left(\sum_{i=1}^{\ell-1} \mathcal{M}_a \cdot \beta_i\right)$$

4 Conclusions

In this paper, a MA controlling mechanism is designed to support for executing the communication-bound tasks of a MU. In such a way, the utilization of the wireless communication can be largely reduced. Different from the past related work, the MA can stay in the most beneficial site, rather than always stays at level 1. The cost equations of a MA are then derived by our developed analysis model. Our ongoing work is to study more complete performance evaluation, and we also further study the MA movement on different network topologies.

References

[CLK02] Chao-Chun Chen, Chiang Lee, and Chih-Horng Ke, "Best Movement of Mobile Agent in Mobile Computing Systems", *Technical Report 91-7-05*, CSIE, NCKU, Taiwan, 2002.

[IB94] T. Imieliński and B. Badrinath, "Mobile Wireless Computing: Challenges in Data Management", *CACM*, Vol. 37, No.10, October 1994.

[MDW99] Dejan Milojicic, Frederick Douglis, and Richard Wheeler, "Mobility – Processes, Computers, and Agents", Addison-Wesley, 1999.

[PS01] Evaggelia Pitoura and George Samaras, "Locating Objects in Mobile Computing", *IEEE Transactions on Knowledge and Data Engineering*, Vol. 13, No. 4, July/August 2001, pp571 -592.

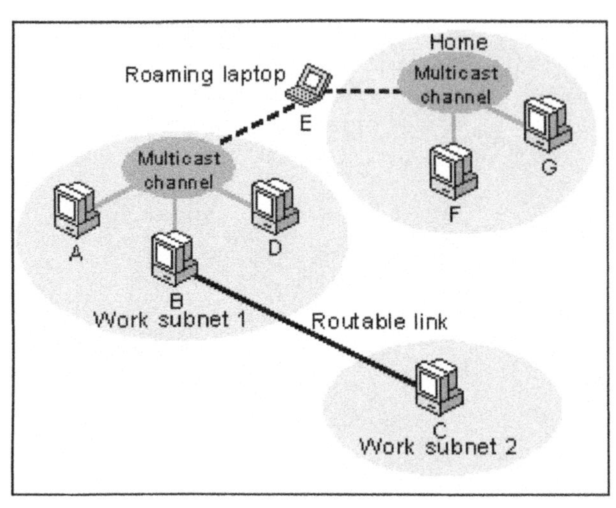
Home
Roaming laptop
Multicast channel
E
Multicast channel
G
F
A
D
B
Work subnet 1
Routable link
C
Work subnet 2

Transactional Peer-to-Peer Information Processing: The AMOR Approach

Klaus Haller[1], Heiko Schuldt[1,2], and Hans-Jörg Schek[1,2]

[1] Database Research Group Institute of Information Systems
Swiss Federal Institute of Technology
CH-8092 Zürich, Switzerland
{haller,schuldt,schek}@inf.ethz.ch
[2] University for Health Informatics and Technology Tyrol (UMIT)
A - 6020 Innsbruck, Austria

Abstract. Mobile agent applications are a promising approach to cope with the ever increasing amount of data and services available in large networks. Users no longer have to manually browse for certain data or services but rather to submit a mobile personal agent that accesses and processes information on her/his behalf. These agents operate on top of a peer-to-peer network spanned by the individual providers of data and services. However, support for the correct concurrent and fault-tolerant execution of multiple agents accessing shared resources is vital to agent-based information processing. This paper addresses this problem and shows how agent-based information processing can be enriched by dedicated transactional semantics — despite of the lack of global control which is an inherent characteristic of peer-to-peer environments — by presenting the AMOR (Agents, MObility, and tRansactions) approach.

1 Introduction

A consequence of the proliferation of Internet technology is the availability of a vast amount of data. Recent trends in accessing this data propose the use of mobile agent technology [CZ98, KG99], because this considerably reduces the amount of data that has to be shipped over the network as well as network latency. Moreover, it allows a user to be off-line while her/his agent program is being executed. Also, by using agent technology, a user no longer has to manually browse for certain data and/or services but rather to submit a mobile personal agent that accesses and processes information on her/his behalf, e.g. places orders or bids in some electronic auctions, or fixes dates. To this end, mobile agents operate on top of a peer-to-peer network spanned by the individual providers of data and services.

Most current mobile agent approaches focus on the support of applications which gather and filter data [BGM+99], [GKP+01]. However, in order to support another practically relevant class of mobile agent applications, the aspect of *information processing* has also to be considered. The extension of the capabilities of mobile agents systems from information gathering towards information processing in a peer-to-peer-based system environment now goes along with additional requirements that users impose on their mobile agent applications. Essentially, mobile agents have to leave the

M.-S. Chen et al. (Eds.): MDM 2003, LNCS 2574, pp. 356–362, 2003.

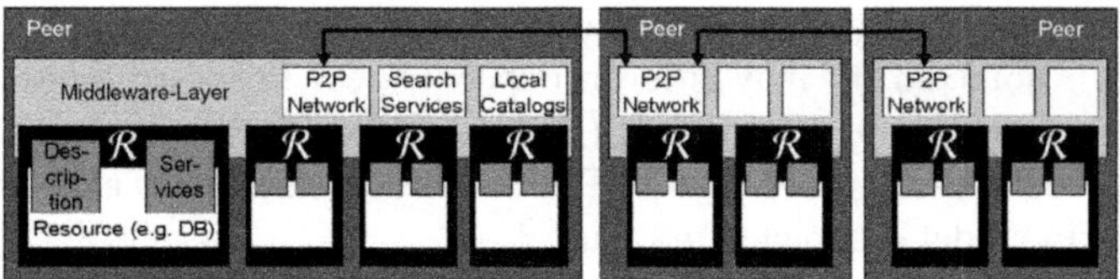

Fig. 1. The AMOR System Model

system in a consistent state, even in the case of failures or when different agents access shared resources and/or services concurrently. This urgently requires support for correct concurrency control and failure handling. Such requirements are well understood in traditional information systems. Essentially, applications are encompassed into transactions and a centralized transaction manager controls their execution, i.e., it enforces their atomic behavior in case of failures and guarantees synchronization in the case of concurrency. However, in the peer-to-peer environments in which mobile agents execute, such a centralized transaction manager does not exist. Yet, the proper failure handling is delegated to the individual agents and the task of synchronization requires communication among groups of agents. Developing an appropriate protocol is one issue of the AMOR project at ETH Zürich, were we want to bring together transaction processing and peer-to-peer environments and to provide the foundations for correct and reliable peer-to-peer information processing.

The main contribution of this paper is a protocol that allows the decentralization of concurrency control and recovery. Yet, it has to be noted that concurrency control and recovery are to be solved jointly and in a global manner; however, by applying sophisticated replication techniques to metadata needed for transaction management, we are able to implement support for transactions in a truly distributed way.

The remainder of the paper is structured as follows: The AMOR system architecture is presented in Section 2. Section 3 discusses the AMOR approach to decentralized transaction management. The AMOR prototype is introduced in Section 4. Section 5 discusses related work and Section 6 concludes.

2 The AMOR System

The bottom layer of the AMOR system consists of a set of distributed and autonomous *resources*, e.g., database systems or flat files. These resources are wrapped by so-called *resource agents* ($\mathcal{R}$), which are responsible for encapsulating resources and for providing unique interfaces to them. Data objects managed by the underlying resources can only be accessed and/or manipulated by the services provided by the corresponding resource agent. Let $\mathcal{R}^*$ denote the universe of all resource agents of the system and $\mathcal{S}^*$ the universe of the services provided by them. We assume the resources wrapped by all $\mathcal{R}_i, \mathcal{R}_k \in \mathcal{R}^*$ to be pairwise disjoint. The independence of resources means that the services of a resource agent only operate on local resources and are not redirected to remote resources.

Resources and therefore also the resource agents are hosted by *peers*. The number of resources is not fixed and may differ from peer to peer. Peers are connected by a middleware layer forming the peer-to-peer (P2P) network. Peers may dynamically decide to join or leave the system such that the P2P network continuously evolves. The overall AMOR system model is depicted in Figure 1.

Applications on top of the AMOR peer-to-peer network are executed by mobile *transactional agents*. Let $\mathcal{T}^*$ be the set of all mobile transactional agents of the system. For the purpose of application execution, each $\mathcal{T}_i$ of $\mathcal{T}^*$ migrates within the network to peers so as to locally invoke services there. Hence, services are used as building blocks for transactional agent applications.

3 Decentralized Transaction Processing in AMOR

Each $\mathcal{T}_i$ implements an individual, independent, and distributed transaction. Traditionally, transaction management in a distributed environment is provided by a dedicated centralized coordinator, e.g., a TP Monitor [BN97]. The coordinator's task is to orchestrate the execution of distributed transactions and to enforce their atomic commitment by using a 2PC protocol [GR93]. Whereas the centralized approach is highly appropriate for small, well-delimited environments with a fixed number of peers, it cannot be applied to large-scale networks with a large number of distributed, heterogeneous, and autonomous peers.

Hence, we must strive for a *distributed implementation* of the functionality of a coordinator. Essentially, following the unified theory of concurrency control and recovery [VHB$^+$98], coordination functionality requires both isolation and atomicity to be provided jointly. In terms of recovery, this means that each failure of some $\mathcal{T}_i$ has to be handled properly: The effects of all services it has invoked so far have to be undone or alternative service invocations leading to a well-defined state have to be enforced. This requires that service invocations are logged which is done by each resource agent mediating transactional agent requests to its resource. In terms of concurrency control, the concurrent invocation of services has to be controlled such that each $\mathcal{T}_i$ is correctly shielded from other transactional agents. Since flow of information is only possible via services manipulating/retrieving data objects, the commutativity behavior of services has to be known (i.e., which service invocations do conflict). Due to the independence of resources, such conflicts may only exist between services wrapped by the same $\mathcal{R}_k$. Hence, conflict detection can also be realized at the level of resource agents. Two service invocations are in *conflict* if their order matters, i.e., if their return values would change when being ordered differently; otherwise, the services *commute*.

Reasoning about global correctness — using information on conflicts — is based on the notion of schedule. A schedule **S** reflects the concurrent execution of a set of transactions. The unified theory addresses both atomicity and isolation by considering regular services and inverse services (which are used to semantically undo the effects of regular ones) within the same framework. Serializability with ordered termination (SOT) has been identified as the correctness criterion that accounts for both problems [AVA$^+$94]. Moreover, it has been shown that the SOT criterion can be checked by means of the acyclicity of the serialization graph of a schedule [AVA$^+$94]. In short, the

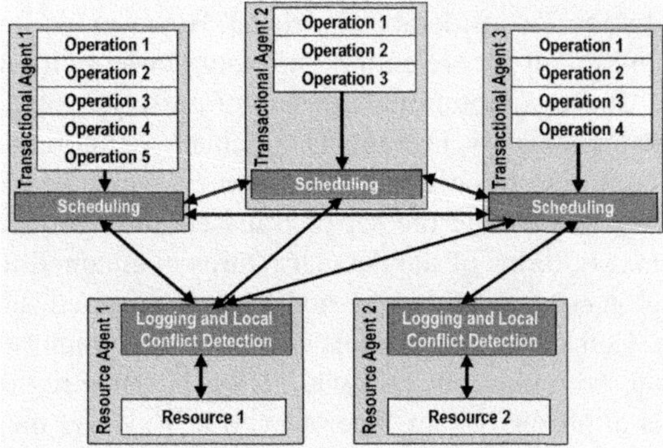

Fig. 2. Decentralized AMOR Approach to Transaction Processing

serialization graph of a schedule **S** is a directed graph where the nodes are the transactions of **S** and the directed edges correspond to the conflicts between transactions. Reasoning about the correctness of a schedule based on a serialization graph requires global information to be available at a central coordinator — the transaction manager. However, due to the absence of a centralized transaction manager in AMOR, conflict information on local service invocations have to be communicated from a resource agent to the corresponding $\mathcal{T}_i$. Then, scheduling can be enforced without centralized coordinator but rather by distributing this conflict information between transactional agents, thereby replicating meta information needed for synchronization purposes. In Figure 2, the distributed AMOR approach to transaction management is illustrated.

AMOR follows an optimistic approach to transaction management. This means that each $\mathcal{T}_i$ executes without determining on the spot whether a service invocation of $\mathcal{T}_i$ is allowed or not. However, at least prior to the commit of $\mathcal{T}_i$, a validation is required which checks whether it has executed correctly and is therefore allowed to commit. This is closely related to well-established optimistic concurrency control protocols like backward-oriented concurrency control, BOCC [KR81] and to serialization graph testing protocols [Cas81]. Following this optimistic protocol, a $\mathcal{T}_i$ is allowed to commit if it does not depend on some active (uncommitted) $\mathcal{T}_j$.

Since conflicts are communicated immediately from a resource agent $\mathcal{R}_k$ to a transactional agent $\mathcal{T}_i$ each time $\mathcal{T}_i$ invokes a service, this information is locally available. Hence, each $\mathcal{T}_i$ is able to locally decide whether or not it is allowed to commit correctly.

While this approach enforces correct concurrency control, it features an important drawback. Essentially, cyclic conflicts and therefore non-serializable executions are not detected until commit time, even when these cyclic conflicts have been imposed much earlier. Thus, detecting such cyclic conflicts at an earlier stage would allow for much less redundancy.

Therefore, each $\mathcal{T}_i$ has to be equipped with metadata reflecting a multi-agent execution. When a copy of the global serialization graph would be available at each transactional agent, cyclic conflicts could be detected and thus resolved immediately. But always maintaining a copy of the complete serialization graph with each $\mathcal{T}_i$ is not practical since it would impose considerable overhead. However, an important observation is that in the type of system we are addressing, more or less closed –albeit not static– communities exist. These communities are sets of closely related resources accessed by the same transactional agents, i.e., agents which aim at addressing the same or similar tasks. Hence, conflicts are only possible between agents invoking services within the same community. We denote the set of transactional agents executing in such a community as *region*. In terms of the data structures maintained for concurrency control purposes, a region corresponds to the nodes of a connected subgraph of the global serialization graph. Consequently, only this connected subgraph (termed *region serialization graph*) has to be replicated among all $\mathcal{T}_i$ of the same region. Obviously, if the serialization graphs of all regions are free of cycles, so is also the global serialization graph. Thus, reasoning about system-wide correctness can be shifted to the universe of region serialization graphs. Therefore, AMOR uses replicas of region serialization graphs maintained by each transactional agent to enforce correct multi-agent executions. This requires that i.) conflicts are communicated from resource agents to the corresponding transactional agents and ii.) the replica of region serialization graphs are kept consistent among all $\mathcal{T}_i$ of the same region. Obviously, requirement i.) is sufficient for correct concurrent execution while ii) allows for an optimization in that incorrect executions can be detected and resolved as early as possible.

However, keeping the region serialization graph up-to-date (requirement ii.) has been an interesting issue during the development of our protocol, because regions are not static. Rather, the composition of regions is affected by the services invoked by the transactional agents of a region. Consider a service invocation of $\mathcal{T}_i$ which imposes a conflict with a service invocation of a $\mathcal{T}_j$ of another region. As a consequence of this conflict, the two originally independent regions have to be merged. Hence, as an additional requirement for consistent metadata management in AMOR, the two region serialization graphs have to be consolidated and finally distributed among all transactional agents of the new region. Similarly, by correctly committing a transactional agent or in case of a rollback, regions may split and so does the corresponding region serialization graph.

4 Prototype

The AMOR prototype implementation is based on the Aglets [LO98] framework which we have extended in several directions. Most importantly, we have added support for agent and place descriptions allowing to search for either places or agents [HS01]. This extension has led to the implementation of two new types: P2P-Agents and Agents Management Agents (AMAs). The latter ones are part of the place infrastructure and responsible for managing the information about places, e.g. which agents are currently residing on that specific place. The former type of agents, P2P-Agents, implements mechanisms for sending messages and for migrating to target places by specifying criteria – in other words, predicates which have to be fulfilled for message recipients or destinations of migration. By this, we follow somehow the idea of autonomous computing [Wla02]: We can specify our transactions (or workflows) using our agents without caring about the system and network configuration, because the agents autonomously adjust to the situation at execution time.

Our P2P-Agents are further refined in transactional agents and resource agents. The latter ones provide an XML-description to support the search for them in the peer-to-peer network. The AMOR prototype encompasses implementations of such resource agents, e.g., one for Oracle8 databases. This resource agent allows to access object-relational tables by wrapping user-defined methods and by providing these services to transactional agents. The other refinement of P2P-Agents is reflected in the concept of transactional agents which implement the protocol introduced in this paper.

5 Related Work

In terms of related work in the context of mobile agents, two directions have to be considered: workflow execution and transactions. The mobile agent approach is commonly agreed to be highly suitable for workflow execution, e.g., [CGN96]. Recent work extends these ideas and deals with the problem of how to schedule mobile agents in case different places are available [XRS01] for a particular service. Other approaches even address the intersection of mobile agent technology and transaction management. First results were achieved by enforcing atomicity and fault tolerance [SP98]. More recent approaches like [SAE01], albeit still concentrating on the atomicity aspect, also provide support for concurrency control. However, this is achieved by combining timestamp-ordering and 2PC, which results in a rather limited degree of concurrency.

6 Summary and Outlook

In this paper, we have presented the ideas of the AMOR project which allows for a novel decentralized implementation of transaction management in a peer-to-peer environment. Applications are embedded within mobile transactional agents. Due to the lack of a central coordinator which is inherent to peer-to-peer systems, the task of enforcing correct concurrency control and failure handling for agent-based applications is shifted to the transactional agents. They receive information about local conflicts each time they invoke a service – information used to update their local view on the relevant portion of global metadata, the region serialization graph. The different local replica of these graphs are kept consistent by means of communication at the agent level.

In our future work, we aim at extending the existing AMOR framework such that unreliable network connections are shielded from the transactional agent. providing support for shielding unreliable network connections.

References

[BGM+99] B. Brewington, et al. *Mobile Agents for Distributed Information Retrieval,* in: M. Klusch: Intelligent Information Agents. Springer, 1999.

[BN97] P. Bernstein and E. Newcomer. *Principles of Transaction Processing*. Morgan Kaufmann, 1997.

[Cas81] M. Casanova. *The Concurrency Control Problem for Database Systems, LNCS 116.* Springer, 1981.

[CGN96] T. Cai, P. Gloor, and S. Nog. *Dartflow: A Workflow Management System on the Web using Transportable Agents.* Technical Report TR96-283, Dartmouth College, 1996.

[CZ98] W. Cockayne and M. Zyda, editors. *Mobile Agents.* Prentice Hall, 1998.

[AVA+94] G. Alonso et al. *Unifying Concurrency Control and Recovery of Transactions.* Information Systems, 1994.

[VHB+98] R. Vingralek et al. *Unifying Concurrency Control and Recovery of Transactions with Semantically Rich Operations.* Theoretical Computer Science, 190(2), 1998.

[GKP+01] R. Gray et al. *Mobile Agent versus Client/Server Performance: Scalability in an Information-Retrieval Task.* 5th Int. Conf. on Mobile Agents, Atlanta, GA, 2001.

[GR93] J. Gray and A. Reuter. *Transaction Processing: Concepts and Techniques.* Morgan Kaufmann, 1993.

[HS01] K. Haller and H. Schuldt. *Using Predicates for Specifying Targets of Migration and Messages in a Peer-to-Peer Mobile Agent Environment.* 5th Int. Conf. on Mobile Agents (MA), Atlanta, GA, 2001.

[KG99] D. Kotz and R. Gray. *Mobile Agents and the Future of the Internet.* Operating Systems Rev., 33(3), 1999.

[KR81] H. Kung and J. Robinson. *On optimistic Methods for Concurrency Control.* ACM Transactions on Database Systems, 6(2), 1981.

[LO98] D. Lange and M. Oshima. *Programming and Deploying Java Mobile Agents with Aglets.* Addison Wesley Longman, 1998.

[SAE01] R. Sher, Y. Aridor, and O. Etzion. *Mobile Transactional Agents.* 21st Int. Conf. on Distributed Computing Systems, Phoenix, AZ, 2001.

[SP98] A. Silva and R. Popescu-Zeletin. *An Approach for Providing Mobile Agent Fault Tolerance.* 2d Int. Workshop on Mobile Agents, Stuttgart, Germany, 1998.

[Wla02] I. Wladawsky-Berger. *Advancing the Internet into the Future.* Talk at the Int. Conf. Shaping the Information Society in Europe, 2002.

[XRS01] R. Xie, D. Rus, and C. Stein. *Scheduling Multi-Task Agents.* 5th Int. Conf. on Mobile Agents (MA), Atlanta, GA, 2001.

Performance Evaluation of Transcoding-Enabled Streaming Media Caching System

Bo Shen, Sung-Ju Lee, and Sujoy Basu

Hewlett-Packard Laboratories
Palo Alto, CA 94304

Abstract. We evaluate the performance of transcoding-enabled streaming media caching proxy system (TeC). TeC is designed for efficient delivery of rich media web contents to heterogeneous network environments and client capabilities. Depending on the connection speed and processing capability of an end user, the proxy transcodes the requested (and possibly cached) video into an appropriate format and delivers it to the user. We study the TeC performance by simulation using real traces derived from enterprise media server logs.

1 Introduction

With the popularity of the Internet and the Web, proxy caching systems have been widely deployed to store frequently requested files close to the end users. Proxy caching reduces the traffic between content origin and the proxies as well as the user perceived latency. Today we find more streaming media contents on the Internet. In order to satisfy various users with different network connection speeds, most streaming content providers encode the video at several different bit-rates. Availability of various bit-rate video is useful as the prosperity of wireless networks brings more heterogeneous client devices (e.g., laptops, PDAs, palm pilots, cell phones, etc.) that have different network connection, processing power, and storage. Popular streaming media files can also be stored in proxy caches for efficient distribution. Traditional caching system treats each client request equally and independently. This policy results in caching various versions with different bit-rates of the same video clip, which is a waste of storage.

We proposed to enable the caching proxy with the transcoding ability so that variants of a video object can be delivered with transcoding the cached content to the appropriate format, instead of accessing the object all the way from the content origin[1]. One of the advantages of having transcoding-enabled proxy is that the content origin servers need not generate different bit-rate videos. Moreover, heterogeneous clients with various network conditions will receive videos that are suited for their capabilities, as content adaptation can easily be done at the network edges. Given that the caching proxy is transcoding-enabled, new adaptive caching systems need to be developed for better utilization of the storage. We proposed a set of caching algorithms that consider the case where variants of the same video object exist in the system at any point of time [1]. We evaluate

M.-S. Chen et al. (Eds.): MDM 2003, LNCS 2574, pp. 363–368, 2003.

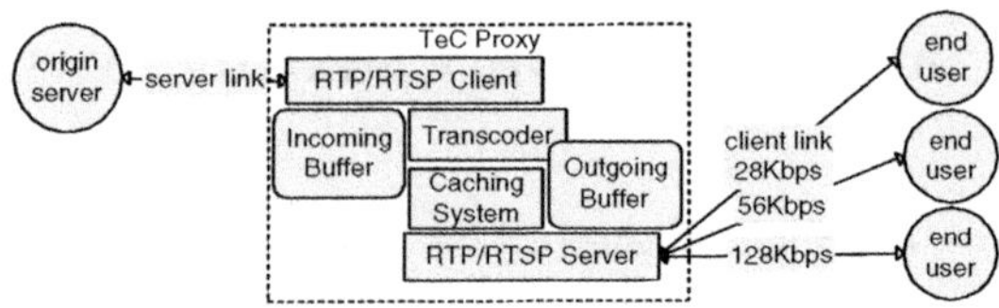

Fig. 1. System and components for transcoding-enabled caching proxy

our proposed TeC (Transcoding-enabled Caching) system using various experiments, including simulations using the proxy trace derived from a real corporate media server log. The trace reveals some degree of heterogeneous access of media objects.

The rest of the paper is organized as follows. We overview the system architecture of TeC in Section 2. Simulation results are reported in Section 3 and we conclude in Section 4.

2 Transcoding-Enabled Caching

2.1 System Architecture

Transcoding-enabled caching (TeC) proxies serve various bit-rate versions of video objects to the end users with different devices or connection profiles. TeC proxy consists of the components shown in Figure 1. The proxy puts the received stream from the origin into the incoming buffer. The transcoder continuously pulls bit streams from the incoming buffer and subsequently pushes the transcoded bits out to the outgoing buffer. The proxy decides to cache the content from the incoming or the outgoing buffer while it is being produced by the transcoder. The data in the outgoing buffer is obtained either from the transcoder or from the caching system.

Given the real time transcoding capability, the TeC proxies dynamically transcode video objects to different variants to satisfy the end users in heterogeneous network environments. Each variant is a version. If version x can be obtained by transcoding from version y, we call version y a *transcodable* version for x. Conversely, version x is the *transcoded* version of y.

2.2 Caching Algorithms

We highlight the TeC caching algorithms in this section. Details of the algorithms can be found in [1]. Let us assume that the origin server has n versions at bit-rates $b_1, b_2, \ldots, b_n$ for each video object. The highest bit-rate version is b_1 and the lowest is b_n, i.e., $b_1 > b_2 > \ldots > b_n$. When version b_i is requested from the end user, and if there is version b_j ($b_j > b_i$, i.e., b_j is a transcodable version for b_i) in cache, the TeC proxy transcodes b_j to b_i instead of fetching b_i from the content origin. Therefore, it is a cache hit even though b_i is not directly available from the cache.

We proposed three different caching algorithms for TeC system. TEC_11 and TEC_12 allow at most one version of a video object to be cached at the proxy at any single time. These two algorithms operate differently when a user requests a version of the video that has lower bit-rate than the one that is already cached. For example, if the client requests version b_i of a video while b_j, where $b_i < b_j$, exists in the cache, b_i is transcoded from b_j and streamed to the user. TEC_11 refreshes the access record of b_j, but TEC_12 removes b_j and caches b_i. On the other hand, TEC_2 may cache multiple versions of the same video object. The goal is to reduce the processing load on the transcoder.

3 Simulations and Results

A stack-based implementation is developed for the performance evaluation simulation study. We conduct simulations based on an enterprise media access trace. The trace logs prolonged media content accesses in mostly a homogeneous network environment. It does however, presented some variant-based media accesses. For performance evaluation in heterogeneous network environments, please refer to [1] in which synthesized traces are used. In simulations here, the TeC proxy uses LRU as the cache replacement algorithm within the TEC caching algorithms. We compare the performance of the TeC proxy with that of a regular caching proxy. The regular caching proxy only serves as an interception proxy using the LRU algorithm without transcoding. This proxy treats each access independently even if the access is to the same media object but a different bit-rate version. We call this regular caching scheme a "reference model."

3.1 Enterprise-Trace Extraction and Analysis

The media server logs provided as input to our simulator are obtained from the servers of HP Corporate Media Solutions. We use log entries from April 1 through May 31, 2001. Since an enterprise media server stores video content at several bit-rates, we believe using the trace driven from this log can help evaluate the TeC performance. Note that the Windows Media Server is not RTP/RTSP-based as we would prefer, but the client access pattern and the video popularity statistics extracted from the server log are nevertheless useful.

To derive proxy traces from primitive server logs, we first partition the server log into separate proxy access logs of four major geographical areas based on the domain names. We then identify multiple bit-rate-variants of the same media object. They are identifiable by suffix of their URL. We find that suffixes such as *28.asf, *56.asf and *110.asf or *112.asf are used. It is verified that these objects are coded at 28Kbps, 56Kbps or above 100Kbps, respectively. We label them as low (b2), mid (b1) and high (b0) bit-rate variants, respectively. Additionally, we look further at the average bandwidth field of each entry in the server logs. We label the access as to a low, mid or high bit-rate variant based on the experienced bandwidth while delivering the object.

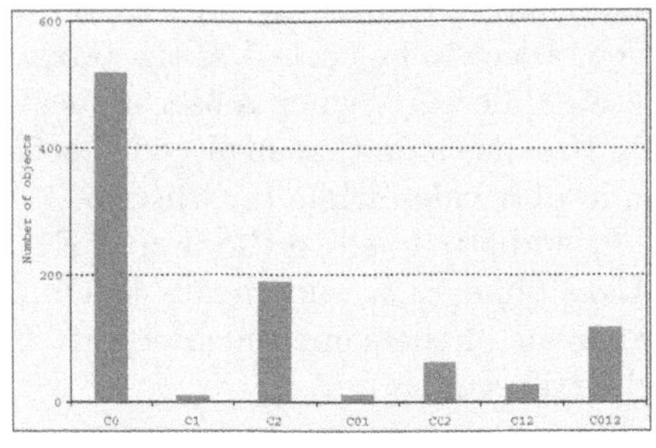

(a) Distribution of variant objects

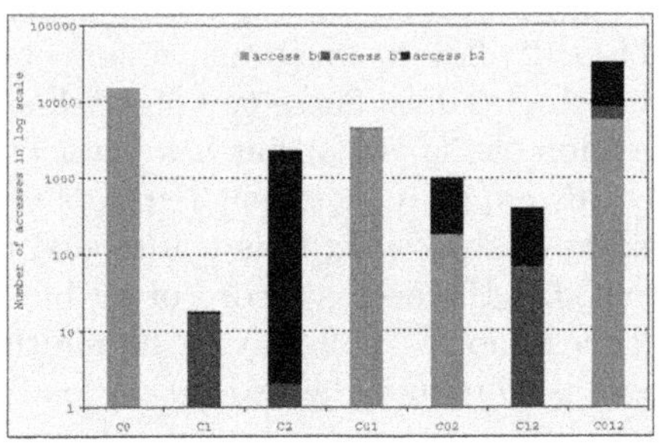

(b) Distribution of accesses

Fig. 2. Workload variance characteristics

To analyze the obtained traces, we define seven variant categories. Categories C0, C1 and C2 contain objects with one version. Categories C01, C02 and C12 contain objects with two variants. Category C012 contains objects with three variants. Figure 2 (a) shows the distribution of the number of objects in each variant category. Since the media server is targeting at corporate intranet users, the majority of the objects are coded in one variant. There are also 115 objects in category C012, and nearly 100 objects coded in various combinations of two variants (in categories C01, C02, and C12). The TeC system is most useful when there are accesses to these objects. Figure 2 (b) shows the number of accesses for each variant categories. Within each category, the number of accesses to different variants is shown. Since the TeC system improves caching performance when there are accesses to bit-rate variants, we focus on C01, C02, C12 and C012 categories. Note that the access to one variant dominates in these categories. In most cases, the highest-bandwidth variant is accessed most often. Note also that the accesses to variants are mainly to objects in C012. The access to variant b0 dominates in C012. The access to other variants is nearly 25% of the total access to objects in the category, and less than 10% of overall total access.

During the simulation, 20 GB of content resides on the origin server and there are total of 27,926 accesses. The accesses are highly temporally located. For contents generated each day, nearly 6 GB of content is accessed at least once more before the end of the measured period. Figure 3 (a) shows the byte hit ratio. The results illustrate the performance improvements gained by the TeC system. TEC_11 provides 4~9% better byte hit ratio over the reference model. Considering that the access to variants is only 10% of the total access, the improvement indicates that 50~90% of the variant-based access can be served from TeC directly. Since the dominant version in enterprise environment is most likely the highest bit-rate version, algorithm TEC_12 that caches lower bit-rate variants produces worse result than TEC_11. TEC_2 achieves less improvement compared with TEC_11. Nevertheless, the transcoding load is significantly reduced

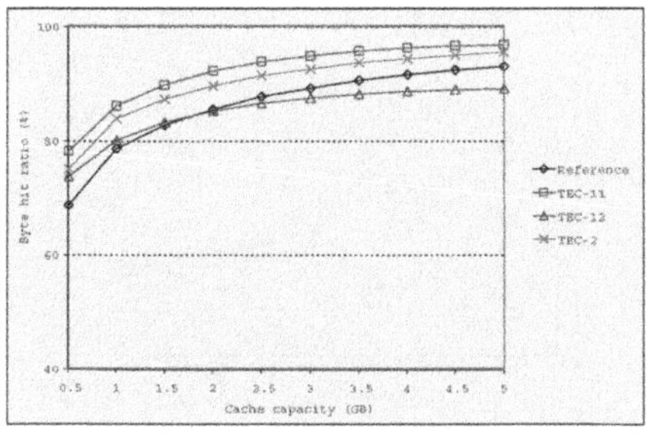

(a) Byte hit ratio

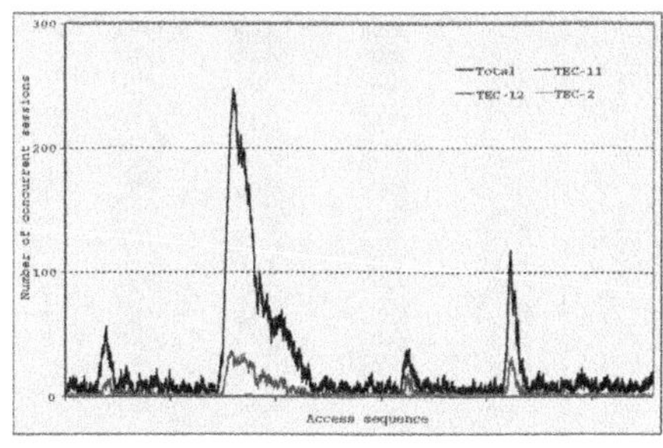

(b) Transcoding load

Fig. 3. Performance Results

in TEC_2 since it caches multiple variants of the same object. This is evident from Figure 3 (b) that shows the number of conconcurrent sessions when each request arrives. For the cache size at 10% of total object size, the average concurrent transcoding session required for TEC_2 is orders of magnitude less than that for TEC_11. The peak transcoding load for TEC_11, TEC_12, and TEC_2 is 36, 36, and 3 concurrent sessions respectively, and on average 3.35, 3.25, 0.13 concurrent sessions respectively.

Bit-rate reduction transcoding often introduces quality degradation less than 1dB. If resolution reduction is considered, the degradation is negligible. Successive transcodings introduce further degradation and this is characterized as a generation loss. Figure 4 (a) shows the distribution of generation losses for the object served from TeC when the cache capacity is 1 GB. For all the TEC algorithms, only 10% of the requests are served with generation losses. There are at most one generation loss for all the requests if TEC_11 is used since it caches variant at higher bit-rate. For TEC_2, only 2% of the requests are served with two generation losses.

We now investigate the startup latency. We define startup delay as the time gap between the instance when the proxy receives a request and the instance when the first packet is served. Assuming a conservative sustained disk I/O bandwidth of 37 MB/s, the startup delay of an exact hit is 37 μsec, which is spent in retrieving of a maximum RTP payload of 1400 bytes. We measured and use the startup delay of 1500 μsec for a transcode hit. A transcode hit is where a cache has a transcodable version of the requested version. Finally, if a miss occurs, the proxy establishes a RTSP/RTP session with the origin server. Assuming an intranet backbone connection (10 Mb/s) between the proxy and the origin, we set the session setup time as 8000 μsec. Using the above values, the simulator calculates the startup delay for each request. Figure 4 (b) shows the results. The startup delay decreases when cache capacity increases since more requests are served directly from the proxy. Since TEC_11 always caches

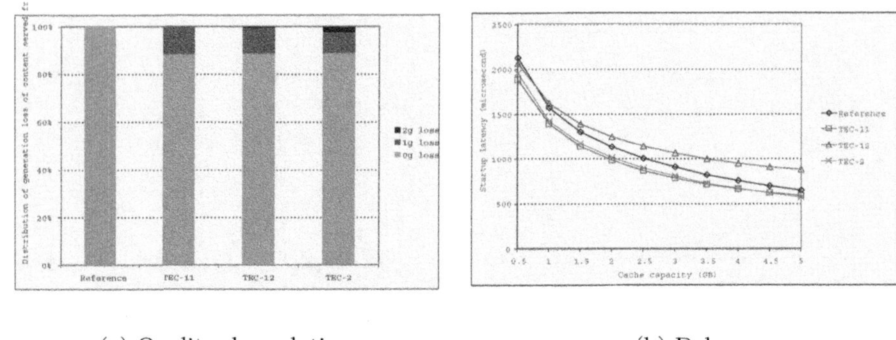

(a) Quality degradation (b) Delay

Fig. 4. Performance Results

the higher bit-rate variant, requests to the lower bit-rate variant of the same object are served with smaller delay as no server access is needed. However, higher bit-rate variant occupies larger space, which reduces the number of exact hits that require the smallest delay. On the other hand, as the startup delay resulted from a transcode hit is much smaller than the delay from a miss, the transcoding-enabled schemes are still beneficial in terms of latency . TEC_11 and TEC_2 generally produces 200 μsec less in startup delay than the reference model. TEC_12 shows the worst delay as most of the user requests is to higher bit-rate variants.

4 Conclusion

We evaluated the TeC performance using the trace derived from an enterprise network. Simulation results showed that TeC system provides better byte hit ratio and less startup latency than traditional caching systems. We also showed that transcoding introduces negligible quality degradation, and requires only small additional processing load on the proxies.

References

1. B. Shen and S.-J. Lee, "Transcoding-Enabled Caching Proxy System for Video Delivery in Heterogeneous Network Environments," Proceedings of IASTED IMSA 2002, Kauai, HI, August 2002.

Adaptive File Cache Management for Mobile Computing

Jiangmei Mei and Rick Bunt

Department of Computer Science
University of Saskatchewan
Saskatoon, SK, Canada

Abstract. By performing file system operations on cached copies of files mobile users reduce their need to access files across the network. Effective cache management strategies will limit use of the network to a small number of read misses and periodic reintegration of changes (updates), and thus can shield the mobile user from changes in bandwidth that will have a significant impact on performance. This paper addresses adaptive approaches to file cache management in mobile environments. The results of a trace-driven simulation study demonstrate that adaptive adjustment of the cache block size can improve performance for mobile users at acceptable cost.

1 Introduction

File systems that support mobile users, such as Coda [3][5] and AFS [2], rely on optimistic file caching to alleviate the impact of mobility. Cache management in such a file system addresses issues such as hoarding files to prepare for mobility, servicing file system requests in cache where possible when mobile, retrieving requested files from the home file server where necessary, and propagating updates to the home file server for timely reintegration. In a mobile environment the bandwidth of the connection to the home file server can change suddenly and dramatically. Since bandwidth has a large impact on the time to complete these file operations, this can have a significant impact on performance. A cache management strategy that assumes that the same bandwidth is available throughout the connection may not perform well in a dynamically changing mobile environment. Cache management in a mobile environment must react dynamically to changes in resource availability – to make effective use of increased resource, and to keep the impact of decreases to a minimum.

The long-term goal of our research is to examine the extent to which cache management in a mobile environment can be designed to adapt to changes in resource availability. This paper extends some earlier work on file cache management for mobile computing [1]. We used the same system model, performance metrics, file system traces and simulator to study the performance benefits of adaptive file cache management in a dynamically changing mobile environment.

M.-S. Chen et al. (Eds.): MDM 2003, LNCS 2574, pp. 369–373, 2003.

2 File Caching for Mobile Clients

Optimistic caching of file replicas at clients is a technique commonly used to improve performance in distributed file systems. This not only reduces latency in servicing file requests, but also avoids traffic on the network.

In a distributed file system that contains mobile computers, caching plays an even more important role in providing mobile users with data availability and performance. When mobile, the user may connect to the home file server through different communication methods at different places (at different times), and the connections may have quite different characteristics. Much of the time, there may be only a "weak" connection to the home file server over a slow and perhaps error-prone link, and sometimes the client may be totally disconnected. Maintaining copies of actively used files in local cache allows the mobile user to continue to work even while the client is disconnected from the file server, and improves the performance when weakly connected since retrievals over the network are less frequent.

3 The Need for Adaptation

Mobility and portability pose significant restrictions on resources such as memory and disk space and it is necessary to cope with these limitations in the design of support software [6]. Mobile clients may also experience considerable variations in connectivity, such as changes in bandwidth and error rate. The quality of service may vary dramatically and unpredictably because of communication noise, congestion, or environmental interference, and disconnections may be frequent. Mobile systems and applications should react to changes dynamically – to take advantage of increased resource availability when that happens, and to avoid the negative impact of decreases when they happen. Being able to adapt successfully to changes in resource availability is widely recognized as a central support issue for mobile computing [6].

Our focus in this short paper is on the amount of data brought into the cache in a block caching implementation – in other words, the cache block size. When bandwidth is limited, fetching smaller blocks conserves bandwidth and reduces access time. When bandwidth is plentiful, it is possible to fetch a large block of data so that the likelihood of satisfying future references in cache is increased.

4 The Experiments

4.1 The Simulated Environment

For this study the mobile environment is defined as follows. After a period of time for hoarding, a mobile user attempts to work at several locations (at different times). While mobile, he/she connects to the home file server over different connection bandwidths and performs work that involves file system activities. The mobile client contacts a single logical home file server for file system requests.

While the file server is always connected to the internet through a fast and reliable link, the mobile client is either weakly connected to the internet when mobile or disconnected.

In the mobile client, the file cache contains copies of file blocks that the user either hoarded prior to going mobile or referenced recently. When notified of any change in the effective bandwidth of the connection, the adaptive cache manager adapts to the changes by modifying its behaviour. For this study, the block size for different bandwidths was predetermined: 16 kB, 8 kB, 4 kB, 2 kB, and 1 kB for bandwidths of 2 Mbps, 512 kbps, 64 kbps, 9.6 kbps, 2 kbps, respectively.

Detailed traces of real user file system activity were collected in the Distributed Systems Performance Laboratory in the Department of Computer Science at the University of Saskatchewan. Each trace is a complete record of an individual user's real day-to-day file system activities in a typical academic research environment over a seven-day period.

The characteristics of the 3 traces we selected for this study are given in Table 1. The number of *Active Hours* in each trace is computed by subtracting the amount of "idle" time in the trace from the time between the first and last references in the trace. *Total Requests* is the number of file system events in the trace, *Unique Files* is the number of different files referenced, and *Unique Bytes* is the total size of those files. *Read Transfers* and *Write Transfers* are the sum of the sizes of all read and write requests in the trace, respectively, and *Write Percentage* is the percentage of all read and write requests that are writes. Finally, *Intervals* is the number of inter-event times of the length of time indicated. Intervals affect assumptions relating to disconnections. If an interval is longer than 15 minutes and shorter than 1 hour, it is assumed that the client is experiencing an unexpected disconnection. If the interval is more than one hour, it is considered to be an expected disconnection. More detail on the trace collection, the post-processing that was done, and the meaning of individual fields is given in [1].

Table 1. Trace Characteristics

Trace		1	2	3
Active Hours		12.9	7.0	8.5
Total Requests		21013	53860	17690
Unique Files		383	1000	443
Unique Bytes (MB)		17	18	22
Read Transfers (MB)		4596	4718	3723
Write Transfers (MB)		4	35	2
Write Requests (%)		38.7	67.9	25.2
Intervals	15-60 (min)	10	6	9
	$>$ 60 (min)	11	5	16

4.2 Results

For this brief presentation performance is assessed primarily through *time expansion* – the ratio of the total time required to process the entire trace in the simulated mobile environment to the time required when strongly connected. It reflects the additional cost to the user of mobile operation and is affected primarily by the additional time required to service read misses over the network and by any interference induced by updates.

The impact of adapting block size is shown in Figure 1. For each trace, adaptation results in reduced time expansion over non-adaptive block caching. In fact, the positive impact on the mobile user is actually even better than what Figure 1 suggests. When only the active hours are considered (i.e. idle time is removed), the time to service actual requests is reduced by 12.5%, 4%, and 5.5% for Trace 1, Trace 2, Trace 3, respectively, at a cache size of 1 MB. With less cache at the client, the reduction is even larger. Full results are available in [4].

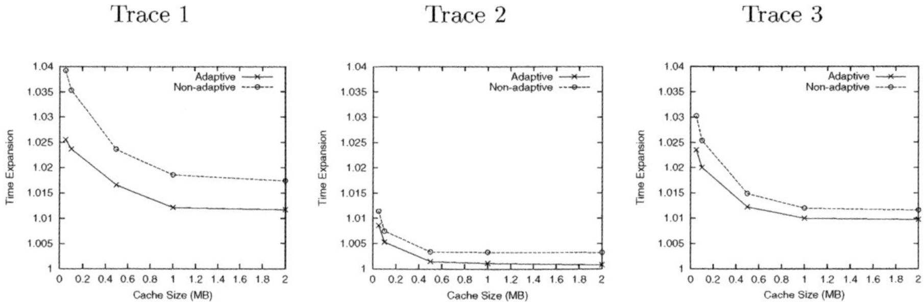

Fig. 1. Time Expansion

Figure 2 shows that adaptive block caching adds only a small amount of extra traffic to the network. This suggests that the benefits of adaptive block caching can be provided at little or no extra cost.

5 Conclusions

The ability to adapt dynamically to environmental changes has been identified as an important success factor for mobile computing. We are experimenting with adaptive approaches to file cache management, seeking to understand the extent to which adaptation can provide improved performance for mobile users.

Adaptation can be beneficial when appropriate adaptation strategies are applied in the constantly changing mobile environment. Adaptive block caching can improve performance over the non-adaptive approaches by dynamically adjusting the amount of data to be fetched as available bandwidth changes. The reduction in time expansion, particularly at small cache sizes, is important when

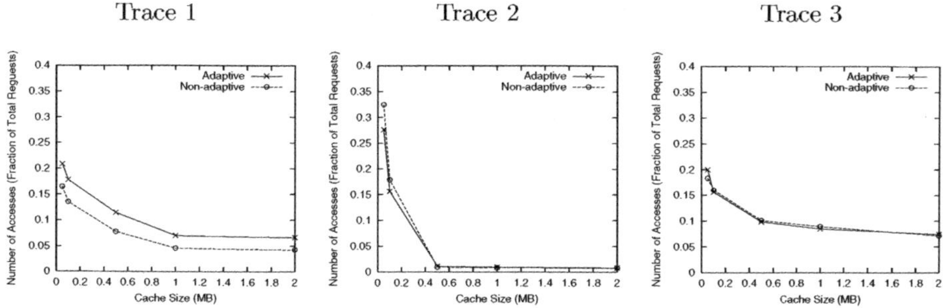

Fig. 2. Number of Network Accesses

the resources at the mobile client are very scarce – as they are for palmtops or PDAs. There may, of course, be negative aspects to adaptation, but our results suggest that these negative impacts remain at acceptable levels.

On balance our adaptive approaches offer performance benefits with very little extra cost. Although this particular study is limited in its scope it adds to the growing list of successful adaptive approaches and provides support for further work in this important area.

6 Acknowledgements

This research was supported by the Natural Sciences and Engineering Research Council of Canada (NSERC) and by TRLabs in Saskatoon.

References

1. K. Froese and R. Bunt. Cache management for mobile file service. *The Computer Journal*, 42(6):442-454, 1999.
2. L. Huston and P. Honeyman. Partially connected operation. In *Proceedings of the Second USENIX Symposium on Mobile and Location-Independent*, pages 91-97, Ann Arbor, MI, April 1995.
3. J. Kistler and M. Satyanarayanan. Disconnected operation in the Coda file system. *ACM Transactions on Computer Systems*, 10(1):3-23, January 1992.
4. J. Mei. Adaptive file cache management for mobile computing. M.Sc. thesis, Department of Computer Science, University of Saskatchewan, Saskatoon, SK, 2002.
5. L. Mummert, M. Ebling, and M. Satyanarayanan. Exploiting weak connectivity for mobile file access. In *Proceedings of the Fifteenth ACM Symposium on Operating Systems Principles*, pages 143-155, Copper Mountain Resort, CO, December 1995.
6. M. Satyanarayanan. Fundamental challenges in mobile computing. In *Proceedings of the Fifteenth ACM Symposium on Principles of Distributed Computing*, pages 1-7, Philadelphia, PA, May 1996.

Adaptive Power-Aware Prefetching Schemes for Mobile Broadcast Environments*

Haibo Hu, Jianliang Xu, and Dik Lun Lee

Department of Computer Science
Hong Kong University of Science and Technology
Clear Water Bay, Hong Kong
{haibo,xujl,dlee}@cs.ust.hk

Abstract. Caching and prefetching are common techniques in mobile broadcast environments. Several prefetching policies have been proposed in the literature to either achieve the minimum access latency or save power as much as possible. However, little work has been carried out to adaptively balance these two metrics in a dynamic environment where the user request arrival patterns change over time. This paper investigates how both *low latency* and *low power consumption* can be achieved for dynamic request patterns. Two adaptive power-aware schemes, $\mathcal{PT}$ with secure region (*PTSR*) and adaptive prefetching with sliding caches (*APSC*), are proposed. Experimental results show the proposed policies, in particular *APSC*, achieve a much lower power consumption while maintaining a similar access latency as the existing policies.

1 Introduction

It has been widely accepted that data broadcasting is an efficient data dissemination technique for the asymmetric and limited-bandwidth wireless links. Although broadcasting has the merit of scalability and low bandwidth consumption, problem is that it may sacrifice a particular user's need to tailor for the majority. Caching data at client-side was proposed to cope with this problem [1]. To fully exploit the cache, prefetching was explored in mobile broadcast environments [2, 5, 3]. In broadcasting, no additional bandwidth is consumed by prefetching, which is an intrinsic merit of it. $\mathcal{PT}$ was the first prefetching scheme for broadcast [2]. Tassiulas and Su improved this idea by looking ahead t time slots so that an optimal prefetching decision is guaranteed for the next t time slots, leading to the *WSLA* policy [5]. While they ignored prefetching cost, i.e., power consumption for downloading data from broadcast, Grassi [3] derived an analytical model for various prefetching policies, from which *PP1* and *PP2* were proposed for minimizing power consumption and latency, respectively.

This paper considers the scenario in which user request patterns, in particular request arrival rates, change over time, which is often observed in reality. For

* Supported by Research Grants Council, Hong Kong SAR under grant HKUST6225/02E .

M.-S. Chen et al. (Eds.): MDM 2003, LNCS 2574, pp. 374–380, 2003.

example, people have a high access rate over stock quotes during trading hours and probably a low rate during after-trading hours. As prefetching data from broadcast comes at the expense of power, a good scheme should emphasize the *effectiveness* of prefetching. Ideally, a prefetching policy should have almost the same access latency as when prefetching is fully utilized and consume power adaptively according to request rates. To this end, we propose two adaptive power-aware prefetching scheme, $\mathcal{PT}$ with secure region (*PTSR*) and adaptive prefetching with sliding caches (*APSC*). Preliminary experimental results show that the proposed policies, in particular *APSC*, achieve almost the same access latency as *WSLA* and $\mathcal{PT}$ with much less power consumption.

The rest of the paper is organized as follows. Section 2 reviews existing prefetching policies. Section 3 identifies problems from them and describes our schemes with performance evaluation. Finally we draw conclusions and discuss possible future work.

2 Prefetching Policies Review

2.1 $\mathcal{PT}$ and *WSLA*

In [2], Acharya et al. proposed a simple yet effective policy for making prefetching decision — $\mathcal{PT}$. It computes the value of a page by taking the product of the access probability (p) of the page and the time (t) that will elapse before that page appears on the broadcast again. This value is called the page's *pt* value. During each broadcasting tick, $\mathcal{PT}$ finds the page in the cache with the lowest *pt* value and replaces it with the currently broadcast page if the latter has a higher *pt* value.

In [5], Tassiulas and Su modeled the data broadcast system in such a way that the aggregate latency reduction (or "prefetching reward" as they called it) from time slot n to $n+t$ was formulated as:$\sum_{n=0}^{t-1}\sum_{i\in C(n)}(\lambda_i\tau_i^f(n)+\frac{\lambda_i}{2})$, They thus came up with an optimal prefetching policy to maximize the above formula, with t as a given parameter. To make the computation feasible for CPU-constrained mobile clients, they yet proposed an approximation substituent, the W-step look-ahead (*WSLA*) policy which instead maximizes the cumulative average reward from slot n up to slot $n+W$. The $W=1$ version of *WSLA* (called *OSLA*) is equivalent to the $\mathcal{PT}$ policy.

2.2 *PP1 and PP2*

The $\mathcal{PT}$ and *WSLA* policies ignored the power consumption issue. In [3], Grassi proposed a general analytical model incorporating energy saving as well as latency reduction. Based on the analytical model, he devised the *PP1* policy to achieve the minimum power consumption. It uses the following as the metric for prefetching decision: $\frac{1-e^{-\lambda u_i d_i}}{d_i}$, where u_i is the request probability for item i, λ is the arrival rate, d_i is the time distance between two consecutive occurrences of item i in the schedule. Similarly, he devised the *PP2* policy to achieve the minimum access latency at an upper bound. The only change from *PP1* to *PP2* is the metric, which is now $u_i d_i$. *PP1* and *PP2* are obviously energy-efficient, for they are "static" policies.

3 Adaptive Power-Aware Prefetching Schemes

Our goal in this paper is to devise prefetching schemes that can achieve good access latency while greatly improve power consumption under dynamic request rates. In the following, we will present insightful observations on the weaknesses of $\mathcal{PT}$ and *PP2*, and propose two remedial schemes, *PTSR* and *ASPC*, respectively.

3.1 Problems with $\mathcal{PT}$ and *WSLA*: $\mathcal{PT}$ with Secure Region

$\mathcal{PT}$ is actually a greedy algorithm that aims at minimizing the average access latency in the next time slot. However, it doesn't care whether the prefetched object will be accessed by the client or not. It is therefore an "aggressive" algorithm which is battery-harmful since each prefetching costs battery power of the mobile device. *WSLA* did not address this problem either.

In order to make $\mathcal{PT}$ adaptive to client requests and power-efficient, we extend the $\mathcal{PT}$ algorithm and devise the *PTSR* scheme, in which the cache space has a special (logical) area called *secure region* (*SR*) where no objects will be victims of cache/prefetch replacement. All the newly prefetched objects are inserted into the *SR*. The *lifetime* of an object residing in this area is a tuning parameter of this algorithm. It is measured by the number of the user's total requests. Suppose the lifetime threshold is set to 5, if after five requests from the user, an object is still not requested, it is moved from the secure region to the *normal region* of the cache where the $\mathcal{PT}$ policy is applied. However, if during this period, the object is requested by the user (prefetching hit), the lifetime of this object is reset to 5. Similarly, if an object residing in the normal region is requested by the user (cache hit), it is moved to the *SR* and its lifetime is set to the threshold.

The scheme increases the utilization ratio of prefetching ($\frac{\#\ prefetch\ hits}{\#\ prefetches}$). When a client is idle, newly prefetched objects are inserted into *SR* and never move out, *SR* expands until it occupies the entire cache and prefetching stops. Thus it is adaptive to client quest patterns. We have a tradeoff between latency and battery consumption: the larger the threshold, the more power saved, but the worse access latency.

3.2 Problems with *PP2*: Adaptive Prefetching with Sliding Caches

PP2 claimed to balance the average access latency and power consumption according to [3]. It is a static policy, so power consumption is extremely low. However, from the result of [3], *OSLA* ($\mathcal{PT}$) always outperforms *PP2* in terms of access latency, for *PP2* is actually a *stateless* policy which ignores the current

Algorithm 1 Adaptive Prefetching with Sliding Caches

Input: t_i: $\mathcal{T}$ value for object i; $\bar{t}_i$: average broadcast interval for object i; p_i: access probability for object i; obj: the object currently being broadcast.

Procedure:

find the minimal value of $p_i t_i$ in DC, denote the object as min;
if obj is currently requested by the client **then**
    replace min with obj;
else
    if $p_{obj} t_{obj} > p_{min} t_{min}$ **then**
        replace min with obj;
    else
        return;
    end if
end if
find the minimal value of $p_i \bar{t}_i$ in SC, denote the object as $min2$;
if $p_{min} \bar{t}_{min} > p_{min2} \bar{t}_{min2}$ **then**
    replace $min2$ with min;
end if

state in the broadcasting, while $\mathcal{PT}$ ($OSLA$) is *stateful*, where the current time slot is essential to make prefetching decisions. Since a stateful policy tries to maximize the benefit of the prefetching for the near future, it is expected to provide a better latency for intensive client accesses.

From latency point of view, when the client's requests are scarce, a stateless algorithm such as *PP2* is efficient enough: since the next request probably won't come after many time slots, it is useless to maximize the prefetching benefit for the recent slots. On the other hand, high request arrival rates favor a stateful algorithm such as $\mathcal{PT}$ and *WSLA*. To save power, we should utilize a stateless policy whenever it performs well enough, and turn to a stateful policy only when it well outperforms stateless counterpart. This motivates our next proposed scheme, adaptive prefetching with sliding caches ($APSC$), which adaptively tunes between the two types of policies according to the client request arrival rate. To be adaptive to various request arrival rates, we split the cache into two caches: *dynamic cache* (DC) and *static cache* (SC).

The new prefetching scheme works in two stages as illustrated in Fig 1 and Algorithm 1: first, the $p_i t_i$ value is the metric for prefetching the current broadcast object and replacing with the victim in the DC; second, if prefetching and replacement actually occurs in stage 1, the victim is not discarded immediately, instead it has the second chance of replacing a victim in the SC in terms of $p_i \bar{t}_i$, where $\bar{t}_i$ is the average broadcast interval of item i. In this manner, we integrate the *static* and *dynamic* prefetching policies. An object with a high value either on pt or $p\bar{t}$ stays in the cache.

In order to make the SC dominate the entire cache when the request arrival rate is low and the DC dominate otherwise, we need to dynamically "slide" the boundary of these two caches (this is where "sliding cache" comes from).

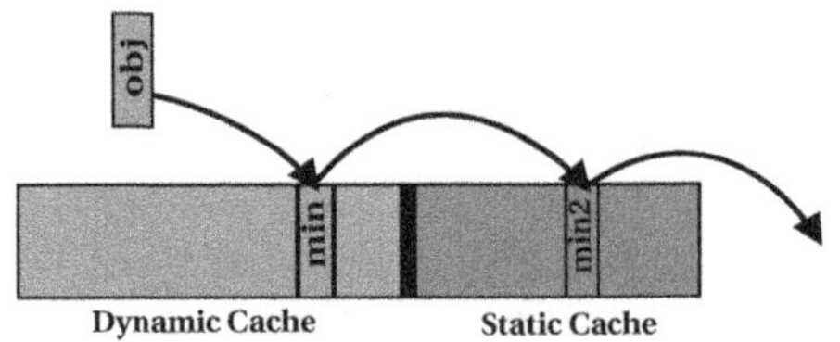

Fig. 1. Prefetch and Replacement of *APSC*

Essentially, we need to intelligently choose when and how much distance to "slide". The client request is a good sign to slide the boundary in favor of the *DC* since the more frequent the user requests, the more dynamic cache we should have in order to fulfill his/her near-future requests. On the other hand, when the client is idle or operates slowly, the boundary should slide toward the *DC* (in favor of *SC*) so that the static cache can dominate. We choose a prefetching operation by the client as a sign to slide the boundary in favor of the *SC*. Sliding algorithm of *APSC* is described in [4].

The next problem is how to define the *relative sliding frequency* of both directions. If we set the sliding frequency toward the *SC* to 1, i.e., for each user request, we will slide the boundary in favor of the *DC* for one cache unit (the size of an object, for simplicity), we then need to decide how often (for how many prefetching operations) the boundary should slide toward the *DC* for one cache unit. This is actually an important tuning parameter of this scheme. The larger the frequency value, the more difficult the boundary sliding toward the *DC*, hence the more the scheme in favor of the *DC* and the more dynamic this scheme is.

3.3 Simulation Results

The simulator of a broadcasting system was implemented using C++ and CSIM. The client has a cache, into which all client requested objects must be fetched before they can be displayed for the user. All prefetched objects are also stored in this cache. The user access arrival interval follows an negative exponential distribution with a mean of $\frac{1}{\lambda}$. Table 1 lists all the parameters for our experiments.

Parameter	Setting	Parameter	Setting
# objects	200	θ	0.5
Total requests	50000	Warmup requests	5000
# disks	5	Δ	5
Cache size	5,10,20,40,50	$\frac{1}{\lambda}$	$25, 50, 100, 200, 500$

Table 1. Parameter Settings

Performance on Various Access Arrival Intervals We evaluate experimental results [1] for the five algorithms as a function of the mean access arrival interval. We can see that while *PTSR* seems a bit unstable, *APSC* has promising performance: a competitively small access latency (better than all the rest four) and its battery consumption outperforms others when the arrival interval is less than 200.

Impact of Dynamic Request Intervals We simulated a dynamic workload in which a client is very active in a certain time period and almost inactive in the others [2]. The experimental results are shown in Fig. 2, the access latency of *APSC* is pretty close to that of $\mathcal{PT}$ and the battery consumption of *APSC* is almost half of the other dynamic prefetching policies. The superiority of *APSC* is hereby fully justified: it can keep a nice balance of access latency and power consumption in a dynamic environment where the user request pattern changes all the time.

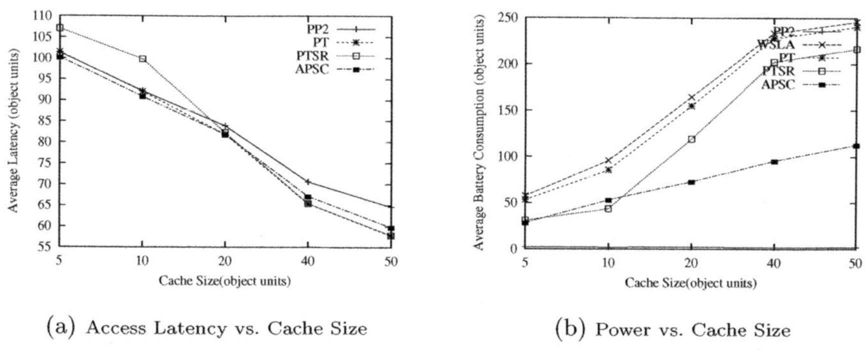

(a) Access Latency vs. Cache Size

(b) Power vs. Cache Size

Fig. 2. Performance under Dynamic User Access Pattern

Tuning the Relative Sliding Frequency of *APSC* The tunable *relative sliding frequency* is essentially the extent to which *APSC* behaves like a dynamic policy. The experimental results [3] shows the *APSC* latency curves as " first go downwards and reach minimum, after that, the latency fluctuates arbitrarily between minimum and a certain value" and "the power goes steadily up until finally it saturates" (reaches almost the same energy consumption as $\mathcal{PT}$). It is

[1] Due to space limitation, please refer to [4] for this experiment settings and statistics.

[2] The mean request arrival interval changes every 500 requests. The request interval ranges from 1 to 500, with a probability of 0.4 being in the range [1, 50], 0.4 in the range [450, 500], and 0.2 to in the rest of the entire range.

[3] Please refer to [4] for detailed statistics.

interesting to see that the two turning (minimal value) points respectively for latency and power are not equal. Therefore, we can pick up a sliding frequency f that has a low access latency as well as a relatively low power consumption.

4 Conclusion and Future Work

Prefetching from broadcast is an important research topic in wireless data dissemination. This paper presented two adaptive schemes to cope with a dynamic user request workload: *PTSR* is an extension of $\mathcal{PT}$, where a secure region idea is appended; *APSC* incorporates the static and dynamic policies in a sliding fashion so that it can still maintain a low access latency while considerably reduce the power consumption compared with the other dynamic policies. Simulation results show that while both of the proposed policies can achieve certain balance between access latency and power consumption, *APSC* is more promising in handling various user access patterns.

As part of our ongoing work, we plan to derive analytical models and investigate optimal settings of the tuning parameters for both schemes (*lifetime* in *PTSR* and *relative sliding frequency* in *APSC*). For future work, we will incorporate data updates.

References

[1] S. Acharya, R. Alonso, M. Franklin, and S. Zdonik. Broadcast disks: Data management for asymmetric communications environments. In *Proceedings of ACM SIGMOD Conference on Management of Data*, pages 199–210, San Jose, CA, USA, May 1995.

[2] S. Acharya, M. Franklin, and S. Zdonik. Prefetching from a broadcast disk. In *Proceedings of the 12th International Conference on Data Engineering (ICDE'96)*, pages 276–285, New Orleans, LA, USA, February 1996.

[3] V. Grassi. Prefetching policies for engery saving and latency reduction in a wireless broadcast data delivery system. In *Proceedings of the 3rd ACM International Work on Modeling, Analysis and Simulation of Wireless and Mobile Systems (MSWiM'00)*, Boston, MA, USA, August 2000.

[4] H.B. Hu, J.L. Xu, and D.L. Lee. Adaptive power-aware prefetching schemes for mobile broadcast environments. Technical report, Department of Computer Science, Hong Kong University of Science and Technology, http://www.cs.ust.hk/haibo/pub/prefetching.pdf, June 2002.

[5] L. Tassiulas and C. J. Su. Optimal memory management strategies for a mobile user in a broadcast data delivery system. *IEEE Journal on Selected Areas in Communications (JSAC)*, 15(7):1226–1238, September 1997.

madrias@umr.edu, yongjian@umr.edu

assourav@ntu.edu.sg

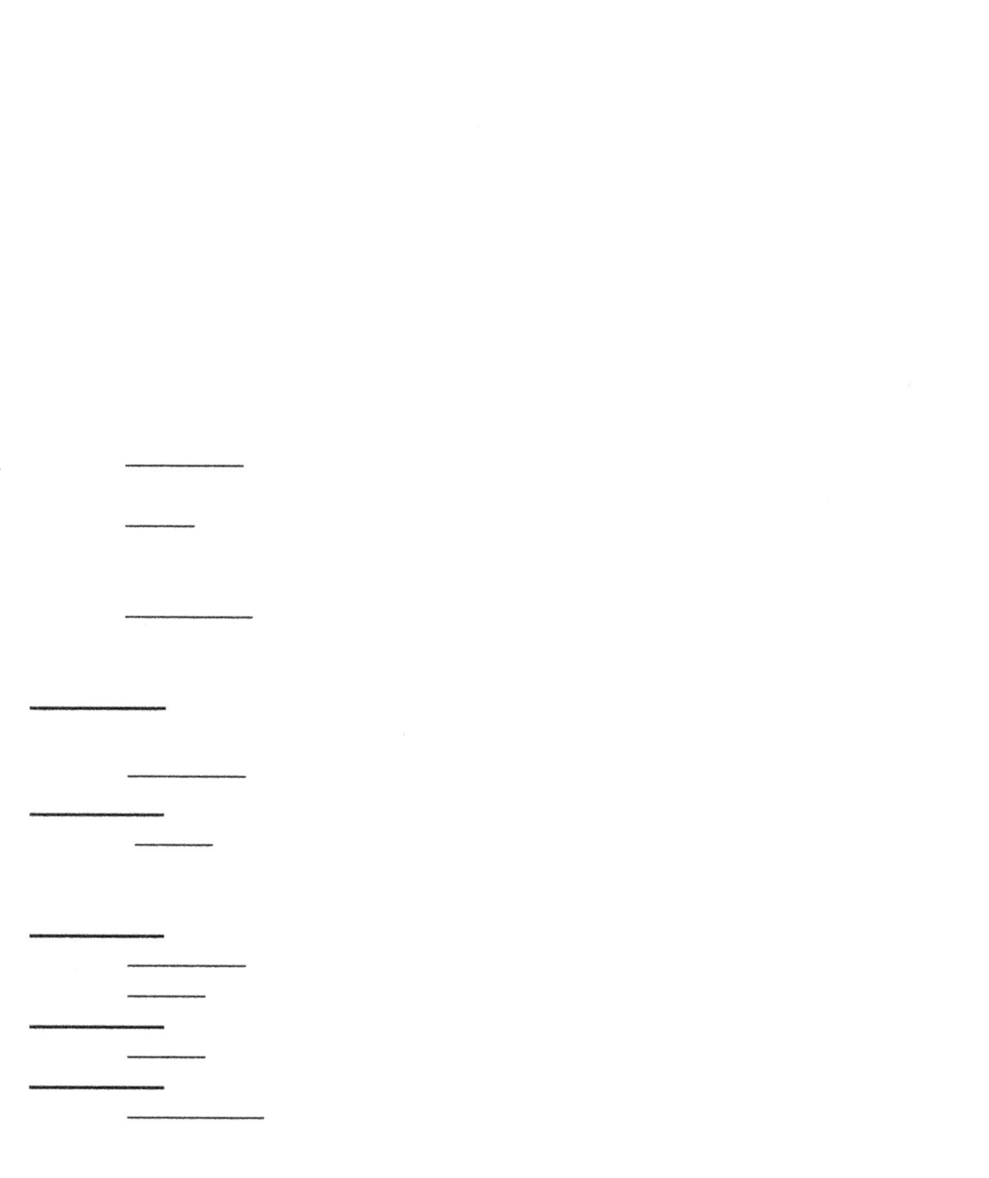

{Francois.Paradis, Francis.Crimmins, Nadine.Ozkan}@csiro.au

```
<task name="archive search">
  <title>Search for an article in the news archives</title>
  <context><device refid="pc"/></context>
  <action><search><service name="P@NOPTIC"/>
     <collection>newsarchive</collection>
     <output>Rank, Title, Summary</output>
</search></action></task>
```

Rank	Document	Document Date
1.	The Age: High density residential plan rejected	*15 March 2001*
2.	The Age: Hulls' illness delays rights bill	*20 March 2001*
3.	The Age: She's our 19th commissioner and she's the first	*17 March 2001*

Rank	Document
1.	Bracks angry at reform agenda block ... Premier Steve **Bracks** yesterday accused the opposition of crippling the government's reform agenda by blocking the upper house. By ADRIAN ROLLINS Tuesday 20 Marc ... 2001 Premier Steve **Bracks** yesterday accused the oppo the government's reform agenda by blocking three key ... McPhee, to advise on reform of the Legislative Council. Th commission will report to Mr **Bracks** with its restructuring plans. ...

olli.sotamaa@uta.fi
http://www.uta.fi/hyper/

veikko.ikonen@vtt.fi
http://www.vtt.fi/tte/research/tte5/tte51/

{takako, katooka, izw}@src.ricoh.co.jp

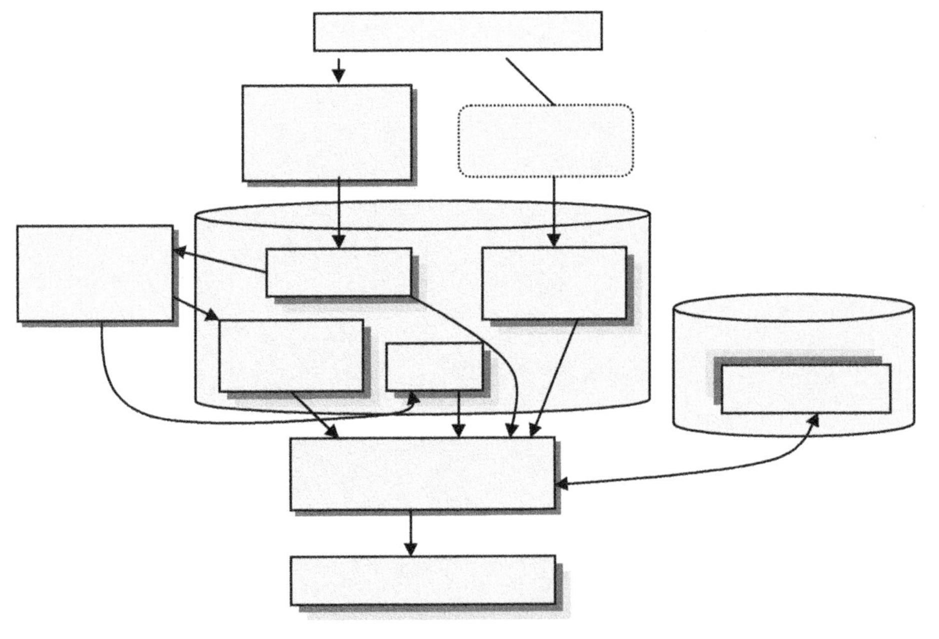

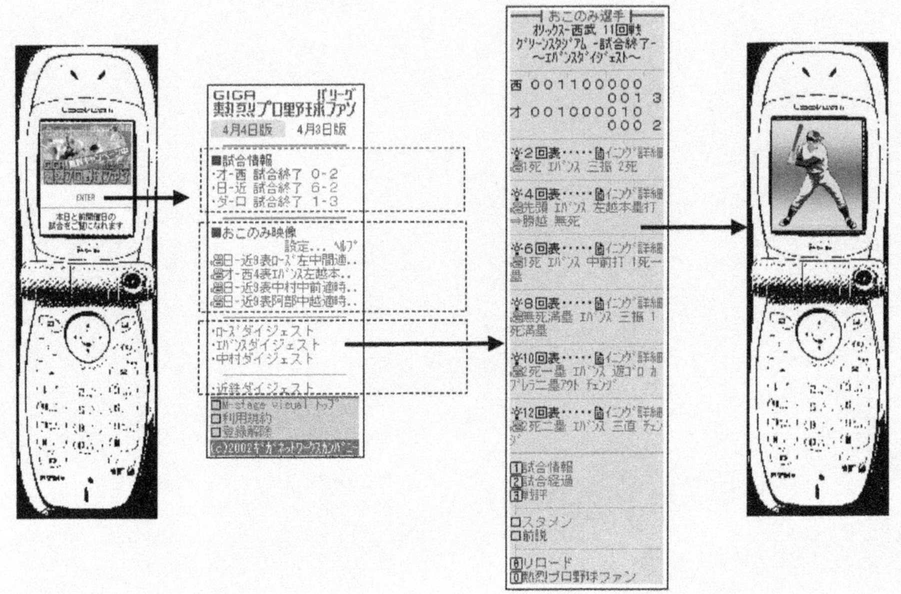
ENTER
本日と前開催日の
試合をご覧になれます
GIGA パリーグ
熱烈プロ野球ファン
4月4日版 4月3日版
■試合情報
・オ-西 試合終了 0-2
・日-近 試合終了 6-2
・ダ-ロ 試合終了 1-3
■おこのみ映像
設定... ヘルプ
日-近9表ローズ左中間適..
オ-西4表エバンス左越本..
日-近9表中村中前適時..
日-近9表阿部中越適時..
・ローズダイジェスト
・エバンスダイジェスト
・中村ダイジェスト
・近鉄ダイジェスト
M-stage visual トップ
利用規約
登録解除
(c)2002ギガネットワークスカンパニー
おこのみ選手
オリックス-西武 11回戦
グリーンスタジアム -試合終了-
～エバンスダイジェスト～
西 001100000
001 3
オ 001000010
000 2
2回表……イニング詳細
1死 エバンス 三振 2死
4回表……イニング詳細
先頭 エバンス 左越本塁打
⇒勝越 無死
6回表……イニング詳細
1死 エバンス 中前打 1死一塁
8回表……イニング詳細
無死満塁 エバンス 三振 1死満塁
10回表……イニング詳細
2死一塁 エバンス 遊ゴロ カブレラ二塁アウト チェンジ
12回表……イニング詳細
2死二塁 エバンス 三直 チェンジ
1試合情報
2試合経過
3戦評
スタメン
前説
8リロード
0熱烈プロ野球ファン

Handling Client Mobility and Intermittent Connectivity in Mobile Web Accesses*

Purushottam Kulkarni, Prashant Shenoy, and Krithi Ramamritham**

Department of Computer Science, University of Massachusetts, Amherst
{purukulk,shenoy,krithi}@cs.umass.edu

Abstract. Wireless devices are being increasingly used to access data on the web. Since intermittent connectivity and client mobility are inherent in such environments, in this paper, we examine the impact of these factors on coherent dissemination of dynamic web data to wireless devices. We introduce the notion of eventual–delta consistency, and in the context of push and pull–based dissemination propose (i) buffering techniques at the proxy to mask the effects of client disconnections and (ii) application–level handoff algorithms to handle client mobility. Our experimental evaluation demonstrates that push is better suited to handle client disconnections due to its lower message overhead and better fidelity, while pull is better suited to handle client mobility due to its lower handoff overheads.

1 Introduction

1.1 Motivation

Recent advances in computing and networking technologies have led to a proliferation of mobile devices with wireless networking capabilities. Like traditional wired hosts, wireless mobile devices are frequently used to access data on the web. These devices differ significantly from their wired counterparts with respect to their capabilities and characteristics—(i) mobile clients can move from one location to another, while the location is fixed for wired hosts and (ii) disconnections and reconnections are frequent in mobile wireless environments, while network failures are an exception in the wired domain. A concurrent trend is the increasing use of dynamic data in today's web. Unlike static web pages that change infrequently, dynamic web data is time–varying and changes frequently (once every few seconds or minutes). In this paper, we address the challenges arising from the intersection of these two trends, namely, the access of dynamic web data using networked mobile wireless devices.

1.2 Problem Formulation and Research Contributions

Consider a mobile client that accesses dynamic data items over a wireless network (see Figure 1(a)). Like in many wired environments, we assume that client web requests are

* This research was supported n part by NSF grants CCR-0098060, EIA-0080119 and CCR-0219520.

** Also with the Indian Institute of Technology Bombay, India.

M.-S. Chen et al. (Eds.): MDM 2003, LNCS 2574, pp. 401–407, 2003.

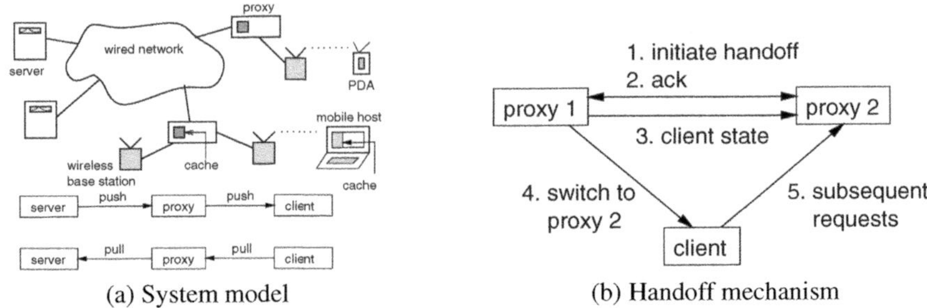

(a) System model (b) Handoff mechanism

Fig. 1. System model and mechanisms for intermittent connectivity and client mobility.

sent to a proxy. The proxy and the server are assumed to communicate over a wired network, while the proxy and the client communicate over a wireless network via a base station(access point); the latter network can be either be a wireless LAN such as 802.11b or a wireless WAN such as a GSM data network. Each base station in the wireless network is assumed to be associated with a proxy.

Intermittent connectivity and client mobility are common in such wireless environments; we discuss each issue in turn. We assume an environment where mobile clients can get disconnected at any time—disconnections occur primarily due to poor network coverage at a particular location or due to the mobile device powering down to save energy. Since the proxy and the server are unreachable during a disconnection, dynamic data items cached at the client can not be refreshed and may become stale. Consequently, to prevent a violation of coherency guarantees, client disconnections should be detected in a timely fashion so that preventive measures can be initiated. Observe that, since the proxy and the server are on a wired network, the proxy can continue to receive updates from the server during a disconnection (but can not propagate these updates to the client). Hence, data cached at a proxy continues to be coherent even during a disconnection—an observation that can be exploited for efficient resynchronization of the client cache upon reconnection. Thus, coherency maintenance in the presence of intermittent connectivity requires: (i) techniques for timely detection of client disconnections and reconnections, and (ii) techniques for efficient resynchronization of cache state at the client upon reconnection.

A second consideration in wireless environments is client mobility. Since a mobile client can move locations, the proxy may decide to hand over responsibility for servicing the client to another proxy in the client's vicinity. Such a *handoff* is desirable since a nearby proxy can reduce network overheads and provide better latency to client requests; the handoff involves a transfer of client–specific cache state from the initiator to the recipient proxy. The design of such application–level handoff mechanisms involves two research issues: (i) the proxy needs to decide *when* to initiate the handoff and (ii) *how* to perform a handoff, the steps involved in such a procedure. These design decisions have implications on the overheads imposed by handoffs and the latency seen by user requests. For instance, the handoff process should be seamless to end-clients—the client should not miss any updates to cached data due to the handoff and all temporal coherency guarantees should be preserved with minimal overheads.

Thus, in the context of intermittent connectivity and client mobility we make three contributions in this paper. First, we propose coherency semantics appropriate for dynamic data access in mobile environments. We then consider two canonical techniques for dynamic data dissemination—push and pull—and show how to adapt these techniques to (i) reduce the overheads of coherency maintenance in the presence of client mobility and (ii) provide temporal coherency guarantees even in the presence of intermittent connectivity.

2 Handling Intermittent Connectivity in Mobile Environments

2.1 Coherency Semantics

Due to the time–varying nature of dynamic data, cached versions of data items need to be *temporally coherent* with the data source. To ensure this property, we define a coherency semantics called *eventual–delta* consistency (Δ_e) for mobile environments. Δ_e semantics provide stricter guarantees when a mobile client is connected and weaken the guarantees upon a disconnection. Formally, Δ_e consistency is defined as follows

$$\Delta_e\text{consistency} \Rightarrow \begin{cases} |S_t - C_t| \leq \Delta & \text{if connected} \\ C_t \rightsquigarrow S_t & \text{if disconnected} \end{cases} \tag{1}$$

where C_t and S_t denote the state of a data item d at the client and the server, respectively, at time t, Δ denotes the desired coherency bound, and $C_t \rightsquigarrow S_t$ indicates that C_t will eventually catch up with S_t on reconnection.

Coherency mechanisms that implement Δ_e consistency can be *buffered* or *unbuffered*. In the unbuffered scenario, each new update from the server overwrites the previously cached version—the proxy cache only maintains the version corresponding to the most recently received update. In the buffered scenario, the proxy buffers recent updates from the server that are yet to be propagated to the client. Thus, updates not yet seen by the client can be buffered at the proxy and delivered next time the client refreshes its cache.

2.2 Maintaining Δ_e Consistency Using Push and Pull

There are two possible techniques to implement Δ_e consistency — *push* and *pull*. In the push approach, cache coherency is maintained by pushing updates to dynamic data items from the server to the proxy and from proxy to the clients. In the pull approach, the onus is on the proxy to poll the server for updates and on the client to poll the proxy to fetch the updates.

We assume that a client registers each data item of interest with the proxy and specifies a coherency tolerance of Δ for each item. In the push approach, the proxy in turn registers these data items with the corresponding server along with the corresponding coherency tolerances. The server then tracks changes to each such data item and pushes all updates that exceed the tolerance Δ to the proxy. The proxy in turn pushes these updates to the client. In the event the client is disconnected, the updates are buffered and propagated to the client upon reconnection. The proxy uses heartbeat messages and client requests as implicit heartbeats to detect client disconnections and reconnections [3]. In the pull approach, the proxy periodically polls the server to fetch updates. To meet client–specified coherency guarantees the proxy should poll the server at a minimum rate of Δ. The time between successive refreshes, *time-to-refresh(TTR)*, for each data item can be statically or dynamically determined. The TTR value has an impact on the fidelity of data at the proxy and the number of polls. The proxy buffers all the updates fetched from the server and on a client poll delivers those that are new since the previous poll. Observe that in the push approach, both the proxy and the server are *stateful*, whereas in the pull approach only the proxy is *stateful*.

3 Handling Client Mobility

As a mobile client moves from one location to another, it switches between base-stations to maintain connectivity. As explained earlier, each mobile client uses a proxy to request web content and it is possible for a mobile client to use the same proxy to service its requests even when it moves locations (since it can always communicate with this proxy so long it remains connected). However, other proxies may be more advantageously located with respect to the new client location. In such a scenario, having the mobile client use a nearby proxy to service its requests may result in improved response times and lower network overhead. To enable such a transition, the proxy needs to hand over responsibility for servicing client requests to the new proxy and also transfer client–specific state to the new proxy. The important steps of such a procedure are: (i) initiating a handoff, (ii) transferring client–state and (iii) committing the handoff (refer to Figure 1(b) and see [3] for details).

Several issues arise in the design of such application–level handoff mechanisms. First, the proxy needs to be aware of the client's physical location so that it can determine whether a handoff is necessary. A second issue is how to ensure that the handoff is seamless and transparent to the end–client. A third issue in the design of handoff algorithms is when and how client–specific state information is transferred from the initiator proxy to the recipient proxy. Depending on the exact mechanisms, following types of handoffs are possible: (i) *Optimistic versus pessimistic*: Objects are transferred lazily in pessimistic handoffs, while they are transferred in an eager fashion in optimistic handoffs and (ii) *Complete versus partial*: In complete handoffs, information about *all* objects accessed by the client is transferred, while in partial handoffs the recipient proxy transfers information only about a fraction of the objects. Figures 2(a) and 2(b) are handoff algorithms for push–based and pull–based mechanisms respectively, detailed explanations of the two algorithms can be found in [3].

4 Experimental Evaluation

We evaluate the efficacy of our approaches using simulations. We use a Fixed-path mobility model [3] to simulate client mobility and an ON–OFF process to simulate client disconnections and reconnections. The simulation environment consists of a number of cells and mobile clients that use the above processes to simulate mobility and intermittent connectivity during web accesses. The workload used for our experiments

1. $P_1 \rightarrow P_2$: initiating handoff for C
2. $P_2 \rightarrow P_1$: ACK
3. $P_1 \rightarrow P_2$: C needs $\{< d_1, \Delta_1, S(d_1) >, \ldots\}$
4. $P_2 \rightarrow P_1$: ACK
5. $P_1 \rightarrow S$: Send updates for object in C to P_2
6. $S \rightarrow P_1$: ACK
7. `if` C `is connected`
   i. $P_1 \rightarrow C$: Switch to P_2
   ii. $C \rightarrow P_1$: ACK
   `else if` C `is disconnected`
   Update record of handoff for C at P_1
8. $P_1 \rightarrow P_2$: Take-over client C $\{< d_1, S(d_1) >, \ldots\}$
9. $P_2 \rightarrow P_1$: ACK
10. $P_1 \rightarrow S$: Update state of objects in C transferred from P_1
11. $S \rightarrow P_1$: ACK

(a) Handoff algorithm for push

1. $P_1 \rightarrow P_2$: initiating handoff for C
2. $P_2 \rightarrow P_1$: ACK
3. $P_1 \rightarrow P_2$: C needs $\{< d_1, ttr_1, S(d_1) >, \ldots\}$
4. $P_2 \rightarrow P_1$: ACK
5. `if` C `is connected`
   i. $P_1 \rightarrow C$: Switch to P_2
   ii. $C \rightarrow P_1$: ACK
   `else if` C `is disconnected`
   Update record of handoff for C at P_1
6. $P_1 \rightarrow P_2$: Take-over client $C\{< d_1, S(d_1) >, \ldots\}$
7. $P_2 \rightarrow P_1$: ACK

(b) Handoff algorithm for pull

Fig. 2. Handoff algorithms

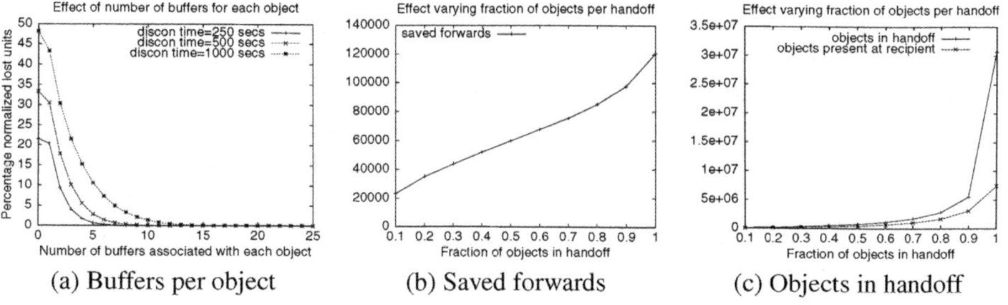

(a) Buffers per object (b) Saved forwards (c) Objects in handoff

Fig. 3. Effect of intermittent connectivity and client mobility in push–based data dissemination.

consisted of the publicly available proxy DEC trace and a synthetic trace (see [3] for details). Updates to objects at the server are simulated by an update process.

To study the effect of intermittent connectivity of clients, we varied three different parameters, one at a time: the maximum disconnection time of clients, the size of the per–object circular buffer and the average time between updates of mutable objects at the server. As the dependent variable, we measured the percentage of *normalized lost units*. Figure 3(a) shows the impact of varying size of the circular buffer associated with each object on the normalized lost units in a push–based approach. We see that, for a given disconnection time, the loss curve has a "knee" beyond which the loss percentage is small. Choosing a buffer size that lies beyond the knee ensures that the proxy can effectively mask the effect of a disconnection (by delivering all updates received during such periods). In general, we find that a small number of buffers (15–20) seem to be sufficient to handle disconnections as large as 15 minutes. As few as 5 buffers can reduce the loss rate from 33% to 2.8% for disconnections of up to 500 seconds.

Figure 3(b) shows the benefit of using application–level handoffs to handle client mobility. The measured metric *saved–forwards* is the number of requests served from

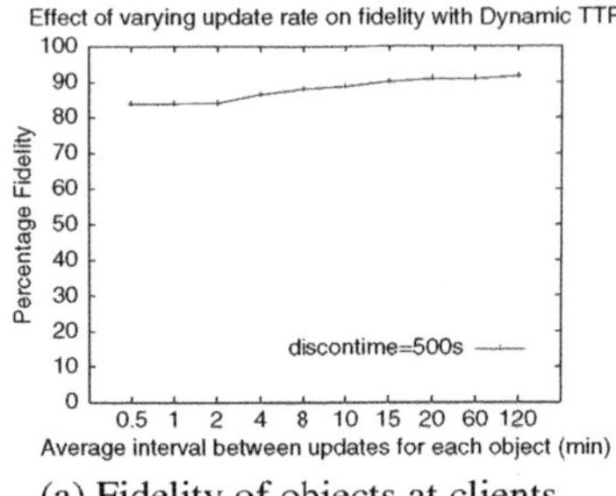

(a) Fidelity of objects at clients

Fraction of objects in handoff	Pull	Push
0.1	5.34	7.14
0.2	9.25	11.02
0.3	14.35	16
0.4	20.65	22.5
0.5	28.82	31.15
0.6	40	44.15
0.7	57.14	64.43
0.8	87.8	102.94
0.9	174	215.64
1	1323.75	2057.85

(b) Handoff overhead

Fig. 4. Comparison of push and pull based dissemination approaches.

the proxy as a result of previous handoffs. Greater the fraction of client–specific state transferred, larger are the savings at an increased message overhead (see Figure 3(c)).

Next we compare the push and pull approaches. Figure 4(a) plots fidelity of data for push and pull in presence of disconnections. As expected, the push–based and the pull with static TTR approaches yield perfect fidelity. Pull with dynamic TTR, which dynamically adjusts TTR values based on observed rate of change of object yields 83.9%–91.7% fidelity. Figure 4(b) compares the average message overhead per handoff in cases of push and pull. We see that pull has lower handoff overhead than push regardless of the fraction of client–state transferred. A detailed experimental evaluation of our techniques is presented in [3].

5 Related Work

File systems such as CODA [2] and distributed systems like Bayou [1] have investigated disconnected operations for mobile clients, techniques for hoarding files and maintaining consistency in weakly connected systems. Several other techniques for maintaining coherency of data in disconnected environments and disseminating data using broadcasts exist and we compare them in [3]. While the previous efforts are for disseminating static web data and for reconciling changes at clients with other hosts on reconnection, our solution is better suited for dynamic data that changes frequently and disseminating data from servers to mobile read–only clients.

6 Conclusions

In this paper, we studied the impact of client mobility and intermittent connectivity on disseminating dynamic web data to mobile hosts. We introduced the notion of eventual–delta consistency and in the context of push–based and pull–based dissemination proposed: (i) proxy–based buffering techniques to mask the effects of client disconnections and (ii) application–level handoff algorithms to handle client mobility. As part of ongoing work, we are implementing these techniques in a prototype proxy for mobile environments.

References

[1] W. K. Edwards, E. D. Mynatt, K. Petersen, M. Spreitzer, D. B. Terry, and M. Theimer. Designing and Implementing Asynchronous Collaborative Applications with Bayou. In *ACM Symposium on User Interface Software and Technology*, pages 119–128, 1997.

[2] J. J. Kistler and M. Satyanarayanan. Disconnected Operation in the Coda File System. In *Thirteenth Symposium on Operating Systems Principles*, volume 25, pages 213–225, 1991.

[3] P. Kulkarni, P. Shenoy, and K. Ramaritham. Handling Client Mobility and Intermittent Connectivity in Mobile Web Access. Technical Report TR02–35, University of Massachusetts, Amherst, 2002.

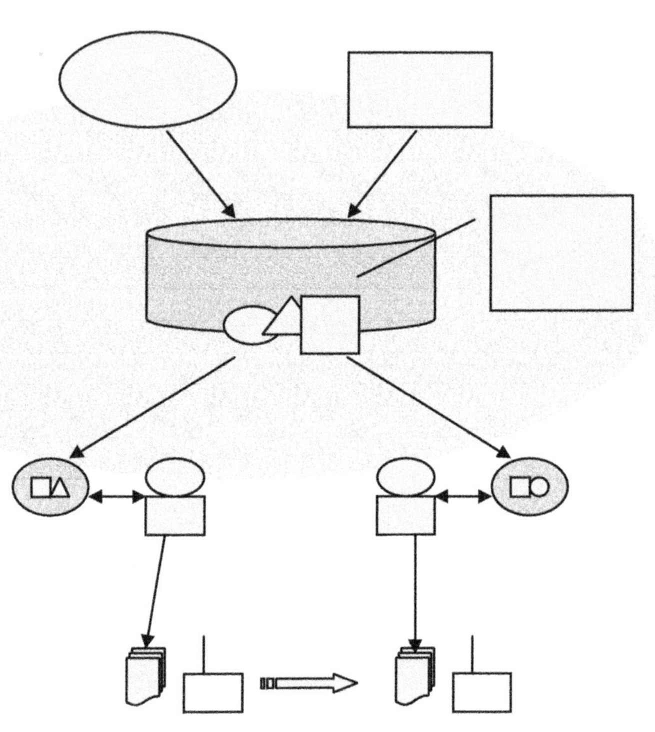

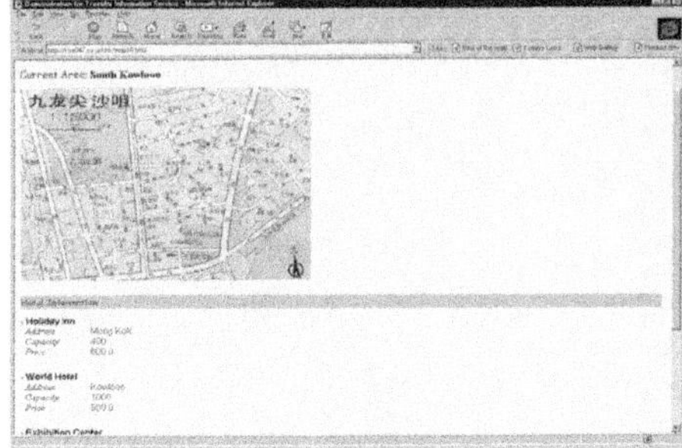
Current Area: South Kowloon
九龙尖沙咀
Hotel Information
Holiday Inn
World Hotel

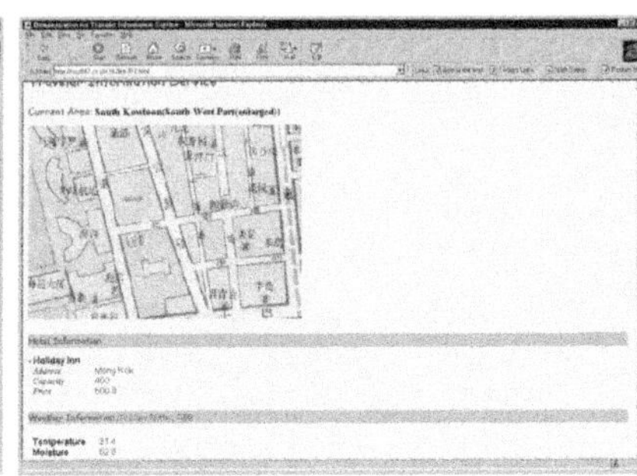
Current Area: South Kowloon(South West Part(enlarged))
Hotel Information
Holiday Inn
Temperature
Moisture

Author Index

Abbadi, Amr El, 165
Agrawal, Divyakant, 165
Akahani, Jun-ichi, 63

Basu, Sujoy, 363
Bellavista, Paolo, 212
Bhowmick, Sourav, 381
Bouguettaya, Athman, 335
Brebner, Gavin, 1
Bubendorfer, Kris, 294
Bunt, Rick, 369

Castro, Paul, 307
Chalon, Denis, 351
Chan, Eddie Y.M., 106
Chen, Chao-Chun, 345
Chen, Ya-Fan, 141
Chen, Ying, 331
Chon, Hae Don, 165
Chuang, John, 45
Connor, Richard, 153
Corradi, Antonio, 212
Crimmins, Francis, 386

Das, Sajal K., 29
Dearle, Alan, 153
Denny, Matthew, 307

Franklin, Michael J., 307
Fu, Yongjian, 381
Funayama, Masataka, 230

Garcia-Molina, Hector, 122
Ghose, Abhishek, 45
Grossklags, Jens, 45

Haller, Klaus, 356
Hashimoto, Takako, 396
He, Weiping, 335
Heer, Johan de, 325
Henricksen, Karen, 247
Hine, John H., 294
Hoi, Ka Kit, 262
Hu, Haibo, 374
Huang, Yongqiang, 122
Hwang, San-Yih, 141

Iizawa, Atsushi, 396
Ikonen, Veikko, 391
Indulska, Jadwiga, 247
Issarny, Valérie, 13

Katooka, Takashi, 396
Kaufman, James, 92
Ke, Chih-Horng, 345
Kim, Myoung Ho, 78
Kirby, Graham, 153
Kulkarni, Purushottam, 401
Kumar, Mohan, 29

Lam, Kwok-Wa, 106
Lankhorst, Marc M., 325
Lee, Chiang, 345
Lee, Dik Lun, 181, 262, 374
Lee, Sung-Ju, 363
Lee, Victor C.S., 106
Lee, Wang-Chien, 181
Li, Shanping, 408
Liu, Dong, 331

Madria, Sanjay Kumar, 381
Majumdar, Ratul kr., 196
McCarthy, Andrew, 153
Mei, Jiangmei, 369
Montanari, Rebecca, 212
Mori, Masakazu, 230
Morrison, Ron, 153
Mullen, Kevin, 153
Myllymaki, Jussi, 92

Nioclais, Donal Mac, 351
Niu, Yongbo, 408

Ozkan, Nadine, 386

Paradis, François, 386
Park, Hyun Kyoo, 78
Peddemors, Arjan J.H., 325
Purakayastha, Apratim, 307

Rakotonirainy, Andry, 247
Ramamritham, Krithi, 196, 401
Rao, Fangyan, 331

Riché, Stéphanie, 1
Richard, Bruno, 351
Robinson, Ricky, 247
Roussopoulos, Nick, 340

Sailhan, Françoise, 13
Satoh, Tetsuji, 63
Schek, Hans-Jörg, 356
Schuldt, Heiko, 356
Shen, Bo, 363
Shen, Huaping, 29
Shenoy, Prashant, 401
Son, Jin Hyun, 78
Song, Zhexuan, 340
Sotamaa, Olli, 391
Stefanelli, Cesare, 212
Strang, Thomas, 279

Takashio, Kazunori, 230
Tokuda, Hideyuki, 230

Wang, Zhijun, 29
Welen, Paula, 153
Wilson, Andy, 153

Xiong, Ming, 196
Xu, Jianliang, 262, 374, 408

Yanagisawa, Yutaka, 63
Yang, Yanyan, 153
Yu, Xiulan, 331

Zheng, Baihua, 181

Lecture Notes in Computer Science

For information about Vols. 1–2484

please contact your bookseller or Springer-Verlag

Vol. 2485: A. Bondavalli, P. Thevenod-Fosse (Eds.), Dependable Computing EDCC-4. Proceedings, 2002. XIII, 283 pages. 2002.

Vol. 2486: M. Marinaro, R. Tagliaferri (Eds.), Neural Nets. Proceedings, 2002. IX, 253 pages. 2002.

Vol. 2487: D. Batory, C. Consel, W. Taha (Eds.), Generative Programming and Component Engineering. Proceedings, 2002. VIII, 335 pages. 2002.

Vol. 2488: T. Dohi, R. Kikinis (Eds), Medical Image Computing and Computer-Assisted Intervention – MICCAI 2002. Proceedings, Part I. XXIX, 807 pages. 2002.

Vol. 2489: T. Dohi, R. Kikinis (Eds), Medical Image Computing and Computer-Assisted Intervention – MICCAI 2002. Proceedings, Part II. XXIX, 693 pages. 2002.

Vol. 2490: A.B. Chaudhri, R. Unland, C. Djeraba, W. Lindner (Eds.), XML-Based Data Management and Multimedia Engineering – EDBT 2002. Proceedings, 2002. XII, 652 pages. 2002.

Vol. 2491: A. Sangiovanni-Vincentelli, J. Sifakis (Eds.), Embedded Software. Proceedings, 2002. IX, 423 pages. 2002.

Vol. 2492: F.J. Perales, E.R. Hancock (Eds.), Articulated Motion and Deformable Objects. Proceedings, 2002. X, 257 pages. 2002.

Vol. 2493: S. Bandini, B. Chopard, M. Tomassini (Eds.), Cellular Automata. Proceedings, 2002. XI, 369 pages. 2002.

Vol. 2495: C. George, H. Miao (Eds.), Formal Methods and Software Engineering. Proceedings, 2002. XI, 626 pages. 2002.

Vol. 2496: K.C. Almeroth, M. Hasan (Eds.), Management of Multimedia in the Internet. Proceedings, 2002. XI, 355 pages. 2002.

Vol. 2497: E. Gregori, G. Anastasi, S. Basagni (Eds.), Advanced Lectures on Networking. XI, 195 pages. 2002.

Vol. 2498: G. Borriello, L.E. Holmquist (Eds.), UbiComp 2002: Ubiquitous Computing. Proceedings, 2002. XV, 380 pages. 2002.

Vol. 2499: S.D. Richardson (Ed.), Machine Translation: From Research to Real Users. Proceedings, 2002. XXI, 254 pages. 2002. (Subseries LNAI).

Vol. 2500: E. Grädel, W. Thomas, T. Wilke (Eds.), Automata Logics, and Infinite Games. VIII, 385 pages. 2002.

Vol. 2501: D. Zheng (Ed.), Advances in Cryptology – ASIACRYPT 2002. Proceedings, 2002. XIII, 578 pages. 2002.

Vol. 2502: D. Gollmann, G. Karjoth, M. Waidner (Eds.), Computer Security – ESORICS 2002. Proceedings, 2002. X, 281 pages. 2002.

Vol. 2503: S. Spaccapietra, S.T. March, Y. Kambayashi (Eds.), Conceptual Modeling – ER 2002. Proceedings, 2002. XX, 480 pages. 2002.

Vol. 2504: M.T. Escrig, F. Toledo, E. Golobardes (Eds.), Topics in Artificial Intelligence. Proceedings, 2002. XI, 432 pages. 2002. (Subseries LNAI).

Vol. 2506: M. Feridun, P. Kropf, G. Babin (Eds.), Management Technologies for E-Commerce and E-Business Applications. Proceedings, 2002. IX, 209 pages. 2002.

Vol. 2507: G. Bittencourt, G.L. Ramalho (Eds.), Advances in Artificial Intelligence. Proceedings, 2002. XIII, 418 pages. 2002. (Subseries LNAI).

Vol. 2508: D. Malkhi (Ed.), Distributed Computing. Proceedings, 2002. X, 371 pages. 2002.

Vol. 2509: C.S. Calude, M.J. Dinneen, F. Peper (Eds.), Unconventional Models in Computation. Proceedings, 2002. VIII, 331 pages. 2002.

Vol. 2510: H. Shafazand, A Min Tjoa (Eds.), EurAsia-ICT 2002: Information and Communication Technology. Proceedings, 2002. XXIII, 1020 pages. 2002.

Vol. 2511: B. Stiller, M. Smirnow, M. Karsten, P. Reichl (Eds.), From QoS Provisioning to QoS Charging. Proceedings, 2002. XIV, 348 pages. 2002.

Vol. 2512: C. Bussler, R. Hull, S. McIlraith, M.E. Orlowska, B. Pernici, J. Yang (Eds.), Web Services, E-Business, and the Semantic Web. Proceedings, 2002. XI, 277 pages. 2002.

Vol. 2513: R. Deng, S. Qing, F. Bao, J. Zhou (Eds.), Information and Communications Security. Proceedings, 2002. XII, 496 pages. 2002.

Vol. 2514: M. Baaz, A. Voronkov (Eds.), Logic for Programming, Artificial Intelligence, and Reasoning. Proceedings, 2002. XIII, 465 pages. 2002. (Subseries LNAI).

Vol. 2515: F. Boavida, E. Monteiro, J. Orvalho (Eds.), Protocols and Systems for Interactive Distributed Multimedia. Proceedings, 2002. XIV, 372 pages. 2002.

Vol. 2516: A. Wespi, G. Vigna, L. Deri (Eds.), Recent Advances in Intrusion Detection. Proceedings, 2002. X, 327 pages. 2002.

Vol. 2517: M.D. Aagaard, J.W. O'Leary (Eds.), Formal Methods in Computer-Aided Design. Proceedings, 2002. XI, 399 pages. 2002.

Vol. 2518: P. Bose, P. Morin (Eds.), Algorithms and Computation. Proceedings, 2002. XIII, 656 pages. 2002.

Vol. 2519: R. Meersman, Z. Tari, et al. (Eds.), On the Move to Meaningful Internet Systems 2002: CoopIS, DOA, and ODBASE. Proceedings, 2002. XXIII, 1367 pages. 2002.

Vol. 2521: A. Karmouch, T. Magedanz, J. Delgado (Eds.), Mobile Agents for Telecommunication Applications. Proceedings, 2002. XII, 317 pages. 2002.

Vol. 2522: T. Andreasen, A. Motro, H. Christiansen, H. Legind Larsen (Eds.), Flexible Query Answering. Proceedings, 2002. XI, 386 pages. 2002. (Subseries LNAI).

Vol. 2525: H.H. Bülthoff, S.-Whan Lee, T.A. Poggio, C. Wallraven (Eds.), Biologically Motivated Computer Vision. Proceedings, 2002. XIV, 662 pages. 2002.

Vol. 2526: A. Colosimo, A. Giuliani, P. Sirabella (Eds.), Medical Data Analysis. Proceedings, 2002. IX, 222 pages. 2002.

Vol. 2527: F.J. Garijo, J.C. Riquelme, M. Toro (Eds.), Advances in Artificial Intelligence – IBERAMIA 2002. Proceedings, 2002. XVIII, 955 pages. 2002. (Subseries LNAI).

Vol. 2528: M.T. Goodrich, S.G. Kobourov (Eds.), Graph Drawing. Proceedings, 2002. XIII, 384 pages. 2002.

Vol. 2529: D.A. Peled, M.Y. Vardi (Eds.), Formal Techniques for Networked and Distributed Sytems – FORTE 2002. Proceedings, 2002. XI, 371 pages. 2002.

Vol. 2531: J. Padget, O. Shehory, D. Parkes, N. Sadeh, W.E. Walsh (Eds.), Agent-Mediated Electronic Commerce IV. Proceedings, 2002. XVII, 341 pages. 2002. (Subseries LNAI).

Vol. 2532: Y.-C. Chen, L.-W. Chang, C.-T. Hsu (Eds.), Advances in Multimedia Information Processing – PCM 2002. Proceedings, 2002. XXI, 1255 pages. 2002.

Vol. 2533: N. Cesa-Bianchi, M. Numao, R. Reischuk (Eds.), Algorithmic Learning Theory. Proceedings, 2002. XI, 415 pages. 2002. (Subseries LNAI).

Vol. 2534: S. Lange, K. Satoh, C.H. Smith (Ed.), Discovery Science. Proceedings, 2002. XIII, 464 pages. 2002.

Vol. 2535: N. Suri (Ed.), Mobile Agents. Proceedings, 2002. X, 203 pages. 2002.

Vol. 2536: M. Parashar (Ed.), Grid Computing – GRID 2002. Proceedings, 2002. XI, 318 pages. 2002.

Vol. 2537: D.G. Feitelson, L. Rudolph, U. Schwiegelshohn (Eds.), Job Scheduling Strategies for Parallel Processing. Proceedings, 2002. VII, 237 pages. 2002.

Vol. 2538: B. König-Ries, K. Makki, S.A.M. Makki, N. Pissinou, P. Scheuermann (Eds.), Developing an Infrastructure for Mobile and Wireless Systems. Proceedings 2001. X, 183 pages. 2002.

Vol. 2539: K. Börner, C. Chen (Eds.), Visual Interfaces to Digital Libraries. X, 233 pages. 2002.

Vol. 2540: W.I. Grosky, F. Plášil (Eds.), SOFSEM 2002: Theory and Practice of Informatics. Proceedings, 2002. X, 289 pages. 2002.

Vol. 2541: T. Barkowsky, Mental Representation and Processing of Geographic Knowledge. X, 174 pages. 2002. (Subseries LNAI).

Vol. 2544: S. Bhalla (Ed.), Databases in Networked Information Systems. Proceedings 2002. X, 285 pages. 2002.

Vol. 2545: P. Forbrig, Q, Limbourg, B. Urban, J. Vanderdonckt (Eds.), Interactive Systems. Proceedings 2002. X, 269 pages. 2002.

Vol. 2546: J. Sterbenz, O. Takada, C. Tschudin, B. Plattner (Eds.), Active Networks. Proceedings, 2002. XIV, 267 pages. 2002.

Vol. 2547: R. Fleischer, B. Moret, E. Meineche Schmidt (Eds.), Experimental Algorithmics. XVII, 279 pages. 2002.

Vol. 2548: J. Hernández, Ana Moreira (Eds.), Object-Oriented Technology. Proceedings, 2002. VIII, 223 pages. 2002.

Vol. 2548: J. Hernández, Ana Moreira (Eds.), Object-Oriented Technology. Proceedings, 2002. VIII, 223 pages. 2002.

Vol. 2549: J. Cortadella, A. Yakovlev, G. Rozenberg (Eds.), Concurrency and Hardware Design. XI, 345 pages. 2002.

Vol. 2550: A. Jean-Marie (Ed.), Advances in Computing Science – ASIAN 2002. Proceedings, 2002. X, 233 pages. 2002.

Vol. 2551: A. Menezes, P. Sarkar (Eds.), Progress in Cryptology – INDOCRYPT 2002. Proceedings, 2002. XI, 437 pages. 2002.

Vol. 2552: S. Sahni, V.K. Prasanna, U. Shukla (Eds.), High Performance Computing – HiPC 2002. Proceedings, 2002. XXI, 735 pages. 2002.

Vol. 2553: B. Andersson, M. Bergholtz, P. Johannesson (Eds.), Natural Language Processing and Information Systems. Proceedings, 2002. X, 241 pages. 2002.

Vol. 2554: M. Beetz, Plan-Based Control of Robotic Agents. XI, 191 pages. 2002. (Subseries LNAI).

Vol. 2555: E.-P. Lim, S. Foo, C. Khoo, H. Chen, E. Fox, S. Urs, T. Costantino (Eds.), Digital Libraries: People, Knowledge, and Technology. Proceedings, 2002. XVII, 535 pages. 2002.

Vol. 2556: M. Agrawal, A. Seth (Eds.), FST TCS 2002: Foundations of Software Technology and Theoretical Computer Science. Proceedings, 2002. XI, 361 pages. 2002.

Vol. 2557: B. McKay, J. Slaney (Eds.), AI 2002: Advances in Artificial Intelligence. Proceedings, 2002. XV, 730 pages. 2002. (Subseries LNAI).

Vol. 2558: P. Perner, Data Mining on Multimedia Data. X, 131 pages. 2002.

Vol. 2559: M. Oivo, S. Komi-Sirviö (Eds.), Product Focused Software Process Improvement. Proceedings, 2002. XV, 646 pages. 2002.

Vol. 2560: S. Goronzy, Robust Adaptation to Non-Native Accents in Automatic Speech Recognition. Proceedings, 2002. XI, 144 pages. 2002. (Subseries LNAI).

Vol. 2561: H.C.M. de Swart (Ed.), Relational Methods in Computer Science. Proceedings, 2001. X, 315 pages. 2002.

Vol. 2562: V. Dahl, P. Wadler (Eds.), Practical Aspects of Declarative Languages. Proceedings, 2003. X, 315 pages. 2002.

Vol. 2566: T.Æ. Mogensen, D.A. Schmidt, I.H. Sudborough (Eds.), The Essence of Computation. XIV, 473 pages. 2002.

Vol. 2567: Y.G. Desmedt (Ed.), Public Key Cryptography – PKC 2003. Proceedings, 2003. XI, 365 pages. 2002.

Vol. 2569: D. Gollmann, G. Karjoth, M. Waidner (Eds.), Computer Security – ESORICS 2002. Proceedings, 2002. XIII, 648 pages. 2002. (Subseries LNAI).

Vol. 2571: S.K. Das, S. Bhattacharya (Eds.), Distributed Computing. Proceedings, 2002. XIV, 354 pages. 2002.

Vol. 2572: D. Calvanese, M. Lenzerini, R. Motwani (Eds.), Database Theory – ICDT 2003. Proceedings, 2003. XI, 455 pages. 2002.

Vol. 2574: M.-S. Chen, P.K. Chrysanthis, M. Sloman, A. Zaslavsky (Eds.), Mobile Data Management. Proceedings, 2003. XII, 414 pages. 2003.

Vol. 2575: L.D. Zuck, P.C. Attie, A. Cortesi, S. Mukhopadhyay (Eds.), Verification, Model Checking, and Abstract Interpretation. Proceedings, 2003. XI, 325 pages. 2003.

GPSR Compliance

The European Union's (EU) General Product Safety Regulation (GPSR) is a set of rules that requires consumer products to be safe and our obligations to ensure this.

If you have any concerns about our products, you can contact us on ProductSafety@springernature.com

In case Publisher is established outside the EU, the EU authorized representative is:

Springer Nature Customer Service Center GmbH
Europaplatz 3
69115 Heidelberg, Germany

Batch number: 09478806

Printed by Printforce, the Netherlands